FINITE MATHEMATICS WITH APPLICATIONS

BY

SHIRLEY O. HOCKETT and MARTIN STERNSTEIN
Professors of Mathematics, Ithaca College

KRIEGER PUBLISHING COMPANY
MALABAR, FLORIDA

Original Edition 1984

Printed and Published by
ROBERT E. KRIEGER PUBLISHING COMPANY, INC.
KRIEGER DRIVE
MALABAR, FLORIDA 32950

Printed in the United States of America

Library of Congress Cataloging in Publication Data

Hockett, Shirley O.
Finite mathematics with applications.

Includes index.
1. Mathematics—1961- I. Sternstein, Martin.
II. Title.
QA39.2.H6224 1984 510 84-960
ISBN 0-89874-674-4

Camera-ready copy prepared on an IBM Selectric by Charles F. Hockett with art work by Valerie B. King.

10 9 8

Preface

Finite Mathematics with Applications is for students in business, economics, or social or biological science. The book is designed for the student to read and use, with explicit attention to difficulties likely to be encountered. Although a course in intermediate algebra is assumed, we have not hesitated to review algebraic techniques or topics as they are needed. Applications are interspersed throughout to show how useful finite mathematics is in a wide variety of fields. Special emphasis has been given to word problems and to ways of attacking them. The presentation of the material is informal and mostly intuitive.

There are six chapters, divided into 33 sections. Each chapter begins with a preview and statement of the relevance of the material to be covered. Each section begins with a description of the specific topic to be introduced; this is intended to motivate, or at least to pique interest. The discussion is in the form of examples, carefully set forth as guides for the reader to follow and study in preparation for the exercises at the end of the section. When possible, the exercises are both graduated in difficulty and also presented in the order in which the topics they touch on have been covered in the preceding discussion. Each section concludes with two mini-tests.

Each chapter ends with a summary of what has been covered, a set of comprehensive Review Exercises, and a Chapter Test aimed at integrating, as much as is useful, the material covered so far.

We have provided answers to the odd-numbered Section Exercises and Chapter Review Exercises, and to all the problems in Section Tests A (often with hints) and in the Chapter Tests.

It is generally agreed that the subject of finite mathematics encompasses a broad range of topics, from combinatorics to game theory, from logic to number theory, from set theory to matrices. In this text the emphasis is on three themes: probability (chapters 1 and 2), linear programming (chapters 3, 4, and 5), and the mathematics of finance (chapter 6). The order in which these themes are presented in the book is by no means the one that must be followed by an instructor, since they are more or less self-contained and can be taught in any arrangement. After carefully considering possible sequences, we chose to begin the text with probability. Many instructors view this as the most difficult material in a course on finite mathematics, but we have found it pedagogically beneficial in our classes to teach probability at the outset, when the students are still fresh, optimistic, and willing to be challenged to learn.

An instructor who chooses a different sequence of topics should note the following:

Chapter 1 (counting techniques) is needed for chapter 2 (probability).

§25 (Markov chains) uses principles of probability developed in chapter 2.

Chapter 3 (linear systems) leads to geometric solutions of linear-programming problems in §26.

Chapter 4 (matrices) is needed for the simplex method in linear programming in §§27-28.

In Appendix 2 we provide two tables. The first is for $(1+r)^n$, with r ranging by increments of 0.005 from 0.005 to 0.80 and with n ranging by units from 1 to 100. The second table is for the present value of an annuity of $1, useful for calculations in problems involving mortgages. Throughout the discussion in chapter 6 (finance) we encourage the use of a calculator; indeed, we have found that most of our students prefer a calculator to the tables.

The text is suitable for a one-semester or two-quarter course.

The authors are grateful to Valerie B. King for the careful figures she has provided and for her intelligent, friendly assistance in final page assembly for the camera. We also thank Randel Morgan of Ithaca College for computing Table 1 of Appendix 2. Finally, we are keenly aware of the vital role Charles F. Hockett has played in bringing this book to fruition. He not only typed the entire manuscript twice, including the superb final camera-ready copy, but also served throughout the project as advisor, mentor, helpful critic, and persistent cheerleader.

SHIRLEY O. HOCKETT
MARTIN STERNSTEIN

Contents

Chapter 1

Counting and sets

One of the major topics we will consider in this text is probability. In order to understand probability you will need to be comfortable with set theory and notation, and to be able to count the objects in a specified set. In this chapter we consider the most basic and useful devices for counting. We will then be ready, in the second chapter, to use these techniques in our study of probability.

A brief summary of the basic concepts and terminology of set theory is available in Appendix 1. It may be helpful to review that material at this time.

SECTION 1. COUNTING WITH VENN DIAGRAMS

Venn diagrams provide a systematic visual method for organizing data. They are therefore extremely useful for certain elementary counting problems, as we will illustrate in this section. We use given data to indicate the known sizes of sets on a diagram, then solve for the sizes of other sets. The procedure is illustrated in the following examples.

Example 1-1

In a small private school with 550 students, 400 are taking English, 100 are taking mathematics, and 60 are taking both subjects. How many students in the school are taking at least one of these two subjects? How many are taking neither?

It is important to recognize that some students are included twice in the data given (see the last topic in Appendix 1). For instance, some of the 400 students taking English are also taking mathematics. Venn diagrams are useful precisely because they enable us to sort out the sizes of the nonoverlapping (*disjoint*) sets.

In Figure 1.1(a) there are two intersecting circles, for the sets

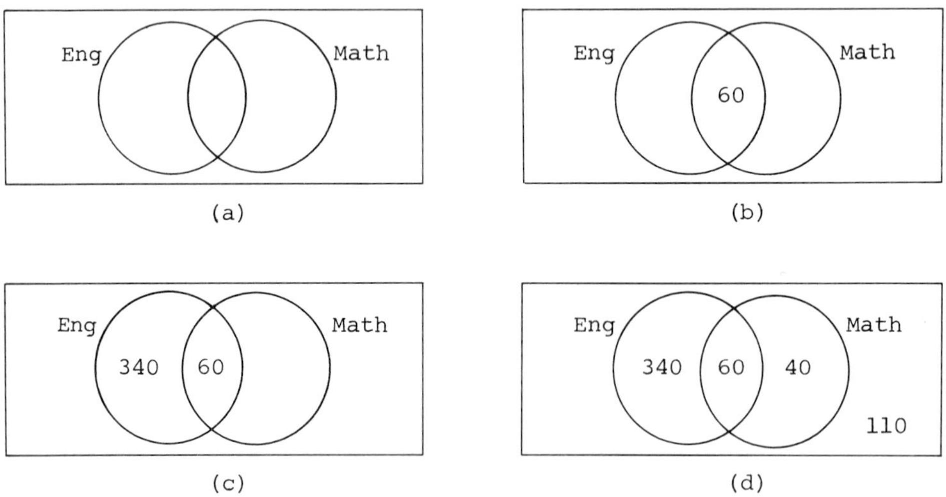

Figure 1.1

of English and mathematics students. In (b) we have entered the number 60 in the intersection of these sets. Since 400 students are taking English, of whom 60 are *also* taking mathematics, the number taking English but *not* mathematics is 400 - 60 or 340; this number is entered in (c). Similarly, the number taking mathematics but not English is 100 - 60 or 40 (d).

We can now answer both questions of Example 1-1. The number taking at least one of the two subjects is 340 + 60 + 40 or 440. Since the universal set in this example is the population of the school, 550 students, the number taking neither English nor mathematics is 550 - 440 or 110 students. Note that the number 110 has been entered outside both circles in Figure 1.1(d).

Example 1-2

A survey of 100 newspaper readers shows that 70 read the morning paper, the *Sun*, and 10 read both the *Sun* and the evening paper, the *Moon*. How many who read the *Moon* do not read the *Sun*? How many in the survey read the *Moon*? (Assume that everyone surveyed reads at least one of the two.)

Since everyone surveyed reads at least one of the two papers, we note that there are 100 readers in the universal set, which is $S \cup M$, where S is the set of *Sun*-readers and M the set of *Moon*-readers. We therefore draw only two intersecting circles and omit the rectangle.

We start (Figure 1.2(a)), by writing the number 10 in the region $S \cap M$. Then, since we know that 10 of the 70 *Sun*-readers also read the *Moon*, it follows that 70 - 10 or 60 read only the *Sun* (b). If we let x be the number who read the *Moon* but not the *Sun*, then, since 100 read one or the other or both, we have the equation

$$100 = 60 + 10 + x$$

and $x = 30$. So 30 people, of those surveyed, read the *Moon* but not the *Sun*, while a total of 10 + 30 or 40 read the *Moon*.

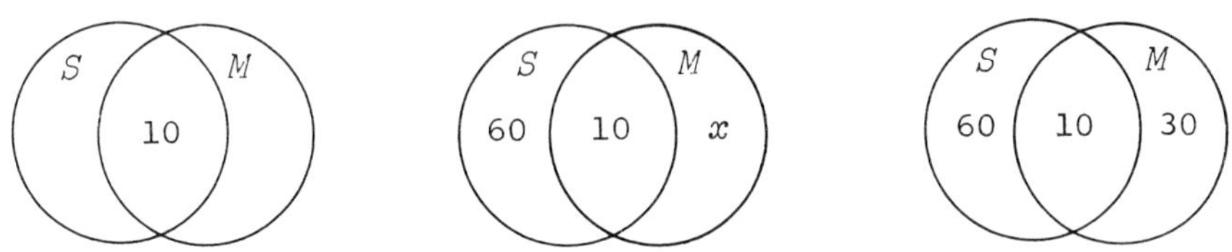

(a) (b) (c)

Figure 1.2

Example 1-3

A gun factory has 500 employees of whom 200 are members of a union. Fifty union members work the night shift, while 180 nonunion employees never work at night. How many employees work the night shift?

In Figure 1.3(a) we show that 50 employees are both union members and night-shift workers. There are therefore 200 - 50 or 150 union members not on the night shift (b). In (c) we show that 180 nonunion employees do not work the night shift. If x is the number of nonunion employees who do work the night shift, then x satisfies the equation

$$500 = 150 + 50 + x + 180$$

and $x = 500 - 380$ or 120 (d). So the total number of employees who work the night shift is 50 + 120 or 170; this is the answer to the question asked.

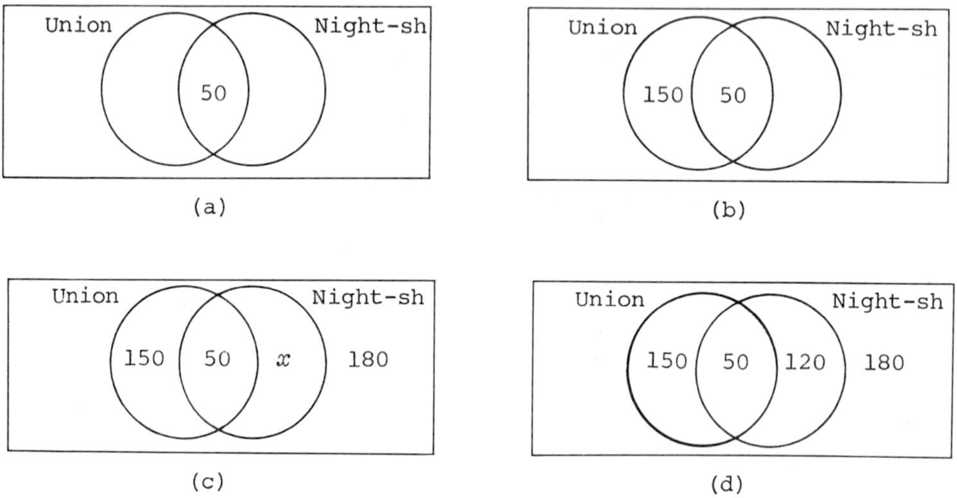

Figure 1.3

Example 1-4

One thousand students in a School of Business passed a course in accounting or a course in calculus last semester. Of these, there were 800 passing grades in accounting and 700 passing grades in calculus. How many of the students passed both courses?

Let A be the set of students who passed accounting and C the set who passed calculus. Then we know that the set $A \cup C$ has 1000 students. We must find x, the number of students in $A \cap C$; see Figure 1.4(a).

Since set A has 800 students, we can represent the number in $A \cap C'$ (the set of students who passed accounting but not calculus) by $800 - x$ ((b) in the figure). Similarly, we can let the number who

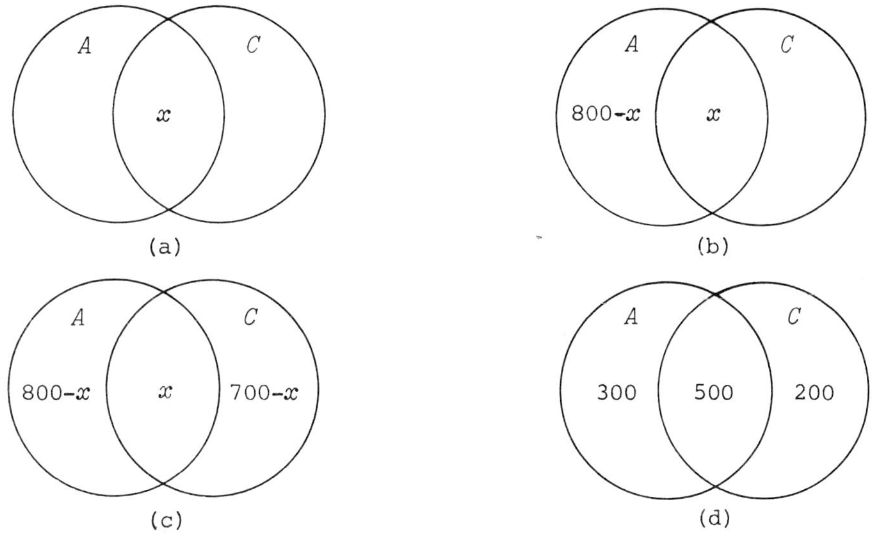

Figure 1.4

passed calculus but not accounting be $700 - x$ (c). To solve for x, we use the fact that the total population under consideration is 1000:

$$
\begin{aligned}
(800 - x) + x + (700 - x) &= 1000 \\
1500 - x &= 1000 \\
x &= 500
\end{aligned}
$$

So 500 students passed both courses. The solution shown in Figure 1.4(d) provides a check, since $300 + 500 + 200 = 1000$.

Example 1-5

A survey of 750 school children showed that

350 had had mumps;
300 had had chicken pox;
100 had had both mumps and chicken pox;
120 had had both measles and chicken pox;
200 had had mumps but neither measles nor chicken pox;
75 had had all three diseases;
50 had had none of the three diseases.

(a) How many of the children surveyed had had measles? (b) How many had had chicken pox but neither mumps nor measles? (c) How many had had both measles and mumps?

The encircled numbers in Figure 1.5 identify the eight nonoverlapping regions involved and show the order in which we determine

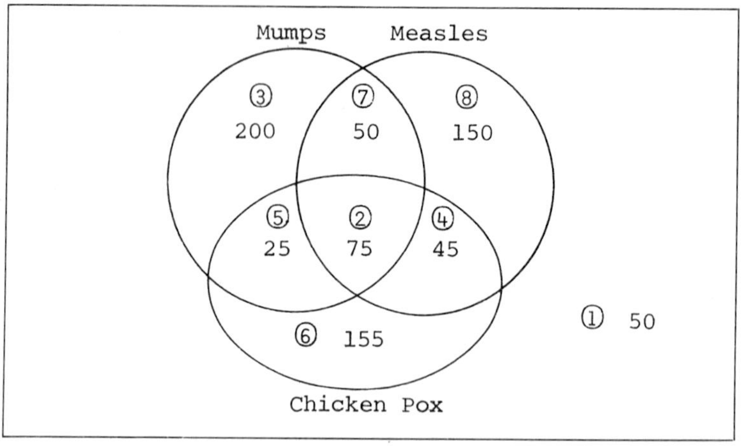

Figure 1.5

the sizes of the disjoint sets. Note especially that we start at the *end* of the list in this example and work backward.

For the sizes of regions ①, ②, and ③, we simply transfer to the Venn diagram the data given for those sets: 50, 75, and 200 respectively. To find the size of ④, we note that 120 children had had both measles and chicken pox (that is, there are 120 altogether in regions ② and ④; we therefore subtract 75 from 120 to get 45 children in region ④. Similarly, to find the size of ⑤, we subtract 75 from 100. For ⑥ we subtract the sum (25 + 75 + 45) from 300; for ⑦ we subtract the sum (200 + 25 + 75) from 350. Finally, to determine the size of the one remaining set, labeled ⑧ on the diagram, we use the fact that the entire population (the universal set) is 750 children, and subtract the sum of the sizes of all the known nonoverlapping sets from that:

$$750 - (200 + 50 + 25 + 75 + 45 + 155 + 50) = 750 - 600 = 150$$

We can now read the answers to the questions asked directly from the Venn diagram: (a) the number who had had measles is

$$50 + 150 + 75 + 45 = 320$$

(b) 155 had had chicken pox but neither mumps nor measles; (c) the number who had had both mumps and measles is 50 + 75 = 125.

EXERCISES

Using Venn diagrams solve each of the following (1 through 11):

1. In a group of businessmen, 80 read the *Wall Street Journal*, 50

read the *New York Times*, and 25 read both of these newspapers. How many of these businessmen read at least one of the two papers?

2. In a nursing home, medications are distributed twice a day with 35 residents receiving pills in the morning and 50 receiving pills in the evening. If 15 patients receive pills twice a day, how many patients are on medications?
3. In a meeting open only to truck drivers who either own their own rigs or belong to the Teamsters' Union, it is noted that, of the 500 present, 200 own their own rigs and 150 both own their own rigs and belong to the Union. How many belong to the union?
4. A survey of 1000 television owners found 800 who own color TV sets and 250 who own both color and black-and-white sets. How many of those surveyed own black-and-white sets?
5. A package of 100 garden-pea seeds contains 40 wrinkled seeds; 25 seeds are both yellow and wrinkled; 35 seeds are neither yellow nor wrinkled. How many of the seeds are yellow?
6. Among 350 applicants for a college teaching position, 300 have a doctoral degree, 200 have both a doctorate and previous teaching experience, and 30 have neither a doctorate nor previous experience. How many of the applicants have previous teaching experience?
7. A grocery store sells only brands of yogurt which are tasty or nutritious. If the store sells 18 brands of which 10 are tasty and 12 are nutritious, how many brands are both tasty and nutritious?
8. Among 135 employees, 40 are executives, 35 make over $25,000, and 75 are neither executives nor earners of more than $25,000. How many executives make over $25,000?
9. Three bond issues were put before the 270 eligible voters of a district with the following results: 20 people voted for all three issues; 30 voted for Bonds 1 and 2; 40 voted for Bonds 2 and 3; 50 voted for Bonds 1 and 3; 100 voted for Bond 1, 80 for Bond 2, and 120 for Bond 3. If each eligible voter voted on each issue, how many people voted against all three issues?
10. In a group of 100 hospital admissions, 50 were scheduled for X rays, 45 for blood tests, 50 for urinalyses, 15 for both X rays and blood tests, 15 for both blood and urine tests, 25 for both X rays and urine tests, and 10 for all three tests. How many admissions, if any, were scheduled for none of these three tests?
11. Human blood is classified according to whether it has none, one, two, or all three of the antigens A, B, and Rh. Figure 1.6 is a Venn diagram for the eight different blood types with the identifying labels shown in the appropriate regions. Note that "O" is used for blood lacking both the A and the B antigens; the blood of a person with the Rh antigen is typed as

positive (+); without, as negative (−) . Thus, a person with type AB+ blood has all three antigens; one with B- blood has only the B antigen; O- blood has none of the antigens; and so on.

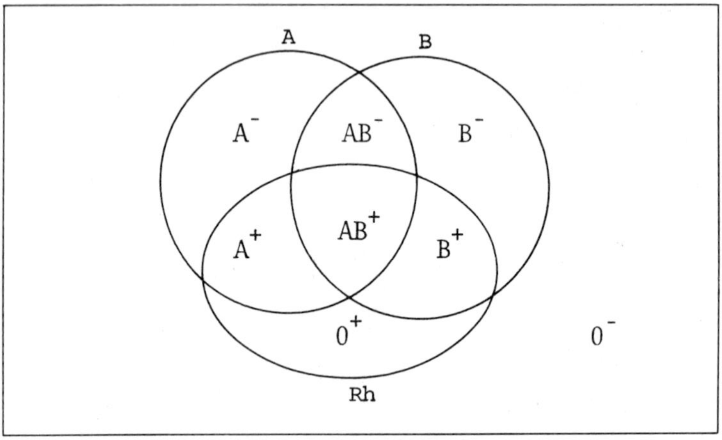

Figure 1.6

The following information was obtained about 84 donors who visited a bloodmobile one weekend:

42 had the B antigen;
50 had the Rh antigen;
26 had both the A and Rh antigens;
36 had both the B and Rh antigens;
25 had both the A and B antigens;
24 had all three antigens;
15 had none of the antigens.

(a) How many donors had type A- blood? (b) How many were O+?

TEST A

1. How many houses in a new development have either a fireplace or more than one bathroom or both, given that 50 houses have fireplaces, 30 have more than one bathroom, and 15 have both of these features?
2. In a set of 60 church buildings, 30 have stained-glass windows, 20 have statues, and 8 have both stained-glass windows and statues. How many of the 60 have neither of these features?

3. In a wine-tasting, 22 wines were commended for their taste, 21 for their aroma, 23 for color, 7 for both taste and aroma, 8 for both taste and color, 9 for color and aroma, and only 3 for all three. How many wines were commended only for taste? Only for aroma? Only for color?

TEST B

1. Of 150 students enrolled in data-processing or in business calculus or both, 100 are taking data-processing and 20 are taking both. How many are enrolled in business calculus?
2. Of 100 houses having either blown-in insulation or rolled batts of insulation or both, 60 have blown-in insulation and 50 have batts. How many have both?
3. In a survey of 1000 college graduates, 200 were familiar with PL1, 150 with FORTRAN, 175 with COBOL, 75 with both PL1 and COBOL, 65 with both FORTRAN and COBOL, 55 with both FORTRAN and PL1, and 25 with all three of these popular programming languages. How many of these graduates had had experience with none of these languages? With at least one of them?

SECTION 2. COUNTING WITH TREE DIAGRAMS

A *tree* diagram provides an efficient routine for listing, counting, and analyzing the possible outcomes in a sequence of experiments or events. This tool, which is also extremely useful in certain kinds of probability problems, is illustrated in the following examples.

Example 2-1

A dinner menu offers a choice of juice or soup as appetizer, of chops, chicken, or fish as main course, and of pie or ice cream as dessert. Construct a tree diagram to show all possible complete dinners.

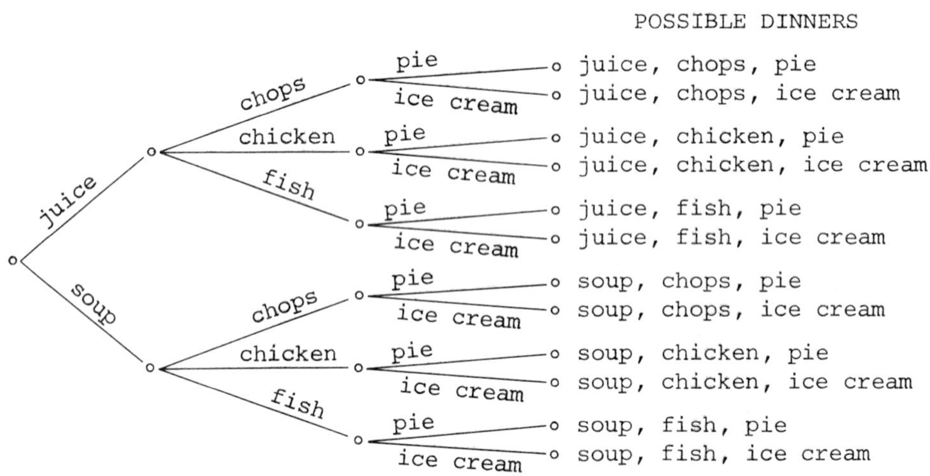

Figure 2.1

Figure 2.1 is called a tree; it shows that there are twelve different complete dinners. Every tree starts from a single point (in this book, always the point furthest to the left). The paths from this initial point to their terminations at the right are called *branches*. The various possible dinners listed at the right in the figure are found by following all the paths from left to right.

Example 2-2

A businessman plans to invest his funds in one and only one of the following three: a savings account, stocks, or bonds. If he invests in stocks, he will choose either preferred or common stocks and only one of two brokers, Mr Sharpe or Mrs Keene. If he selects bonds, he will invest all his funds in one of the three types A, AA, or AAA. Show his investment choices in a tree diagram.

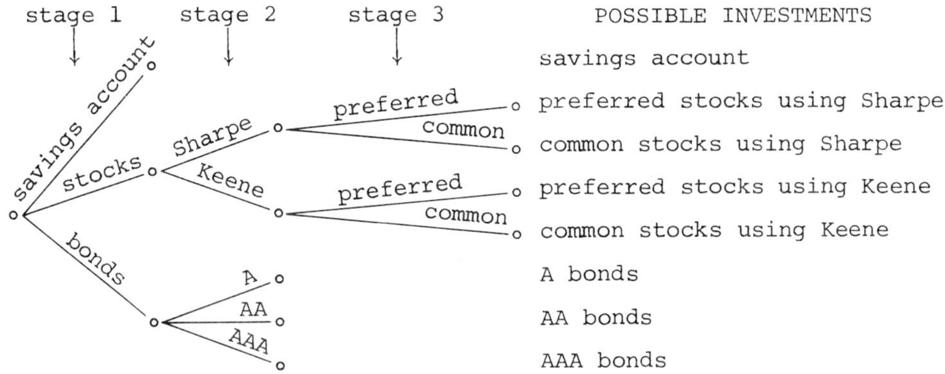

Figure 2.2

In this example (Figure 2.2) we see that different branches end at different *stages*: some branches are shorter than others. If a savings account is chosen there is nothing more to be decided, so the uppermost branch has only one line segment and ends after the first stage. For stocks, there are still two decisions to be made (broker and kind of stock); each of these branches thus involves three stages (or *subbranches*). For bonds, there remains only one decision to be made (on the bond-rating desired), so there are two stages.

Note that for the branches leading from the choice of stocks instead of savings account or bonds, we could also have considered the kind of stock at the second stage and the choice of broker at the third. Draw a new tree, using this order, to show that we end up with the same eight investment possibilities.

Example 2-3

Draw a tree diagram to illustrate all possible outcomes from tossing three coins. (a) How many of these have exactly two heads showing? (b) How many have at least one tail showing?

In Figure 2.3 (top of page 12), where "H" stands for heads and "T" for tails, we have for convenience placed the symbols for the outcome at the branching points. Two heads appear in each of the

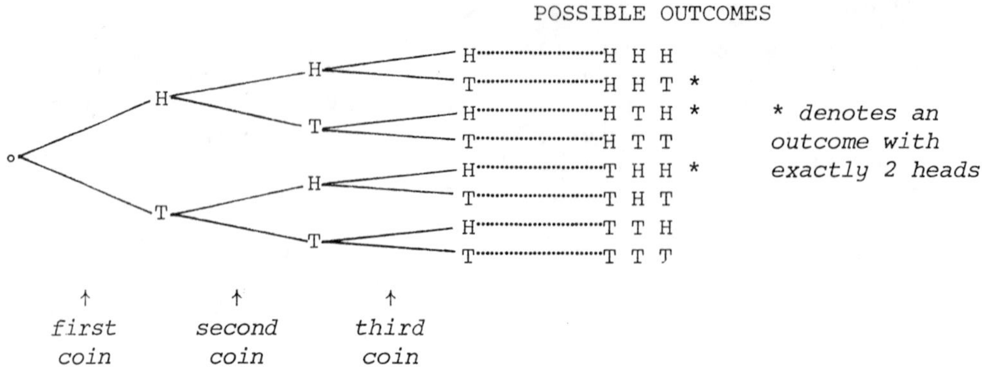

Figure 2.3

outcomes labeled with an asterisk; counting these as different we have three as the answer to part (a). Since at least one tail shows in each outcome except HHH, the answer to part (b) is seven.

Example 2-4

Suppose baseball teams A and B are in the midst of the world series and team A has won the first two games. Draw a tree to show the possible outcomes, using the fact that the team that first wins four games wins the series. How many different ways can the series be completed? How many of these ways require seven games to be played? How many require only five games?

See Figure 2.4. A branch terminates in this tree when a team has won four games (and the series). We have indicated this by a black dot by the winner. Since there are fifteen dots, there are fifteen different outcomes for the series, assuming team A has won the first

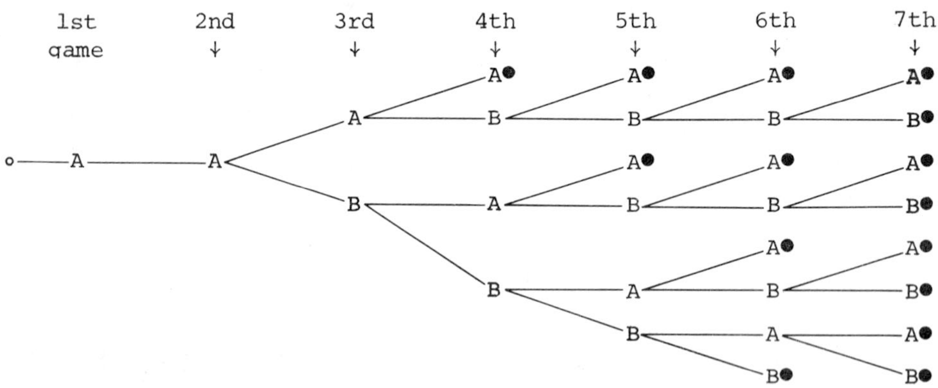

Figure 2.4

two games. Eight of these are seven-game series; only two end after the fifth game, and for both of these team A is the champion.

Example 2-5

Color blindness is a recessive sex-linked hereditary trait for which the gene is located on the X-chromosome. Since males carry one X- and one Y-chromosome, any male with an X^B gene ("B" for color blindness) is himself color-blind. Since females have two X-chromosomes, a color-blind female must have two doses of the gene: that is, she must be X^BX^B. We will call an X^BX female (one with only one dose of the gene) a carrier, since she may transmit color blindness to her offspring although she herself has normal vision.

(a) Draw a tree to show the possible outcomes for the vision of a child born to a color-blind man and a carrier woman. (b) Draw a tree for the child of a carrier woman and a man with normal vision. (c) Is it possible for a woman who is XX to have a color-blind daughter? a color-blind son?

The answers to (a) and (b) are shown in Figure 2.5 (a) and (b) respectively.

(c) It is not possible for a woman who is neither color-blind nor a carrier to have color-blind offspring. Since the mother is XX, a daughter could inherit at most a single dose of the color-blindness gene (from her father), whereas a son must inherit the Y-chromosome from his father and an X-chromosome from his mother. So anyone whose mother is XX has normal vision.

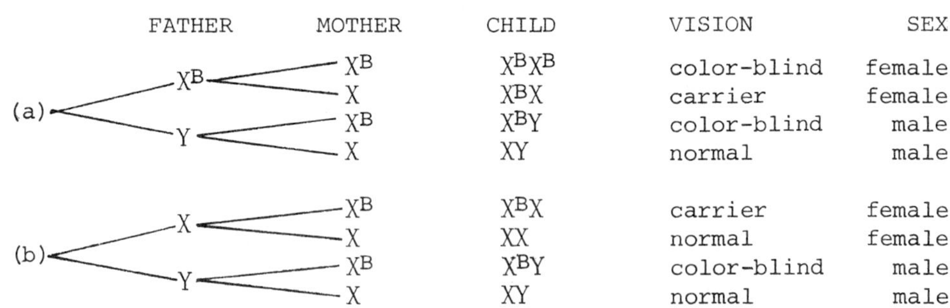

Figure 2.5

EXERCISES

Draw a tree diagram for each of the exercises 1 through 9.

1. Select one of the three countries Alpha, Omega, and Ultra for foreign aid; then choose between military and economic aid for that country.

2. A card is selected at random from a playing-card deck and its suit is noted; a second card is then chosen and its color is noted.
3. A student must take a mathematics or a science course. If he chooses math, he can take either calculus or finite math; if he chooses science, he can take biology, chemistry, or physics.
4. Each of two models of a car is available in three different colors and in two different body styles.
5. Tom or Mary is to be chosen president; Alice, John, or Ellen is to be named vice-president; and Mike or Meg is to be selected treasurer.
6. Show all possible outcomes for the sexes of children in a family with three children. How many of these have at most one girl?
7. Show all possible outcomes of tossing four coins.
8. One of three brands of coffee is to be chosen for Abe's Restaurant I and one of the two remaining brands for Abe's Restaurant II.
9. Three people board a bus having four vacant seats. Show all possible seating arrangements involving the empty seats. How many are there?
10. The offices for a company's first, second, third, and fourth vice-presidents are to be distributed among Brown, Smith, Jones, and Adams. Illustrate the ways this may be done, using two different tree diagrams.
11. A bag contains three pennies, two nickels, and one dime. Draw a tree for all possible outcomes if you choose two coins in succession without replacing the first. How many different *amounts* of money can you obtain? What if you replace the first coin before drawing the second?
12. The disease hemophilia involves abnormal blood clotting, often resulting in uncontrollable bleeding. Like color blindness (Example 2-5, p. 13), hemophilia is a hereditary recessive sex-linked trait, carried by a gene on the X-chromosome (denoted by X^H). X^HX^H females and X^HY males are "bleeders"; they often die before they are able to reproduce; few of the survivors have children. Draw a tree to show the possible outcomes for a child born to a normal man (XY) and a hemophiliac woman. What dire conclusion must be reached?
13. Suppose teams S and T are in the midst of a baseball world series in which S has won the first game and T has won both the second and the third games. Draw a tree to show all the different ways the series may come out, and tell how many ways there are. In how many of these is team T the champion? In how many ways can team T win the series in six or fewer games? (Remember, the first team to win four games wins the series.)

14. A rat may go through a maze (shown at the right), starting at A and ending at I, by moving along one segment at a time either to the right or upward. Show on a tree diagram all the possible paths the rat may follow.

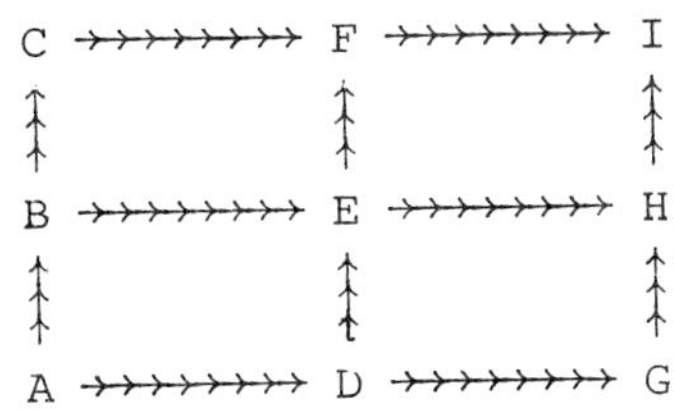

TEST A

Draw a tree diagram for each of the following (1 through 3).

1. All the ways of answering a test consisting of four true-false questions, assuming every question is answered.
2. All the ways you can buy a hand calculator from one of three companies, where the first and second companies each have two models and the third has four models.
3. All the ways of arranging four different books in a row on a shelf.

TEST B

Draw a tree diagram for each of the following (1 through 3).

1. All the ways to answer a test consisting of two multiple-choice questions, each with the three possible answers A, B, and C, assuming every question is answered.
2. All the ways of choosing a secretary and a treasurer from among Baker, Fromm, Kines, and Jackson.
3. All words, actual or nonsense, using just once each of the letters A, B, C, and D.

SECTION 3. THE MULTIPLICATIVE RULE

The multiplicative rule is a basic principle of counting that is actually an application of trees. It says:

> If a first event can happen in p ways, and if for each of these a second event can happen in q ways, then the total number of ways these two events can happen together is p times q.

Of course we can replace "event can happen" by "operation can be performed," "task can be done," "decision (or choice) can be made," and so on. Figure 3.1 shows how this rule can be derived from a tree diagram.

The tree in Figure 3.1 has two stages, one for each event. It is clear that if a sequence of three events can occur in p, q, and r ways respectively, the total number of ways all three events can happen in the order given is $p \times q \times r$. It is not hard to visualize a three-stage tree for such a situation, but trees are often more trouble than they are worth when the number of events or of the ways in which the events can occur is large. It is easy to see how the

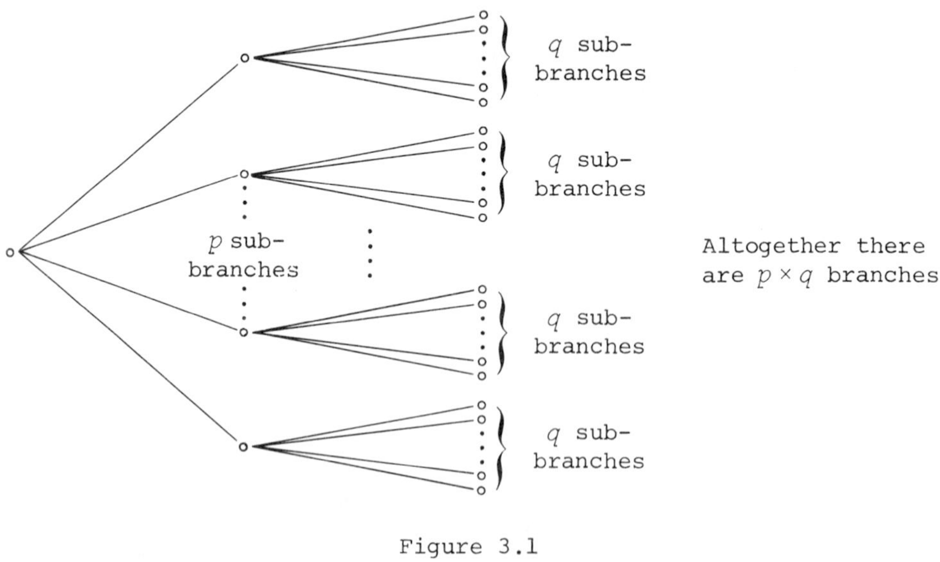

Figure 3.1

multiplicative rule can be extended to more than three events. The rule, illustrated in the examples that follow, is at the heart of the counting procedures discussed in the remainder of this chapter.

Example 3-1

If a clothing manufacturer makes shirts in three colors, in six neck sizes, and in five sleeve lengths, then the total number of different shirts offered is $3 \times 6 \times 5 = 90$.

Example 3-2

Johann Gregor Mendel (1822-1884), the founder of the science of genetics, noted the following characteristics of garden peas: the seeds are either smooth or wrinkled, and are also either yellow or green; the flowers are either red or white and either axial or terminal; the pods are either inflated or constricted and either green or yellow; and the stems are either short or long. So the number of distinguishable varieties is $2 \times 2 \times 2 \times 2 \times 2 \times 2 \times 2 = 128$.

DISCUSSION

In many applications, the numbers that are to be multiplied are not so obvious as are those in the preceding examples. In such cases we must first determine which events are to occur (or which decisions are to be made), then specify, for each, the number of alternatives involved.

Example 3-3

Four people enter a theater in which there are nine vacant seats. In how many different ways can they be seated?

As shown in the upper part of Figure 3.2, there are four decisions to be made, corresponding to the four questions stated in the boxes. The first person can choose a seat in nine different ways. The second person then has only eight alternatives available. Similarly, the third and fourth people have seven and six seats respec-

How many different seats can the first person choose from?	How many different seats can the second person choose from?	How many different seats can the third person choose from?	How many different seats can the fourth person choose from?
9	8	7	6

Figure 3.2

tively to choose from. In the lower part of the figure we place these answers in the boxes. So the four people can be seated in a total of $9 \times 8 \times 7 \times 6 = 3024$ different ways.

Example 3-4

Three company representatives and two union leaders are all scheduled to speak at a bargaining session. (a) In how many different orders can they speak? (b) In how many different orders can they speak if they must alternate, beginning with a company representative?

(a) How many choices are there for

first speaker?	second?	third?	fourth?	fifth?
5	4	3	2	1

The number of different orders is thus $5 \times 4 \times 3 \times 2 \times 1 = 120$.

(b) Note especially the difference in conditions between the preceding and this case. Here, how many choices are there for

first company speaker?	first union speaker?	second company speaker?	second union speaker?	third company speaker?
3	2	2	1	1

Thus here the number of different orders is $3 \times 2 \times 2 \times 1 \times 1 = 12$.

Example 3-5

Find the total number of six-character license plates where the first two characters are capital letters (from our 26-letter alphabet) and the remaining four are digits (0, 1, ..., 9), if (a) repetitions are permitted; and if (b) repetitions are not permitted.

	first letter	second letter	first digit	second digit	third digit	fourth digit
(a)	26	26	10	10	10	10
(b)	26	25	10	9	8	7

As (a) shows, the number of possible plates allowing repetitions is $26 \times 26 \times 10 \times 10 \times 10 \times 10 = 6,760,000$. But if repetitions are not allowed, then, as shown in (b), the number is much smaller: there are only $26 \times 25 \times 10 \times 9 \times 8 \times 7 = 3{,}276{,}000$ possible plates.

EXERCISES

1. If a woman owns ten blouses, eight skirts, and twelve pairs of shoes, and if she pays no attention to the dictates of fashion (except that her shoes always match), how many different outfits can she choose from?
2. If a company has three applicants for a junior executive position, seven applicants for a secretarial position, and five for a janitorial position, in how many ways can the three positions be filled?
3. In how many ways can a true-false examination with ten questions be answered, assuming all questions are answered?
4. In how many ways can a multiple-choice questionnaire be filled out if there are five questions each with three possible answers?
5. A survey asks four questions each with two possible answers, and three questions each with four possible answers. In how many different ways can the survey be answered?
6. (a) How many different four-letter radio-station names are possible if all 26 letters are available and letters may be repeated? (b) How many if repetitions are prohibited? (c) How many without repetitions if the first letter must be W?
7. Find the number of different six-character license plates if the first three characters must be letters and the last three must be digits, (a) without repetitions, and (b) if repetitions are allowed.
8. (a) In how many ways can two of six different books be chosen and arranged on a shelf? (b) Three out of six?
9. (a) If six people board a plane in which there are ten vacant seats, in how many ways can they be seated? (b) How about six people and only six seats?
10. (a) If seven companies bid on a project, in how many ways can the lowest, second lowest, and third lowest bids come in (assuming no information in advance)? (b) What if only three companies bid?

TEST A

1. A student must take four courses as follows: one of two available mathematics courses, one of four science courses, one of three humanities courses, and one of three social-science courses. How many different programs of study are possible?
2. How many ways are there to answer an examination consisting of ten questions if five are multiple-choice, each with four pos-

sible answers, and five are true-false questions? (Assume every question is answered.)

3. In how many different conceivable orders can five horses finish a race?
4. In how many ways can a board president and vice-president be chosen arbitrarily from twenty equally qualified board members?

TEST B

1. A manufacturer must choose one of three warehouses, one of two assembly plants, and one of five trucking firms. In how many ways can the combination of choices be made?
2. (a) How many three-letter secret codes can be made from the letters in the word CODES if repetitions are allowed? (b) If repetitions are not allowed?
3. A biologist is about to test five species of animals so she can rank them according to intelligence. How many different ways can she rank them *before* the test, assuming she has no preliminary information?

SECTION 4. PERMUTATIONS

A *permutation* of a set of objects is any arrangement of the objects in a definite linear order. For example, there are six permutations of the three letters A, B, C; they are:

ABC ACB BAC BCA CAB CBA

To count the number of permutations of a set of objects, we can draw a tree or use the multiplicative rule. In the example just given, the first position can be filled by any of three letters; there are then only two choices of letter for the second position (since one letter has already been used); the third and last position must then be filled by the one remaining letter. So the number of permutations of the three letters is

$$3 \times 2 \times 1 = 6$$

Example 4-1

In how many ways can a dispatcher send out five trucks, one to each of five warehouses?

The dispatcher has five choices for the first truck; four choices are then left for the second truck; and so on. The multiplicative rule yields

$$5 \times 4 \times 3 \times 2 \times 1 = 120 \text{ ways}$$

FACTORIAL NOTATION

The answer to Example 4-1 can also be written

$$5!$$

read "five factorial." So $5! = 5 \cdot 4 \cdot 3 \cdot 2 \cdot 1$. Note that

$$\begin{aligned} 1! &= 1 \\ 2! &= 2 \cdot 1 \\ 3! &= 3 \cdot 2 \cdot 1 \\ 4! &= 4 \cdot 3 \cdot 2 \cdot 1 \end{aligned}$$

In general, if n is a positive integer then $n!$ is the product of the first n positive integers:

$$n! = n \cdot (n-1) \cdot (n-2) \cdots 3 \cdot 2 \cdot 1$$

Note also that, for any n, $n! = n \times (n-1)!$:

$$
\begin{aligned}
3! &= 3\cdot 2\cdot 1 &&= 3\cdot 2!\\
7! &= 7\cdot 6\cdot 5\cdot 4\cdot 3\cdot 2\cdot 1 &&= 7\cdot 6!\\
n! &= n\cdot(n-1)\cdot(n-2)\cdots 3\cdot 2\cdot 1 &&= n\cdot(n-1)!
\end{aligned}
$$

If, in this last equation, we replace n by 1, we get

$$1! = 1\cdot 0!$$

--which is meaningless because 0! has not been defined. But in order to make this statement true we *define* zero factorial to be equal to one:

$$0! = 1$$

PERMUTATIONS OF n OBJECTS TAKEN n AT A TIME

The number of permutations of three letters using all three--that is, if they are "taken three at a time"--is 3!. Similarly, the number of permutations of four objects, taken four at a time, is

$$4! = 4\cdot 3\cdot 2\cdot 1 = 24$$

The latter is the answer to numerous questions like the following:

(a) How many ways can four different letters be arranged, without allowing repetitions, in four-letter "words"?
(b) In how many ways can we name a president, a vice-president, a secretary, and a treasurer from a group of four people?
(c) How many different four-digit dog tags can be formed from the set of digits {2, 3, 5, 9}, using each digit only once?

We can now generalize. The number of permutations (*ordered* arrangements) of n objects, taken n at a time, will be denoted by the symbol

$$P(n,n)$$

We see, then, that

$$P(n,n) = n! = n(n-1)(n-2)\cdots 3\cdot 2\cdot 1$$

Example 4-2

In how many different orders can the seven questions for a survey be arranged?

The answer is $P(7,7)$, or 7!, which equals $7\cdot 6\cdot 5\cdot 4\cdot 3\cdot 2\cdot 1$ or 5040.

PERMUTATIONS OF n OBJECTS TAKEN r AT A TIME

In the next two examples we are concerned with the permutations of a set of objects in which some of the objects are set aside and not used.

Example 4-3

In how many ways can a panel of judges award a blue ribbon, a red ribbon, and an honorable mention to ten presumably equally qualified chefs who are competing in an edible-art contest?

Any of the ten chefs is (theoretically) eligible for the blue ribbon; any of the remaining nine chefs can then get the red ribbon; and then there is an eight-way choice from the remaining chefs for the honorable mention. So there are $10 \cdot 9 \cdot 8 = 720$ ways of choosing three chefs out of ten for the awards. (This is in advance of the actual evaluation of the chefs' entries: the task faced by the panel is to eliminate all but one of these 720 different theoretically possible ways of making the awards.)

Example 4-4

Find the number of ways in which each of four bears can choose a private cave in which to hibernate, from among six caves that are equally attractive (to bears, that is).

The first bear has six choices. After it has chosen its cave, the second bear can choose from among the five remaining caves. The third then has four alternatives, and the fourth has three. The number of ways of choosing four caves out of six is obtained by multiplying:

$$6 \times 5 \times 4 \times 3 = 360$$

Notation and Discussion

There are several common ways to symbolize the number of permutations of n objects taken r at a time:

$$P(n,r) \qquad {}_nP_r \qquad P_{n,r}$$

In this book we shall use only the first of these.

To arrive at a formula for $P(n,r)$ we take another look at the last two examples:

Example 4-3	Example 4-4
(choosing among)	(choosing among)
10 chefs for 3 prizes	6 caves for 4 bears
$n = 10,\ r = 3$	$n = 6,\ r = 4$
The number of permutations of 10 things taken 3 at a time is $P(10,3) = 10 \cdot 9 \cdot 8$	The number of permutations of 6 things taken 4 at a time is $P(6,4) = 6 \cdot 5 \cdot 4 \cdot 3$
$= 10 \cdot 9 \cdot 8 \cdot \frac{7 \cdot 6 \cdot 5 \cdot 4 \cdot 3 \cdot 2 \cdot 1}{7 \cdot 6 \cdot 5 \cdot 4 \cdot 3 \cdot 2 \cdot 1}$	$= 6 \cdot 5 \cdot 4 \cdot 3 \cdot \frac{2 \cdot 1}{2 \cdot 1}$
So $P(10,3) = \frac{10!}{7!} = \frac{10!}{(10-3)!}$	So $P(6,4) = \frac{6!}{2!} = \frac{6!}{(6-4)!}$

In Example 4-3 we multiplied by 1 in the form $\frac{7\cdot6\cdot5\cdot4\cdot3\cdot2\cdot1}{7\cdot6\cdot5\cdot4\cdot3\cdot2\cdot1}$ to obtain 10! in the numerator; in Example 4-4 we multiplied by 1 in the form $\frac{2\cdot1}{2\cdot1}$ to obtain 6! in the numerator.

Here is the general principle suggested by these examples: The number of permutations of n objects taken r at a time is

$$P(n,r) = \frac{n!}{(n-r)!}$$

It is very important to note that we find $P(n,r)$, using this formula, only when

(1) the n objects under consideration are all different;
(2) repetitions are not allowed;
(3) the order of the objects matters.

A study of the following examples suggests how to compute $P(n,r)$:

$P(20,5) = 20\cdot19\cdot18\cdot17\cdot16$; $P(20,5)$ has 5 factors
$P(9,6) = 9\cdot8\cdot7\cdot6\cdot5\cdot4$; $P(9,6)$ has 6 factors
$P(12,9) = 12\cdot11\cdot10\cdot9\cdot8\cdot7\cdot6\cdot5\cdot4$; $P(12,9)$ has 9 factors

We see that in a computation of $P(n,r)$ we write down factors starting with n and decreasing by 1 until we have r factors altogether, thus:

$$\begin{aligned} P(n,r) &= \overbrace{n(n-1)(n-2)\cdots[n-(r-1)]}^{r \text{ factors}} \\ &= n(n-1)(n-2)\cdots(n-r+1) \end{aligned}$$

It is easy to show that the formula for $P(n,n)$ is a special case of that for $P(n,r)$ when $r = n$. Recall first that we have defined 0! to be equal to one (page 22). Now, since

$$P(n,r) = \frac{n!}{(n-r)!}$$

we have

$$P(n,n) = \frac{n!}{(n-n)!} = \frac{n!}{0!} = n!$$

This shows us another reason why it is convenient to define 0! as we have.

Example 4-5

A company plans to hire two new trainees, one in finance, the other in advertising. Ten presumably equally qualified college graduates apply for the positions. In how many possible ways can the choices be made?

Since repetitions are not allowed (that is, the same candidate cannot be given both jobs) and order is important (choosing Smith for finance and Jones for advertising is different from choosing Jones for finance and Smith for advertising), we conclude that the problem is one of permutations. In the conditions stated we have the permutations of ten objects taken two at a time:

$$P(10,2) = \frac{10!}{8!} = 10 \cdot 9 = 90$$

Example 4-6

Suppose ten horses are running in the first race and eight are entered in the second. With payoffs for first, second, and third places in each race, how many different results are possible?

In the first race the number of possible finishes is

$$P(10,3) = \frac{10!}{7!} = 10 \cdot 9 \cdot 8 = 720$$

Similarly, in the second race the number of possible finishes is

$$P(8,3) = \frac{8!}{5!} = 8 \cdot 7 \cdot 6 = 336$$

We now apply the multiplicative rule, since for *each* of the possible results in the first race there are 336 possible results in the second. So there are $720 \times 336 = 241{,}920$ different overall results.

Example 4-7

If ten people are running for the positions of first and second vice-president, five others are running for treasurer, and a third group, numbering twelve, are running for the positions of first, second, and third secretary, how many different outcomes are possible?

The final answer again calls for the multiplicative rule:

$$P(10,2) \times P(5,1) \times P(12,3) = \frac{10!}{8!} \times 5 \times \frac{12!}{9!} = 90 \times 5 \times 1320 = 594{,}000$$

Example 4-8

A town has been notified by the state government that its new automobile license plates may bear either three different letters, chosen from our 26-letter alphabet, or four different digits, from the set $\{0,1,2,3,4,5,6,7,8,9\}$. If there are 20,000 licensed drivers in town, will there be enough license plates to go around?

There are $P(26,3)$ licenses involving letters and $P(10,4)$ involving digits. Since a plate has either letters or digits, but not both, the total number is found by adding:

$$P(26,3) + P(10,4) = \frac{26!}{23!} + \frac{10!}{6!} = 26 \cdot 25 \cdot 24 + 10 \cdot 9 \cdot 8 \cdot 7$$

which comes out to 15,600 + 5040 = 20,640. The answer to the question posed is "Probably not!" While not every licensed driver has a car, some undoubtedly own two or more, and businesses may even own several trucks, and so on.

EXERCISES

1. Evaluate (a) 6! (b) 7! (c) 8! (d) $\frac{20!}{16!}$ (e) $\frac{10!}{5!5!}$
2. Evaluate (a) $P(6,6)$ (b) $P(10,1)$ (c) $P(47,1)$ (d) $\frac{P(8,3)}{3!}$ (e) $\frac{P(10,4)}{4!}$
3. Evaluate (a) $\frac{n!}{(n-1)!}$ (b) $\frac{n!}{(n-2)!}$ (c) $\frac{n(n-1)!}{n!}$ (d) $\frac{n!}{0!}$
4. Show that (a) $9 \cdot 8 \cdot 7 = \frac{9!}{6!}$ (b) $12 \cdot 11 \cdot 10 \cdot 9 \cdot 8 = \frac{12!}{7!}$ (c) $n(n-1)(n-2)\cdots(n-r+1) = \frac{n!}{(n-r)!}$
5. If there are four lots on which to build new supermarkets, in how many different ways can four grocery chains split up the lots, one per grocery chain?
6. There are six new areas to be considered for shopping plazas. In how many ways can six survey teams choose one each to evaluate?
7. In how many ways can a farmer plant five crops in five different fields, one crop to a field?
8. In how many ways can two couples choose one house each from among five houses up for sale?
9. In how many possible ways can four investment companies choose to support one invention apiece from among nine which have been submitted for consideration?
10. How many three-letter code-words can be made from the letters of the English alphabet if repetitions are not allowed?
11. A team of big-game hunters has captured five lions and six tigers. They have agreed to send one lion to each of two zoos, in Chicago and in New York respectively, and to send one tiger to each of three zoos, in Washington, Miami, and San Diego. How many different options are available to the hunters?
12. A baseball scout must select and then rank the best three out of ten infielders, the best four out of fifteen outfielders, and the best two out of five pitchers. How many possible ways can he do this?
13. For a particular assignment, a professional taster will have the option either of choosing the three best-tasting wines from a group of seven and then ranking them, or of selecting the three most flavorful beers from a group of six and then ranking them.

How many possibilities are there altogether?

14. A company is buying enough advertising space in a newspaper either to list in order five of its best products, to be selected from a group of 20, or to list in order three of its most popular services, to be chosen from among 10 different available services. How many different ads are possible?

TEST A

1. As part of a test of relative pain relief, identical-looking capsules of aspirin, acetaminophen, demerol, and a placebo are given one to each of four cancer patients. How many ways can this be done?
2. In how many ways can a national book club choose a winter, spring, summer, and fall offering from a set of ten books?
3. A designer is choosing carpeting and drapes for both a living room and a dining room from a group of seven carpet designs and five drapery patterns, no one of the latter matching any of the former. If he decides that no design or pattern is to be used in more than one room, how many choices does he have altogether?

TEST B

1. In how many different orders can five dignitaries be seated in a row of five chairs on a podium?
2. In how many ways can a theater-goer indicate his first, second, and third choices of dates to see a ballet which will be playing for seven consecutive evenings?
3. To open a bank safe, the correct sequence of three different numbers must be dialed on each of two combination locks. If one of the locks goes from 0 to 49 and the other from 0 to 99, how many different unlocking sequences are possible?

SECTION 5. COMBINATIONS

In §4 we noted (page 21) that a permutation of a set of objects is an *ordered* arrangement of the objects. A *combination* is a selection rather than an arrangement; in a combination *order does not matter*. If, for example, we consider subsets of two letters taken from the set of three letters {A, B, C}, we have

PERMUTATIONS	COMBINATIONS
AB, BA	AB
AC, CA	AC
BC, CB	BC

There are six permutations of the set of three letters taken two at a time; there are only three combinations.

A combination of elements drawn from some set is simply *a subset of that set*. A permutation of elements drawn from some set is a subset of that set *with its elements in a specific order*. When we are listing the elements of a combination (or of a set, or of a subset) the order in which we list them is irrelevant; thus {A,B} and {B,A} are the same combination (or subset). But in listing the elements of a permutation the order counts: thus AB and BA are different permutations.

Permutations and combinations are closely related, and both are important in the study of probability. In this section we will see in more detail how they are related and how they differ.

COMBINATIONS VERSUS PERMUTATIONS

The following examples are designed to show how permutations and combinations differ. The answer to each question below is a number: in the column at the left it is the number of permutations, whereas at the right it is the number of combinations.

In §4 (page 23) we introduced the notation "$P(n,r)$" to denote the number of permutations of n objects taken r at a time. For the number of combinations of n objects taken r at a time several notations are common:

$$C(n,r) \qquad {}_nC_r \qquad C_{n,r} \qquad \binom{n}{r}$$

We will use the first of these, uniform in style with our notation for the number of permutations.

Examples

5-1

(a) In how many ways can a president and a vice-president be named from a group of 10 people?
Ans: $P(10,2)$.

(b) In how many ways can a two-person committee be selected from a group of 10 people?
Ans: $C(10,2)$.

5-2

(a) In how many ways can a judge award first, second, and third prizes to three artists out of a competing group of twenty?
Ans: $P(20,3)$.

(b) In how many ways can a judge award prizes of $25 each to three artists out of a competing group of twenty?
Ans: $C(20,3)$.

5-3

(a) A company wants to advertise four of its seven products, one each of TV, radio, newspaper, and magazine. In how many different ways can this be done?
Ans: $P(7,4)$.

(b) A company can afford to advertise four of its seven products. How many different selections can be made?
Ans: $C(7,4)$.

5-4

(a) A TV network has ten soap operas from which to choose shows to broadcast at noon, 2 PM, and 3 PM. How many different schedules are possible?
Ans: $P(10,3)$

(b) In how many ways can a TV network choose three of ten available soap operas to show during the afternoon?
Ans: $C(10,3)$.

FINDING $C(n,r)$

In Example 5-1 (b), suppose that one of the two-person committees consists of Kirk and Young. This single combination "supplies" two different permutations for Example 5-1 (a): Kirk as president and Young as vice-president; and Young as president, Kirk as vice-president. The same is true of *every* two-person committee selected from the group of ten. In fact,

$$P(10,2) = C(10,2) \times 2!$$

where the factor 2! "acts to permute" each two-person committee.

Similarly, to obtain the answer to Example 5-2 (a) from that to 5-2 (b), we multiply by 3! (the "permuting" factor). Thus:

$$P(20,3) = C(20,3) \times 3!$$

This equation is correct because there are 3! different ways of assigning first, second, and third prizes to an unordered set (a combination) of three artists, say {Anna, Beth, Carol}.

In Examples 5-3 and 5-4 we conclude similarly that

$$P(7,4) = C(7,4) \times 4!$$
$$P(10,3) = C(10,3) \times 3!$$

From these examples we infer that to obtain $P(n,r)$, the number of permutations of n objects taken r at a time, we can: FIRST, find $C(n,r)$, the number of (unordered) combinations of n objects taken r at a time; and THEN multiply that by $r!$:

$$P(n,r) = C(n,r) \times r!$$

From this it follows that

$$C(n,r) = \frac{P(n,r)}{r!} = \frac{n!}{r!\,(n-r)!}$$

where, in the second step, we simply replace $P(n,r)$ by the formula already obtained for it in §4 (page 24).

We now use the formula for $C(n,r)$ to evaluate the answers to the four examples in the right-hand column, 5-1 through 5-4:

5-1. The number of two-person committees that can be selected from a group of ten, or, equivalently, the number of combinations of ten (people) taken two at a time, is

$$C(10,2) = \frac{10!}{2!8!} = \frac{10\cdot 9\cdot \cancel{8!}}{2!\cancel{8!}} = \boxed{\frac{10\cdot 9}{2\cdot 1}} = 45$$

5-2. The number of ways of awarding prizes of \$25 each to three artists chosen from a group of twenty, or, equivalently, the number of combinations of twenty (artists) taken three at a time, is

$$C(20,3) = \frac{20!}{3!17!} = \frac{20\cdot 19\cdot 18\cdot \cancel{17!}}{3!\cancel{17!}} = \boxed{\frac{20\cdot 19\cdot 18}{3\cdot 2\cdot 1}} = 1140$$

5-3. The number of selections of four products from seven, or, equivalently, the number of combinations of seven (products) taken four at a time, is

$$C(7,4) = \frac{7!}{4!3!} = \frac{7\cdot 6\cdot 5\cdot 4\cdot \cancel{3!}}{4!\cancel{3!}} = \boxed{\frac{7\cdot 6\cdot 5\cdot 4}{4\cdot 3\cdot 2\cdot 1}} = 35$$

5-4. The number of ways of choosing three TV shows out of ten, or, equivalently, the number of combinations of ten (shows) taken three at a time, is

$$C(10,3) = \frac{10!}{3!7!} = \frac{10\cdot 9\cdot 8\cdot \cancel{7!}}{3!\cancel{7!}} = \boxed{\frac{10\cdot 9\cdot 8}{3\cdot 2\cdot 1}} = 120$$

EVALUATING $C(n,r)$

Look again at the fractions enclosed in boxes above:

$$C(10,2) = \frac{\overbrace{10\cdot 9}^{2\text{ factors}}}{\underbrace{2\cdot 1}_{2!}} \qquad C(20,3) = \frac{\overbrace{20\cdot 19\cdot 18}^{3\text{ factors}}}{\underbrace{3\cdot 2\cdot 1}_{3!}}$$

$$C(7,4) = \frac{\overbrace{7\cdot 6\cdot 5\cdot 4}^{4\text{ factors}}}{\underbrace{4\cdot 3\cdot 2\cdot 1}_{4!}} \qquad C(10,3) = \frac{\overbrace{10\cdot 9\cdot 8}^{3\text{ factors}}}{\underbrace{3\cdot 2\cdot 1}_{3!}}$$

We see that to evaluate $C(n,r)$ we write down a fraction whose numerator is a product starting with n and with successive factors decreasing by one until there are r factors altogether, and whose denominator is $r!$. Thus, for example

$$C(11,4) = \frac{11\cdot 10\cdot 9\cdot 8}{4\cdot 3\cdot 2\cdot 1} = 330 \qquad C(13,5) = \frac{13\cdot 12\cdot 11\cdot 10\cdot 9}{5\cdot 4\cdot 3\cdot 2\cdot 1} = 1287$$

From this computational rule we can verify the following:

$$C(7,5) = \frac{7\cdot 6\cdot \cancel{5\cdot 4\cdot 3}}{\cancel{5\cdot 4\cdot 3}\cdot 2\cdot 1} = \frac{7\cdot 6}{2\cdot 1} = C(7,2)$$

$$C(10,6) = \frac{10\cdot 9\cdot 8\cdot 7\cdot \cancel{6\cdot 5}}{\cancel{6\cdot 5}\cdot 4\cdot 3\cdot 2\cdot 1} = \frac{10\cdot 9\cdot 8\cdot 7}{4\cdot 3\cdot 2\cdot 1} = C(10,4)$$

and, in general,

$$C(n,r) = C(n,n-r)$$

This formula simplifies the computation of $C(n,r)$ when r is larger than half of n. For example

$$C(15,12) = C(15,3) = \frac{15\cdot 14\cdot 13}{3\cdot 2\cdot 1} = 455$$

$$C(100,98) = C(100,2) = \frac{100\cdot 99}{2\cdot 1} = 4950$$

$$C(8,7) = C(8,1) = \frac{8}{1} = 8$$

The following is a plausible explanation for the equality of, say, $C(7,5)$ and $C(7,2)$. There are $C(7,5)$ different five-person committees that can be formed from a group of seven people. Each such five-person committee leaves a different two-person "residue" (consisting of the two not on that particular five-person committee). So there are just exactly as many two-person residues as there are five-person committees, showing that $C(7,5) = C(7,2)$.

Example 5-5

How many ways can a committee be formed with two representatives from five fraternity members and two representatives from ten dormitory residents?

The two fraternity representatives can be chosen in $C(5,2)$ ways and the two dorm representatives in $C(10,2)$ ways. Since the multiplicative rule prevails here, the final answer is

$$C(5,2) \times C(10,2) = \frac{5\cdot 4}{2\cdot 1} \times \frac{10\cdot 9}{2\cdot 1} = 450$$

Example 5-6

A bin contains ten good calculators and four defective ones. How many ways can three be chosen at random (that is, all equally likely to be selected) so that (a) all three are good? (b) all three are defective? (c) exactly one is defective?

(a) We can choose three of the ten good calculators in

$$C(10,3) = \frac{10\cdot 9\cdot 8}{3\cdot 2\cdot 1}$$

or 120 ways.

(b) We can choose three of the four defective calculators in

$$C(4,3) = C(4,1) = \frac{4}{1} = 4$$

ways.

(c) If, when choosing three calculators, exactly one is defective, then the other two must be good. The answer is therefore

$$C(4,1) \times C(10,2) = \frac{4}{1} \times \frac{10\cdot 9}{2\cdot 1} = 4 \times 45 = 180$$

Example 5-7

A company can afford to hire either two executives, for which there are five applicants, or four junior executives, for which there are seven applicants. How many different decisions are possible?

This is an "either-or" situation similar to that dealt with in Example 4-8 (page 25). The multiplicative rule does not apply. Instead, we add:

$$C(5,2) + C(7,4) = C(5,2) + C(7,3) = \frac{5\cdot 4}{2\cdot 1} + \frac{7\cdot 6\cdot 5}{3\cdot 2\cdot 1} = 10 + 35 = 45$$

Example 5-8

In how many ways can three grades of A, five of B, ten of C, and two of F be assigned arbitrarily to a class of twenty students?

Three of the twenty students can be chosen for A's in $C(20,3)$ ways.

Of the remaining seventeen students, five can be chosen for B's in $C(17,5)$ ways.

Then ten of the remaining twelve students can be selected for C's in $C(12,10)$ ways.

Finally, the two remaining students can be given F's in $C(2,2)$ or only one way.

To get the answer we of course multiply:

$$C(20,3) \times C(17,5) \times C(12,10) \times C(2,2)$$

$$= \frac{20 \cdot 19 \cdot 18}{3!} \times \frac{17 \cdot 16 \cdot 15 \cdot 14 \cdot 13}{5!} \times \frac{12 \cdot 11 \cdot 10 \cdot 9 \cdot 8 \cdot 7 \cdot 6 \cdot 5 \cdot 4 \cdot 3}{10!} \times \frac{2 \cdot 1}{2!}$$

But this can be written more elegantly by retaining factorials:

$$\frac{20!}{3!5!10!2!}$$

Example 5-9

In how many ways can exactly five heads appear in a toss of nine coins?

Each of the nine coins shows heads or tails. We want the number of ways in which five of the nine coins show heads. The answer is $C(9,5)$.

Note that whenever there are exactly five heads showing when nine coins have been tossed, there are also exactly four tails. This is a reaffirmation that $C(9,5) = C(9,4)$.

EXERCISES

1. Evaluate (a) $C(8,2)$ (b) $C(6,1)$ (c) $C(10,7)$ (d) $C(4,4)$
2. Evaluate $C(n,n)$ and justify your answer.
3. Prove that $C(n,r) = C(n,n-r)$ using the formula on page 30.
4. (a) How many combinations are there of the letters in the set {P, Q, R, S, T} taken three at a time? (b) How many combinations are there of these five letters taken two at a time? (c) List the combinations in (a) and in (b) matching them up in pairs so that each pair exhausts the five-letter set; for example, we have the three-letter combination {Q, S, T} in (a) to match with the two-letter combination {P, R} in (b).
5. (a) How many ways can two of 50 employees be chosen for equal merit raises? (b) What if one employee is to get a $500 bonus, the other only a $300 bonus?
6. How many different poker hands are possible? That is, how many ways can a hand of five cards be dealt from a deck of 52 cards?
7. (a) How many ways can three winners of 15-inch color TV sets be chosen in a sweepstakes contest in which 1000 people participate? (b) How many ways if the prizes are a grandfather clock, a dishwasher, and a stereo set?
8. (a) How many different bridge hands are possible? That is, how

many ways can a thirteen-card hand be dealt from a deck of fifty-two cards? (b) How many thirteen-card bridge hands have no clubs?

9. How many ways can a committee of three faculty members and four administrators be selected from a pool of 30 faculty members and ten administrators?
10. How many ways can a panel of coaches honor three football players, three basketball players, and three baseball players at a sports banquet if the selections are to be made from starting teams of 22 football players, five basketball players, and nine baseball players.
11. An investor is considering two options: he can buy three of five secure dividend-producing stocks or five of ten risky but high-growth-potential stocks. How many different decisions are possible?
12. A health official is trying to find the cause of a fast-spreading epidemic. During the first day of his investigation he has enough time to check only two out of five possible bacterial sources or one of three possible viral sources. How many different ways can he proceed?
13. In how many ways can an instructor arbitrarily assign 20 S's and five U's to a class of 25 students who are taking the course on the satisfactory-unsatisfactory basis?
14. If ten coins are tossed, in how many ways can exactly seven of them show heads?
15. How many ways can raises be given to ten employees if two are to get 7% raises, three to get 6% raises, and five to get 4%?
16. How many ways can 100 students be assigned to four homerooms if each is to have 25 students?
17. How many ways can the word "NUMBER" be misspelled using all its letters?
18. In a congressional district seven candidates are running for two seats in the House of Representatives. (a) How many different possible election outcomes are there? (b) How many ways can a person vote for two, one, or no candidates?
19. A brokerage firm advertises positions for four programmers, two stenotypists, and three junior executives. In an attempt to comply with affirmative-action practices, the firm wants to hire only women as programmers, only men as stenotypists, but either sex as junior executives. If 12 women and eight men apply and all are qualified for all three types of position, in how many ways can the positions be filled?
20. (a) There are 14 teams in the National Conference of the National Football League. (i) How many games must be played during a season if each pair of teams in this Conference plays exactly once? (ii) How many games must each team play? (b) There are 28 teams in the National Football League. Answer (i) and (ii) above replacing "Conference" by "League."

21. To study sibling interaction, a psychologist is working with a family of six children. (a) If she decides to observe each pair separately for an hour-long session, how many hours must she schedule? (b) Suppose two of the children are twins and she wishes to spend an extra hour with any pair that includes one or both twins. How long will it take in all?
22. Show that for Example 5-8 (page 32) you get the same answer if you assign the grades in a different order; for example, first the five B's, then the two F's, then the three A's, and finally the ten C's.

TEST A

1. In how many ways can three of 20 products be chosen for a special promotional campaign?
2. In how many ways can four of ten classical selections and three of ten pop pieces be chosen for a new record album?
3. In how many ways can 12 new cars be sent, three to each of four distributors?

TEST B

1. In how many ways can two of ten patients be chosen to try out an experimental cancer drug?
2. In how many ways can three of eight Whites and three of seven Blacks be selected to serve on a biracial committee dealing with inner-city problems?
3. The housing office on a college campus needs to assign six freshmen to three two-person rooms. How many different options are available?

SUMMARY OF §§1-5

You should now be able to

(1) count using Venn diagrams;
(2) count using tree diagrams;
(3) count using the multiplicative rule;
(4) understand and use factorial notation;
(5) determine when order matters and when it does not matter in a counting problem;
(6) distinguish between permutations and combinations of a set of objects;
(7) use appropriate formulas to evaluate the number of permutations or of combinations of a set of objects;
(8) apply counting procedures correctly in applications where repetitions may or may not be permitted;
(9) use addition (rather than multiplication) in appropriate "either-or" situations.

REVIEW EXERCISES

1. Interviews with 180 men in their sixties shown that 130 had high blood pressure and 100 consumed alcoholic beverages. If 30 men of those interviewed neither had high blood pressure nor used alcohol, how many of those with high blood pressure also used alcohol?
2. A survey of 800 college freshmen on their extracurricular activities in high school revealed that

   180 had participated in dramatic activities;
   260 had played or sung in musical groups;
   45 had participated in both drama and music;
   50 had been in sports and music;
   125 had been involved in drama but in neither music nor sports;
   20 had participated in all three activities;
   190 had not participated in any of the three activities.

   Use a Venn diagram to answer the following questions:
   (a) How many of the freshmen surveyed had participated in sports but in neither drama nor music?
   (b) How many had been involved in both drama and sports?
3. (a) Draw a tree diagram that shows all the possible three-digit numbers that can be formed from the digits {0, 2, 8} without repetitions. How many such numbers are there?
   (b) How many numbers are there if repetitions are permitted?
4. (a) There are four different prizes in a drawing, won by York, Burns, Allen, and Wertz. Only York and Wertz are eligible for first prize. Draw a tree to show the different ways the prizes can be awarded.
   (b) How many ways are there if any winner may win any prize?
   (c) How many ways are there if all four prizes are identical?
5. (a) How many ways can four Congressmen be chosen for a committee from a group of 12?
   (b) How many ways if the Speaker of the House is one of the 12 and must be a member of the committee?
6. How many ways can three Congressmen and two Senators be selected from a group of six Congressmen and four Senators?
7. The fifth edition of Grove's *Dictionary of Music and Musicians* is in ten volumes. How many ways can they been arranged on a shelf so that they are out of order?
8. An ordinary die is tossed four times.
   (a) How many different sequences are possible?

(b) In how many of these sequences does exactly one "3" show?
(c) In how many sequences does no "3" show?

9. The United States Postal Service Zip Code is a five-digit number in which each digit is chosen from the set $\{0,1,2,3,4,5,6,7,8,9\}$, with repetitions permitted.
(a) How many different Zip Codes are possible?
(b) The first three digits of a Zip Code identify a major city or distribution center. How many different Zip Codes are available altogether within a major city or within the area served by a distribution center?

10. A basketball team has five players: one center, two guards, and two forwards. If 15 people try out for the team, four of whom can play only center, six only guard, and the rest only forward, in how many different ways can a team be formed?

11. (a) How many ways can two Aces, two Kings, and three Queens be chosen from an ordinary deck of 52 cards (that is, one with four of each face card and four Aces)?
(b) How many different 13-card hands dealt from a standard deck contain two Aces, two Kings, and two Queens?

12. In how many ways can you leave a tip for a waiter if you have only four coins: a dime, a quarter, a half-dollar, and an Anthony dollar?

13. A ten-question test has seven true-false questions and three four-way multiple-choice questions.
(a) How many different sets of answers are there?
(b) How many of the ways of answering yield a perfect score on the test?
(c) How many sets of answers yield a score of zero?

14. For a dinner in a Chinese restaurant, you may select one of two soups, either three dishes from the eight in column A or two dishes from the six in column B, and one of three desserts. How many different dinners are available?

15. At our favorite pizzeria we can order a plain pizza or one with as many of the following additional ingredients as we want: mushrooms, pepperoni, black olives, onions, peppers, bacon.
(a) How many different kinds of pizza are available?
(b) Express the answer to (a) in terms of combinations.
(c) Express the answer to (a) as a power of two, then justify this form of the answer.

CHAPTER TEST

1. In a group of 415 sophomores:

   217 are taking accounting;
   170 are taking mathematics;
   131 are taking biology;
   48 are taking math and biology;
   25 are taking accounting and biology;
   82 are taking math and accounting;
   10 are taking all three subjects.

   Draw a Venn diagram, then answer the following questions:
   (a) How many in the group are taking none of the three subjects?
   (b) How many are taking accounting but neither math nor biology?
2. A just-married couple who would like to have two daughters agree that they will continue to have children until two girls are born or until they have four children. Draw a tree diagram and tell how many different outcomes are possible. (Assume no multiple births.)
3. Draw a tree diagram that shows all possible outcomes in tossing three coins.
   (a) How many of these outcomes show two or more tails?
   (b) Express the answer to (a) in terms of combinations.
4. Patients admitted to a hospital are classified according to sex, blood type (there are eight types), age (under 20, between 20 and 50, over 50), and racial extraction (White, Black, Far Eastern, and Other). How many different classes are there in all?
5. How many ways can a chairperson, a secretary, and a treasurer be chosen from a group of eleven people?
6. A student must sign up for five courses: two of six humanities courses, one of four in science, and two of seven electives. How many different options are available?
7. A telephone area code has three digits. The first and last digits are selected from the set {2,3,4,5,6,7,8,9}; the middle digit is either 0 or 1. How many different area codes are possible?
8. (a) How many ways can you choose three of five books to read next month?
   (b) How many ways can you arrange three of five books on a shelf?
9. (a) How many ways can you choose a five-person committee from a group of ten men and six women?
   (b) How many ways are there if the five-person committee is to

consist of either all women or all men?

10. In a class on public speaking, an instructor may choose either three of ten students to give 15-minute presentations or four of eight students to give ten-minute presentations. How many choices does the instructor have?

Chapter 2

Probability

A *random experiment* is one that can be repeated and which has a definite outcome each time it is performed, but whose actual outcome on any particular performance cannot be predicted with certainty. Some examples of random experiments are the tossing of a coin, the rolling of a die, the birth of a baby (which can be of either sex), the attempt to catch a fish (which succeeds or fails), the administering of a test for tuberculosis (which comes out positive or negative).

Although we may not be able to predict the specific outcome of any particular performance of a random experiment, we can often make reliable predictions about the relative frequencies of different outcomes when the experiment is to be repeated a sufficient number of times. Our study of probability will involve just such repeated experiments.

HISTORICAL NOTE. The theory of probability had its origins in the seventeenth century when mathematical problems arising from games of chance were examined by the French mathematicians Blaise Pascal (1623-1662) and Pierre de Fermat (1601?-1665). It is interesting to note that many of the simple examples we still use today to illustrate the rules and properties of probability (cards, dice, roulette, coin tossing) are a survival from those times. But probability now plays an essential role in a wide range of physical, biological, and behavioral disciplines, from weather prediction and nuclear physics to life insurance and public-opinion polls.

§6. Basic concepts of probability.
§7. Equally likely outcomes
§8. Conditional probability
§9. Independent events
§10. Bayes' law
§11. Finite stochastic processes
§12. Binomial probabilities
§13. Expected value
Summary
Review exercises
Chapter test

SECTION 6. BASIC CONCEPTS OF PROBABILITY

In this section we present the essentials of probability theory. In particular, we define a sample space for an experiment, then discuss the probabilities of the various outcomes of the experiment.

INTUITIVE DEFINITION OF PROBABILITY

To specify the probability of an outcome of an experiment is to make a mathematical statement about the likelihood of occurrence of that outcome. Since "probability" and "likelihood" are near-synonyms, this definition is not only intuitive but also circular. Nevertheless, we would no doubt all agree that (1) if a "fair" coin is tossed, the probability that heads will show is $\frac{1}{2}$; that (2) if a "fair" die is rolled, the probability that a five will show (on the uppermost face) is $\frac{1}{6}$; and that (3) if a card is chosen "at random" from a deck of 52 cards, the probability that it will be the Queen of Diamonds is $\frac{1}{52}$. Our agreement in the first case is based on the conviction that if we toss a well-balanced coin (that is, a fair one) a good many times, about half of those times heads will show. Indeed, the more tosses, the more likely that the relative occurrence, or frequency, of heads will be close to $\frac{1}{2}$ (otherwise we might wonder whether the coin is really fair). Similar statements can be made about experiments (2) and (3).

We will very shortly lay down some rules that must be followed in assigning probabilities to the outcome of an experiment. These will make the definition of probability precise.

SAMPLE SPACE

A set S of all possible outcomes of an experiment is called a *sample space* for the experiment.

For example, if the experiment consists of tossing a coin to see which face shows, the sample space is {heads, tails} (or {H,T}). If the experiment is choosing a family from the set of all families with three children and we are interested in the number of boys, an appropriate sample space is the set {0,1,2,3}. In the first case, the sample space has two outcomes; in the second, there are four outcomes.

An *event* is any subset of S; in particular, an event containing only one element of S, that is, a single possible outcome of the experiment, is called a *simple event*.

In this text we consider only finite sample spaces. A *finite* set has n elements, where n is a positive integer.

It is often possible to define many different sample spaces for the same experiment. The choice depends on the purpose of the experiment and on the specific events in which we are interested. In the examples that follow, note that each outcome corresponds to exactly one element in the sample space.

Example 6-1

Suppose we roll a die. If we are interested in the number that shows on the uppermost face, then the set $S_1 = \{1,2,3,4,5,6\}$ is the appropriate sample space; it has six simple events. If we are interested in whether the number showing is even or odd, then an appropriate sample space is $S_2 = \{\text{even, odd}\}$, which has two simple events.

Note that $S_3 = \{<5, \geqslant 5\}$, where "$<5$" denotes one of the numbers 1, 2, 3, or 4 and "$\geqslant 5$" denotes 5 or 6, is also a possible sample space for rolling a die. However, neither of the following two sets is possible: $\{1,3,4,6\}$; $\{<5, \geqslant 2\}$. (Why not?)

Example 6-2

Suppose we draw a card from a well-shuffled deck. A standard deck of cards has four suits (S = spades, H = hearts, D = diamonds, C = clubs), and each suit has thirteen cards (A = ace, K = king, Q = queen, J = jack, 10, 9, ..., 2). If we are interested in the precise card chosen, then

$$S_1 = \{SA,SK,SQ,\ldots,S4,S3,S2,HA,\ldots,H2,DA,\ldots,D2,CA,\ldots,C2\}$$

is a sample space, with 52 simple events.

If we let E_1 be the event that the card drawn is a 3 or a 5, then

$$E_1 = \{S3,H3,D3,C3,S5,H5,D5,C5\}$$

Suppose E_2 is the event that the card drawn is a spade or a king. Then

$$E_2 = \{SA,SK,SQ,\ldots,S3,S2,HK,DK,CK\}$$

Note that event E_2 "happens" if any of (only) 16 different outcomes occurs; we were careful not to count the outcome SK (king of spades) twice.

If our interest when drawing a card is only in the suit, then a suitable sample space is

$$S_2 = \{S,H,D,C\}$$

Example 6-3

One of four applicants is to be selected for a new position. Two possible sample spaces are

$$S_1 = \{\text{person } 1, \text{ person } 2, \text{ person } 3, \text{ person } 4\}$$
$$S_2 = \{\text{male, female}\}$$

The personnel manager may be more interested in S_1, while a government investigator concerned with fair employment practices may be more interested in S_2.

PROBABILITY ASSIGNMENT: RULES

Given a sample space S for an experiment, with n simple events, we assign to each simple event e_i in S a number $p(e_i)$, called its *probability*, such that

$$(1) \quad 0 \leqslant p(e_i) \leqslant 1$$
$$(2) \quad p(e_1) + p(e_2) + \ldots + p(e_n) = 1$$

Note that (1) says that the probability of each simple event is not less than zero and not greater than one; and (2) says that the sum of the probabilities of all the simple events is one.

If an outcome is certain to occur, we say that its probability is one. If an outcome is certain *not* to occur, we say it has probability zero. If, for example, we are interested in the heights of a group of collegiate basketball players, we might use the sample space $S = \{e_1, e_2\}$, where e_1 is the outcome that a player is less than ten feet tall, e_2 the outcome that a player is ten feet tall or taller. Then

$$p(e_1) = 1 \quad \text{and} \quad p(e_2) = 0$$

For a specified sample space, any probability assignment that satisfies rules (1) and (2) is *valid*; any that violates either (1) or (2) is not valid.

Example 6-4

Let $S = \{e_1, e_2, e_3, e_4\}$ be a sample space. Then

$$p(e_1) = p(e_2) = p(e_3) = p(e_4) = \tfrac{1}{4}$$

is a valid probability assignment. So is each of the following:

$$p(e_1) = \tfrac{1}{2} \quad p(e_2) = 0 \quad p(e_3) = \tfrac{1}{6} \quad p(e_4) = \tfrac{1}{3}$$
$$p(e_1) = 1 \quad p(e_2) = 0 \quad p(e_3) = 0 \quad p(e_4) = 0$$

But none of the following is a valid probability assignment:

$$p(e_1) = \tfrac{1}{4} \quad p(e_2) = \tfrac{1}{4} \quad p(e_3) = \tfrac{1}{4} \quad p(e_4) = \tfrac{1}{5}$$

$$p(e_1) = \tfrac{1}{4} \quad p(e_2) = 0 \quad p(e_3) = \tfrac{1}{3} \quad p(e_4) = 0$$

$$p(e_1) = \tfrac{1}{2} \quad p(e_2) = \tfrac{3}{4} \quad p(e_3) = -\tfrac{1}{8} \quad p(e_4) = -\tfrac{1}{3}$$

(Why not?)

PROBABILITY ASSIGNMENT: JUSTIFICATION

We would of course like to assign probabilities to simple events in a meaningful and nonarbitrary way. If an experiment is repeated many times, then probability can be given the "relative frequency" interpretation mentioned earlier (page 42). The toss of a fair coin or the roll of a fair die is an example of such an experiment. If a gold-mining company that has explored a good many sites has made a worthwhile find in roughly one of every three sites examined, then it may be reasonable to say that the probability that a similar site will pay off is $\frac{1}{3}$.

However, we are sometimes called on to assign probabilities to events which cannot be repeated exactly, or that occur only once, such as that a given student will pass a course, that a specific baseball team will win next year's World Series, or that there will be enough natural snow for the next Winter Olympics. Such nonexperimental probabilities, which are frequently based on personal convictions, will generally not concern us here.

It is important to realize that the assigning of probabilities is not a trivial task. Still, there are a few basic guidelines which should be carefully followed. If, for instance, two outcomes are equally likely to occur (for example, heads and tails on the toss of a fair coin), then we assign them the same probability. Also, if one outcome, judging from experience, is more likely to occur than another, then we assign a greater probability to the first than to the second.

Example 6-5

Roll a fair die and observe the number on the uppermost face. Here $S = \{1,2,3,4,5,6\}$. Since the six outcomes are equally likely and the probabilities must sum to one, we assign to each outcome the probability $\frac{1}{6}$.

Example 6-6

Given a class of 100 students from which a student is to be chosen at random, we assign the probability $\frac{1}{100}$ to each of the 100 equally likely outcomes.

In §7 we will consider in further detail experiments all of whose outcomes have the same probability.

Example 6-7

Consider a sample space $S = \{e_1, e_2, e_3\}$. Suppose we know that $p(e_2) = 2p(e_1)$ and that $p(e_3) = \frac{1}{3}p(e_1)$. What are the probability assignments?

Letting $p(e_1) = x$, we have $p(e_2) = 2x$ and $p(e_3) = \frac{1}{3}x$. Since the probabilities must add up to one, we have

$$x + 2x + \tfrac{1}{3}x = 1$$

from which $x = \frac{3}{10}$. So

$$p(e_1) = \tfrac{3}{10} \qquad p(e_2) = 2(\tfrac{3}{10}) = \tfrac{3}{5} \qquad p(e_3) = \tfrac{1}{3}(\tfrac{3}{10}) = \tfrac{1}{10}$$

Example 6-8

A store manager notes that if he keeps five checkout lines open, a customer is twice as likely to find an empty line as not to find one. What are the respective probabilities?

If x and y are the probabilities respectively of finding and of not finding an empty line, then $x = 2y$ and $x + y = 1$. We solve simultaneously:

$$2y + y = 1$$

so

$$y = \tfrac{1}{3} \qquad x = \tfrac{2}{3}$$

PROBABILITY OF AN EVENT

If an event E is the union of two or more simple events in a sample space S, then the probability of E, $p(E)$, is the sum of the probabilities of those simple events. It follows that $p(E)$ is also between zero and one inclusive. For completeness, we note that since the empty set $\emptyset$ is a subset of every set (see Appendix 1), it is an event for *every* sample space. Since the set $\emptyset$ has no elements, we say that the event $\emptyset$ is *impossible*. Also, since S is the union of all the simple events, one of the outcomes in S MUST occur. So we have

$$p(\emptyset) = 0 \quad \text{and} \quad p(S) = 1$$

Example 6-9

Suppose there are five horses, A, B, C, D, and E, in a race and that the sample space for the race is $\{a,b,c,d,e\}$, where a is the event that horse A wins, and so on. Assume that there are no ties and that

$$p(a) = \tfrac{1}{5} \quad p(b) = \tfrac{1}{3} \quad p(c) = \tfrac{2}{15} \quad p(d) = \tfrac{1}{6} \quad p(e) = \tfrac{1}{6}$$

Let W be the event that A, B, or C wins. Then $W = \{a,b,c\}$. Find $p(W)$.

We have $p(W) = p(a) + p(b) + p(c)$; so

$$p(W) = \frac{1}{5} + \frac{1}{3} + \frac{2}{15} = \frac{2}{3}$$

PROBABILITY OF A UNION OF EVENTS

At the end of Appendix 1 we state the following theorem for the number of elements in the union of two sets A and B:

$$n(A \cup B) = n(A) + n(B) - n(A \cap B)$$

This theorem enables us to compute the probability of the union of two events. To find $p(X \cup Y)$, where X and Y are events of a given sample space, we consider two cases.

CASE 1. $X \cap Y = \emptyset$. Events X and Y are *mutually exclusive*; that is, they cannot occur simultaneously. Here $X \cup Y$ consists of the simple events in X together with the simple events in Y. The events X and Y have no simple events in common. Therefore

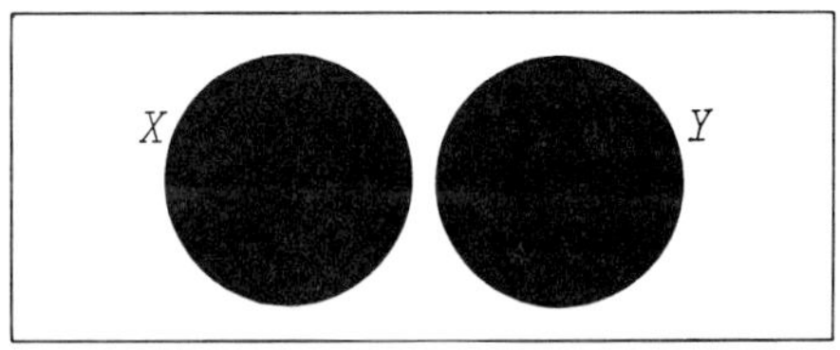

$$p(X \cup Y) = p(X) + p(Y)$$

Example 6-9 above illustrates this case since we assumed no ties in the horse race.

CASE 2. $X \cap Y \neq \emptyset$. In this case the events X and Y share one or more simple events. Here $p(X) + p(Y)$ is too large an answer, since it "counts" the probability of any shared event twice. The figure below answers the question "How much too large?" Since the proba-

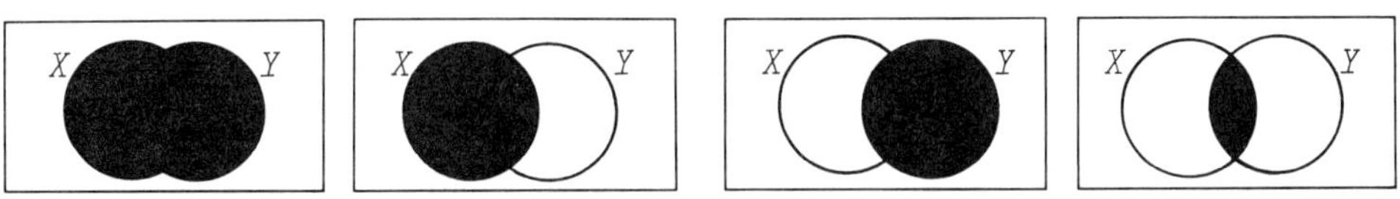

$p(X \cup Y)$ = $p(X)$ + $p(Y)$ $p(X \cap Y)$

bility $p(X \cap Y)$ of the intersection $X \cap Y$ is counted *twice* when we add $p(X)$ and $p(Y)$, our formula subtracts $p(X \cap Y)$ to yield the correct result:

$$p(X \cup Y) = p(X) + p(Y) - p(X \cap Y)$$

This, known as the *addition formula*, actually holds for *any* pair of events X and Y. For if X and Y are mutually exclusive (as in Case 1), then $X \cap Y = \emptyset$ and $p(X \cap Y) = 0$.

Note especially that if events X and Y are not mutually exclusive --that is, if they share one or more simple events--then the word "or," as in "in X or Y," will mean "in X or in Y or in both." This is precisely the definition of the union of the two sets, $X \cup Y$.

Example 6-10

Suppose $X \cap Y = \emptyset$, $p(X) = \frac{1}{4}$, $p(Y) = \frac{1}{3}$. Evaluate $p(X \cup Y)$.
Since $X \cap Y = \emptyset$, we have

$$p(X \cup Y) = p(X) + p(Y) = \frac{1}{4} + \frac{1}{3} = \frac{7}{12}$$

Example 6-11

The probability that a construction company will be awarded a certain contract is $\frac{1}{4}$, the probability that it will be awarded a second contract is $\frac{1}{5}$, and the probability that it will get both contracts is $\frac{1}{10}$. Find the probability that the company will be awarded at least one of the two contracts.
We have

$$p(X \cup Y) = p(X) + p(Y) - p(X \cap Y) = \frac{1}{4} + \frac{1}{5} - \frac{1}{10} = \frac{7}{20}$$

PROBABILITY OF A COMPLEMENTARY EVENT

The complement E' of a set E is the set of all outcomes in the sample space S which do not belong to E (Appendix 1). Since $E \cup E' = S$ and $E \cap E' = \emptyset$ (see the figure) we note that $p(E \cup E') = 1$ and that $p(E \cap E') = 0$. Therefore the addition formula given above yields

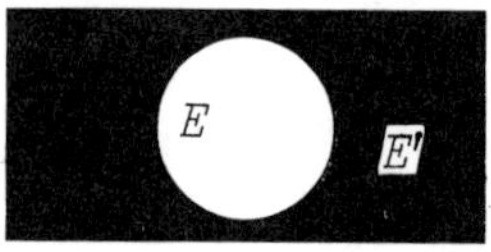

$$p(E \cup E') = p(E) + p(E') - p(E \cap E')$$
$$1 = p(E) + p(E') - 0$$

so that

$$p(E') = 1 - p(E)$$

Example 6-12

The probability that a baby born in the United States will be a boy is 0.512. Find the probability that a newborn baby will be a girl.
Since the events "girl baby" and "boy baby" are complementary, the probability of a girl is $1 - 0.512 = 0.488$.

VENN DIAGRAMS FOR PROBABILITY

In Chapter 1 we illustrated the usefulness of the Venn diagram as an organizational tool in problems of counting. It is often helpful to enter known numerical probabilities directly on a Venn diagram, as in the following example.

Example 6-13

Given that $p(A) = \frac{1}{2}$, $p(B) = \frac{1}{3}$, and $p(A \cap B) = \frac{1}{12}$, find $p(A \cap B')$, $p(A' \cap B)$, and $p(A \cup B)$.

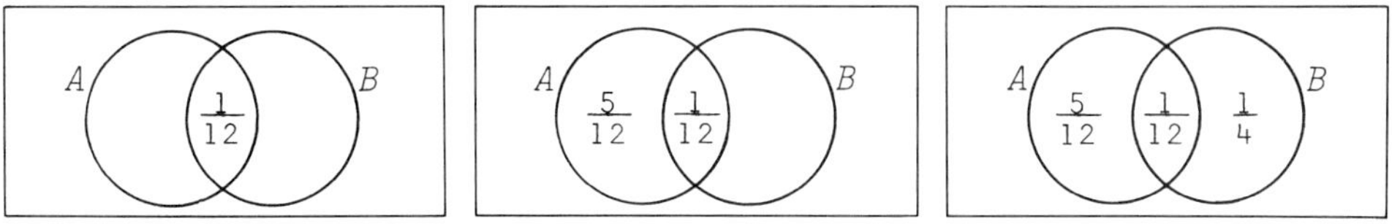

$p(A \cap B) = \frac{1}{12}$ $\qquad p(A \cap B') = \frac{1}{2} - \frac{1}{12} = \frac{5}{12}$ $\qquad p(A' \cap B) = \frac{1}{3} - \frac{1}{12} = \frac{1}{4}$

In the figure we proceed as we did in counting problems using Venn diagrams in §1, but now recording not the numbers of elements but the probabilities. Note the order in which we enter the given data on the diagrams. We see from the figure that

$$p(A \cup B) = \frac{5}{12} + \frac{1}{12} + \frac{1}{4} = \frac{9}{12} = \frac{3}{4}$$

We can check by using the given data and the addition formula:

$$p(A \cup B) = p(A) + p(B) - p(A \cap B) = \frac{1}{2} + \frac{1}{3} - \frac{1}{12} = \frac{9}{12} = \frac{3}{4}$$

EXERCISES

1. Consider the experiment of tossing a coin three times. (a) Describe a suitable sample space if one is concerned only with how many heads come up. (b) Describe a suitable sample space if one's concern is the order in which heads and tails appear.
2. Consider choosing at random a family that has three children. (a) Describe a suitable sample space if one is concerned with the number of boys. (b) Describe a suitable sample space if one is concerned with the order of male and female births.
3. Redo Exercise 1 with four tosses of the coin instead of three.
4. Redo Exercise 2 with families of four children instead of three.
5. Consider rolling a pair of dice. Describe two suitable sample spaces: (a) supposing one's concern is only the sum of the spots on the uppermost faces; (b) supposing one's concern is the ordered pair of numbers showing respectively on the first and the second die.
6. An experiment consists of tossing a coin and then rolling a die. Describe a sample space for the possible results. (For a multistage experiment like this one, drawing a tree diagram--or just visualizing one--is often helpful.)
7. A sample space for an experiment has three simple events, e_1, e_2, and e_3. Which rows in the following table show valid probability assignments?

ROW	$p(e_1)$	$p(e_2)$	$p(e_3)$	ROW	$p(e_1)$	$p(e_2)$	$p(e_3)$
1	0	1	0	5	0	2	-1
2	$\frac{1}{4}$	$\frac{1}{3}$	$\frac{1}{2}$	6	$\frac{1}{4}$	$\frac{1}{4}$	$\frac{1}{4}$
3	$\frac{1}{3}$	$\frac{1}{3}$	$\frac{1}{3}$	7	$\frac{1}{2}$	$\frac{3}{4}$	$-\frac{1}{4}$
4	$\frac{1}{2}$	$\frac{1}{3}$	$\frac{1}{6}$	8	$\frac{12}{35}$	$\frac{1}{7}$	$\frac{1}{5}$

8. Suppose a sample space consists of three events. Given that $p(e_1) = 3p(e_2)$ and that $p(e_3) = 2p(e_2)$, find the probabilities $p(e_1)$, $p(e_2)$, and $p(e_3)$.
9. A sample space consists of four simple events. Find the probability of each if

$$\text{(a)}\quad p(e_1) = 2p(e_2) = 3p(e_3) = 4p(e_4)$$
$$\text{(b)}\quad p(e_1) = \tfrac{1}{2}p(e_2) = \tfrac{1}{3}p(e_3) = \tfrac{1}{4}p(e_4)$$

10. Suppose we toss a weighted coin for which heads is three times as likely to show as tails. What is the probability of heads? of tails?
11. Suppose we roll a weighted die for which the 2-spot is twice as likely to land face up as is the 1-spot, the 3-spot three times as likely as the 1-spot, and so on. Find the probabilities of obtaining a 1, a 2, a 3, a 4, a 5, and a 6.
12. Suppose a fair die is rolled--that is, one for which all of the six outcomes are equally likely. What is the probability of obtaining: (a) a 4? (b) an even number? (c) an odd number? (d) a number less than 5? (e) a number greater than 6? (f) a number between 1 and 6 inclusive?
13. Find $p(X \cup Y)$ given that $p(X) = \frac{1}{10}$, $p(Y) = \frac{1}{5}$, and $X \cap Y = \emptyset$.
14. Find $p(X \cup Y)$ given that $p(X) = \frac{1}{2}$, $p(Y) = \frac{1}{4}$, and $p(X \cap Y) = \frac{1}{8}$.
15. Find $p(X)$ given that $p(X \cup Y) = \frac{3}{4}$, $p(Y) = \frac{1}{2}$, and $p(X \cap Y) = \frac{1}{8}$.
16. Find $p(X \cap Y)$ given that $p(X \cup Y) = \frac{1}{2}$, $p(X) = \frac{1}{3}$, and $p(Y) = \frac{1}{4}$.
17. Find $p(X')$ given that $p(X) = \frac{3}{4}$.
18. Find $p[(X \cup Y)']$ given that $p(X) = \frac{1}{7}$, $p(Y) = \frac{1}{14}$, and $X \cap Y = \emptyset$.
19. To be in good standing at a certain college a freshman must pass a mathematics or an English course during the first semester. If the probability of passing math is $\frac{3}{4}$, of passing English is $\frac{7}{8}$, and of passing both is $\frac{11}{16}$, what is the probability that the freshman will be in good standing?
20. It is known that the probability of striking oil in a certain region on land is 0.3, while the probability of striking oil in a certain region at sea is 0.2. If a company spokesman says that the probability of striking oil in both places is 0.06, what does that imply as to the probability of striking oil in at least one of the two places?
21. During a tropical storm that passes near the coast of Florida, a weather forecaster asserts that the probability that winds in excess of 60 mph will hit the coast is 0.5 and that the probability of a tsunami ("tidal wave") over ten feet is 0.2. What is the probability of both high wind and tsunami if the probability of having at least one of the two is 0.6?

22. The manager of a certain restaurant knows that nine out of ten customers will order dinner, five out of ten will order a cocktail, and nine out of twenty will order both a cocktail and dinner. What is the probability that someone will order dinner but no cocktail? a cocktail but no dinner? neither a cocktail nor dinner?
23. As noted in §2 (page 13), color blindness is a sex-linked genetic trait. Suppose that 6% of a certain population is color-blind, 5.7% both male and color-blind, and 51.3% male or color-blind (or both). What percentage of the population is male? female? female and color-blind? male but not color-blind? (Hint: Use a Venn diagram.)

TEST A

1. Consider an impending vote for or against each of three bond issues. Describe two sample spaces: (a) with reference to how many issues pass; (b) with reference to which issues pass and which do not.
2. Suppose a sample space consists of three events. Find the probabilities, given that $p(e_1) = 2p(e_2)$ and $p(e_2) = \frac{1}{2}p(e_3)$.
3. Find $p[(X \cap Y)']$, given that $p(X \cup Y) = 0.5$, $p(X) = 0.3$, and $p(Y) = 0.4$.
4. Suppose that the probability that the sun will shine in the morning on St. Croix, Virgin Islands, is 0.9, that it will shine in the afternoon is 0.95, and that it will shine both morning and afternoon is 0.88. (a) What is the probability that the sun will shine in the morning or the afternoon (or both)? (b) in the morning but not the afternoon? (c) in neither morning nor afternoon?

TEST B

1. Republican and Democratic candidates are seeking each of two elective offices. Describe two sample spaces: (a) each event is a pair of numbers, indicating respectively the number of Republican and of Democratic winners; (b) each event is the winning of a particular office by a specific party.
2. In a tennis match one player is thought to be $1\frac{1}{2}$ times as likely to win as the other. What are their probabilities of winning?
3. Find $p(Y)$ if $p(X \cup Y) = 0.75$, $p(X \cap Y) = 0.4$, and $p(X) = 0.5$.
4. For an experiment involving a parachute jump, the probability that the main chute will not open is 0.01, that the reserve chute will not open is 0.05, and that neither will open is 0.0005. What is the probability that at least one chute will open? that the main chute will fail the reserve one open?

SECTION 7. EQUALLY LIKELY OUTCOMES

In this section we apply the principles outlined in §6 to the special case in which the outcomes of an experiment are all equally likely.

EQUIPROBABLE SAMPLE SPACES

In §6 we constructed several sample spaces in which the simple events were equally likely; for example, for the toss of a *fair* coin,

$$S_1 = \{H,T\}$$

for the roll of a *fair* die,

$$S_2 = \{1,2,3,4,5,6\}$$

or, for the drawing of a card *at random* from a full deck,

$$S_3 = \{AS,\ldots,2S,AH,\ldots,2H,AD,\ldots,2D,AC,\ldots,2C\}$$

The words "fair" and "at random" are in italics above precisely to emphasize that in each of the three experiments any outcome is as likely as any other.

Since the sum of probabilities of the simple events in any sample space must equal one, it is clear that in S_1 we assign probability $\frac{1}{2}$ to each simple event, that we assign probability $\frac{1}{6}$ to each element in S_2, and that each element in S_3 is assigned probability $\frac{1}{52}$. Such sample spaces are called *equiprobable*.

In general, if an equiprobable sample space S has n simple events then the probability of each is $\frac{1}{n}$.

Furthermore: if an event E in an equiprobable sample space S consists of m simple events, then the probability of E is

$$\overbrace{\frac{1}{n}+\frac{1}{n}+\ldots+\frac{1}{n}}^{m\text{ addends}} = \frac{m}{n}$$

If we regard any outcome in event E as "favorable," then we have the rule

$$p(E) = \frac{\text{no. of elements in } E}{\text{no. of elements in } S} = \frac{\text{no. of favorable outcomes}}{\text{total no. of (possible) outcomes}}$$

Example 7-1

Of 20 mice in a cage, 12 are female. If one mouse chosen at random is to be used in an experiment, what is the probability that it will be female? male?

Since the mice are equally likely to be chosen, the probability of choosing a particular one is $\frac{1}{20}$, the probability of choosing a female is $\frac{12}{20} = \frac{3}{5}$, and the probability of choosing a male is $\frac{8}{20} = \frac{2}{5}$.

Example 7-2

Four cards are to be dealt from a well-shuffled deck. What is the probability that they will all be aces?

We choose as the sample space the set of all four-card "hands" that can be dealt from a deck of 52 cards. The number of such different four-card hands is

$$C(52,4) = \frac{52!}{48!4!} = 270{,}725$$

Since a hand of four aces is only one simple event in this equiprobable space, the probability of getting four aces is $\frac{1}{270{,}725}$.

Choosing Equiprobable Sample Spaces

When considering a sample space we must be careful in deciding whether the simple events are equally likely. For example, suppose we are interested in the number of heads that appear when we toss a nickel and a dime. Here are two possible sample spaces:

(1) $S_1 = \{0,1,2\}$, where each simple event is the number of heads.

(2) $S_2 = \{HH, HT, TH, TT\}$, where each simple event is the sides showing respectively on the nickel and on the dime (see the figure).

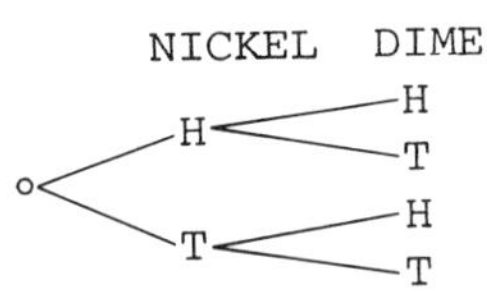

$= \{HH, HT, TH, TT\}$

Which of these two sample spaces is equiprobable? Certainly not both, for then the probability of two heads (for example) would be both $\frac{1}{3}$ and $\frac{1}{4}$, which is impossible.

If the coins are fair then S_1 cannot be equiprobable, since one head is more likely to occur than two: two heads can occur in only one of four ways (both coins must show heads) whereas one head can be obtained in two of four ways.

Furthermore, if we actually toss a fair nickel and a fair dime a great many times we expect each of the four simple events of S_2 to occur in about $\frac{1}{4}$ of the tosses. Indeed, extensive actual experiments bear this out. So, empirically (that is, based on experiment), S_2 is the sample space to select for equally likely outcomes.

Another good experiment that shows the need for care in selecting a sample space is the rolling of a pair of dice, say one yellow and one red, noting the sum of the numbers showing. Here are two possible sample spaces:

(1) If we take each possible sum as a simple event we have

$$S_1 = \{2,3,4,\ldots,12\}$$

(2) Taking as simple events each possible ordered pair of numbers on the yellow and on the red die respectively, we have

$$\begin{aligned} S_2 = \{&(1,1),\ (1,2),\ (1,3),\ (1,4),\ (1,5),\ (1,6),\\ &(2,1),\ (2,2),\ (2,3),\ (2,4),\ (2,5),\ (2,6),\\ &(3,1),\ (3,2),\ (3,3),\ (3,4),\ (3,5),\ (3,6),\\ &(4,1),\ (4,2),\ (4,3),\ (4,4),\ (4,5),\ (4,6),\\ &(5,1),\ (5,2),\ (5,3),\ (5,4),\ (5,5),\ (5,6),\\ &(6,1),\ (6,2),\ (6,3),\ (6,4),\ (6,5),\ (6,6)\} \end{aligned}$$

So S_1 has 11 simple events, while S_2 has 36.

Again we ask: assuming the dice are fair, which sample space has equiprobable simple events? Again we answer: certainly not both! The probability of obtaining the sum 7, for instance, cannot be both $\frac{1}{11}$ and $\frac{6}{36} = \frac{1}{6}$.

We can see from the display for S_2 that there are six ways, out of a total of 36, to get the sum 7: (1,6), (2,5), (3,4), (4,3), (5,2), and (6,1). And extensive experimentation in rolling a pair of fair dice affords grounds for assigning equal probabilities to the 36 simple events in S_2. So S_2, not S_1, is equiprobable.

We can readily obtain probabilities of other events by using formulas from §6 for the probability of a union, a complement, and so forth, as shown in the following examples.

Example 7-3

A pair of fair dice is rolled. Find the probability (a) that the sum will be 9; (b) that the sum will be less than 11; (c) that the two dice will show the same number (a double).

For (a) we have

$$p(\text{sum is } 9) = p\{(3,6),\ (4,5),\ (5,4),\ (6,3)\} = \frac{1}{36}+\frac{1}{36}+\frac{1}{36}+\frac{1}{36} = \frac{1}{9}$$

For (b),

$$\begin{aligned} p(\text{sum} < 11) &= 1 - p(\text{sum is 11 or 12})\\ &= 1 - p\{(5,6),\ (6,5),\ (6,6)\}\\ &= 1 - \frac{3}{36} = \frac{33}{36} = \frac{11}{12} \end{aligned}$$

And for (c),

$$p(\text{a double}) = p\{(1,1),\ (2,2),\ (3,3),\ (4,4),\ (5,5),\ (6,6)\} = \frac{6}{36} = \frac{1}{6}$$

Example 7-4

Again we roll a pair of dice. Let X be the event "the sum is even" and Y the event "the sum is greater than 7." Find $p(X \cup Y)$.

$$\begin{aligned} X = \{&(1,1),\ (1,3),\ (1,5),\ (2,2),\ (2,4),\ (2,6),\\ &(3,1),\ (3,3),\ (3,5),\ (4,2),\ (4,4),\ (4,6),\\ &(5,1),\ (5,3),\ (5,5),\ (6,2),\ (6,4),\ (6,6)\} \end{aligned}$$

$$\begin{aligned} Y = \{&(2,6),\ (3,5),\ (3,6),\ (4,4),\ (4,5),\\ &(4,6),\ (5,3),\ (5,4),\ (5,5),\ (5,6),\\ &(6,2),\ (6,3),\ (6,4),\ (6,5),\ (6,6)\} \end{aligned}$$

$$\begin{aligned} X \cap Y = \{&(2,6),\ (3,5),\ (4,4),\ (4,6),\ (5,3),\\ &(5,5),\ (6,2),\ (6,4),\ (6,6)\} \end{aligned}$$

Since there are 36 equally likely simple events in the sample space, each has probability $\frac{1}{36}$. We use the addition formula:

$$\begin{aligned} p(X \cup Y) &= p(X) + p(Y) - p(X \cap Y)\\ &= \tfrac{18}{36} + \tfrac{15}{36} - \tfrac{9}{36} = \tfrac{24}{36} = \tfrac{2}{3} \end{aligned}$$

Example 7-5

Assume that a person chosen at random is equally likely to have been born in any month of the year. What is the probability that (a) a person chosen at random was born in August? (b) two people chosen at random were both born in August? (c) two people chosen at random were born in the same month? (d) the first person chosen of two was born in August, the second in February? (e) of two chosen at random, one was born in August and one in February?

(a) Assuming equiprobability, since there are 12 months in the year the probability that a person chosen at random was born in August is $\frac{1}{12}$.

(b) For two people chosen at random there are 144 equally likely possibilities (this follows from the multiplicative rule in §3, page 16). Since "both in August" is one of these 144 possibilities, the probability of this event is $\frac{1}{144}$.

(c) The event of having been born in the same month consists of 12 of the 144 possibilities (the 12 "favorable" outcomes); the probability is therefore $\frac{12}{144} = \frac{1}{12}$.

(d) That the first was born in August, the second in February, is, like both being born in August, a simple event, with probability $\frac{1}{144}$.

(e) The probability that one is August and one February is $\frac{1}{144} + \frac{1}{144}$ or $\frac{1}{72}$. (Why? Visualizing a tree diagram for this sample space may make your explanation easier.)

Example 7-6

Assume equal probabilities for boy babies (B) and girl babies (G).

If a three-child family is selected at random, what is the probability that it has (a) no boys? (b) exactly one boy? (c) exactly two boys? (d) three boys?

Note the similarity between this example and the illustration, given earlier, of tossing a nickel and a dime and being concerned only with the number of heads. So {0,1,2,3}, where the simple event is the number of boys, is *not* an equiprobable sample space. The equiprobable sample space heeds the order of birth and the sex of each: {BBB,BBG,BGB,BGG,GBB,GBG,GGB,GGG}. The probability (a) of no boys is $\frac{1}{8}$; (b) of exactly one boy is $\frac{3}{8}$; (c) of exactly two boys is $\frac{3}{8}$; and (d) of three boys is $\frac{1}{8}$.

USING COMBINATIONS TO EVALUATE PROBABILITIES

The examples that follow use the counting techniques of §3 and §5. In each problem we assume that the outcomes are all equally likely.

Example 7-7

A shipment of ten radios is known to contain four defective ones. If a sample of three is chosen at random, what is the probability that exactly two are defective?

From §5 we know that there are $C(10,3) = 120$ ways of choosing three radios from ten. There are $C(4,2) = 6$ ways of selecting two defective radios from four defective ones, and $C(6,1) = 6$ ways of choosing one nondefective radio from among six that are okay. We use the multiplicative rule to find the number of ways of selecting exactly two defective radios and one nondefective one:

$$C(4,2) \times C(6,1) = 6 \times 6 = 36$$

The required probability is therefore

$$\frac{C(4,2) \times C(6,1)}{C(10,3)} = \frac{36}{120} = \frac{3}{10}$$

Example 7-8

Suppose it is known that there are three defective light bulbs in a sample of 20. What is the probability that (a) a bulb chosen at random will be good? (b) two bulbs chosen at random will both be good? (c) three selected at random will all be good?

(a) The probability that a single one will be good is $\frac{17}{20}$.

(b) There are $C(20,2) = 190$ ways of selecting two bulbs out of 20, and $C(17,2) = 136$ ways of choosing two good bulbs out of 17. So the probability that two bulbs chosen at random will both be good is

$$\frac{136}{190} = \frac{68}{95}$$

(c) Similarly, the probability that three drawn at random from the sample of 20 will all be good is

$$\frac{C(17,3)}{C(20,3)} = \frac{34}{57}$$

Example 7-9

If a committee of six is chosen at random from a group consisting of five supervisors and ten employees, what is the probability that the committee will have (a) exactly four employees? (b) at least four employees?

(a) For there to be exactly four employees on the projected six-person committee, there must also be exactly two supervisors. The number of ways of doing this is $C(10,4) \times C(5,2) = 210 \times 10$. Since the number of ways of selecting six people from a group of 15 is $C(15,6) = 5005$, the answer is

$$\frac{C(10,4) \times C(5,2)}{C(15,6)} = \frac{210 \times 10}{5005} = \frac{60}{143}$$

(b) To find the probability that the committee will include at least four employees, we add the probabilities for four, five, and six employees:

$$\frac{60}{143} + \frac{C(10,5) \times C(5,1)}{C(15,6)} + \frac{C(10,6) \times C(5,0)}{C(15,6)} = \frac{102}{143}$$

Example 7-10

Eleven patients suffering from leukemia are being treated with a drug called cytoxan. Six of the patients have improved. If a medical researcher selects four of the 11 randomly in connection with a cancer study, what is the probability that (a) exactly two of those chosen will have improved? (b) at least one of them will have improved?

(a) The number of ways of selecting four patients from a group of 11 is $C(11,4)$. If two of the patients selected are to have improved then two must not have improved. The number of ways of selecting two of six (who have improved) and also two of five (who have not) is $C(6,2) \times C(5,2)$. The probability called for is therefore

$$\frac{C(6,2) \times C(5,2)}{C(11,4)} = \frac{15 \times 10}{330} = \frac{5}{11}$$

(b) The simple way to compute the probability that at least one of the four chosen will have improved is to find the probability of the complementary event:

$$p(\text{no. impr} \geqslant 1) = 1 - p(\text{no. impr} < 1) = 1 - p(\text{none impr})$$

Since the probability that no patient in the sample group will have improved is

$$\frac{C(6,0) \times C(5,4)}{C(11,4)} = \frac{1 \times 5}{330} = \frac{1}{66}$$

the answer is

$$1 - \frac{1}{66} = \frac{65}{66}$$

Of course, the following much lengthier calculation also yields the correct answer:

$$\frac{C(6,1) \times C(5,3)}{C(11,4)} + \frac{C(6,2) \times C(5,2)}{C(11,4)} + \frac{C(6,3) \times C(5,1)}{C(11,4)} + \frac{C(6,4) \times C(5,0)}{C(11,4)}$$

Justify the last statement.

General Rule for Equally Likely Outcomes

Note that the three examples just given illustrate, once again, the general principle stated at the beginning of this section:

> If the outcomes in a sample space are equally likely, then the probability of an event is the ratio of the number of ways it can occur to the total number of ways that are possible.

EXERCISES

1. For each of the following, describe, first, a sample space in which the simple events are equiprobable, and then a sample space for which the simple events are not equiprobable.
   (a) Tossing two fair coins.
   (b) Tossing three fair coins.
   (c) Tossing four fair coins.
   (d) Rolling a fair die.
   (e) Rolling a pair of fair dice.
   (f) Guessing at all questions on a four-question true-false test.
   (g) Guessing on a two-question multiple-choice test in which the first question has the possible answers A, B, and C and the second has the possible answers A and B.
2. For each of the sample spaces with equiprobable simple events in Exercise 1, give the probability of a simple event.
3. We roll a pair of fair dice and note the sum. Find the probability that the sum will be (a) 7; (b) 5, 6, or 7; (c) not 12; (d) not 7 or 11; (e) 13.
4. What is the probability of being dealt a specific bridge hand; that is, of being dealt a specified set of 13 cards out of a deck of 52?

5. Suppose a roulette wheel has the numbers 0, 1, 2, ..., 36, with 0 green, half the numbers from 1 to 36 red, the other half black. The wheel is spun and a ball is rolled. If the ball is as likely to land on any number as on any other, find the probability that it will land on (a) 25; (b) 0; (c) a number other than 0; (d) a red number; (e) a black number; (f) a number between 1 and 12 inclusive.
6. Suppose a newborn baby is as likely to be a girl as a boy. If a four-child family is chosen at random, what is the probability of (a) no girls? (b) exactly two girls? (c) at least one boy?
7. An inspection procedure calls for the random selection of a box of 12 units followed by the random choice of four units from the box. If all four are good, the whole load is approved. (a) Suppose the box chosen happens to contain two defective units and ten good ones, what is the probability that the load will be approved? (b) Suppose, instead, that the box chosen contains five defective units; what is the probability of approval?
8. Assume that people are equally likely to have been born on any of the seven days of the week. What is the probability that (a) a random choice will hit on a person born on Monday? (b) two people chosen at random will both have been born on Monday? (c) two people chosen at random will have been born on the same day of the week? (d) of two chosen at random, the first will have been born on Monday and the second on Tuesday? (e) of two chosen at random, one will have been born on Monday and the other on Tuesday?
9. In a town meeting of 20 people, 12 are Democrats and the rest Republicans. A committee of three is chosen at random. What is the probability that (a) all three will be Democrats? (b) all three will be Republicans?
10. Two senators are chosen at random from among the 100 in the Senate. What is the probability that (a) both are from New York? (b) neither is from New York? (c) exactly one is from New York? [Hint: solve (c) using your answers to (a) and (b).]
11. Suppose there are five defective rifles in a shipment of a dozen. If four are randomly chosen, what is the probability that (a) exactly two are defective? (b) at least one is defective?
12. A committee of five is selected at random from a group composed of seven House members and eight Senate members. What is the probability that (a) exactly three House members will be on the committee? (b) a majority of the committee will be from the Senate?
13. Three of ten qualified applicants for a job are women. If two of the applicants are selected at random, what is the probability that (a) both are women? (b) exactly one is a woman? (c) both are men?

14. Five city council members favor a particular bond issue, while the other three members are opposed. If three of the council members are interviewed at random, what is the probability that (a) a majority of those interviewed favor the issue? (b) a majority are opposed?
15. Six of nine mice have been trained to run a maze. If four of the nine are chosen at random, what is the probability that (a) all chosen can run the maze? (b) at least one can? (c) only one can?

TEST A

1. The numbers 1, 2, 3, and 4 are written on slips of paper. Two of these slips are drawn at random without replacement, and the sum of the numbers is noted.
   (a) Construct a sample space for which the simple events are equiprobable.
   (b) Find the probability that the sum will be two.
   (c) Find the probability that the sum will be five.
2. Suppose there are three green marbles and five red marbles in a box and two are randomly chosen. Find the probability that (a) both are green; (b) both are red.
3. A retail store has ten hand calculators on sale, three of which are programmable. If a teacher randomly picks two to purchase, what is the probability that at least one of those picked is programmable?

TEST B

1. A president and a vice-president are to be chosen as follows: the names of four candidates, Brown, Jones, Smith, and Fields, are placed in a hat; the first slip drawn determines the presidency; the second slip drawn determines the vice-presidency. (a) Construct a sample space for which the events are equiprobable. (b) Find the probability that Brown will become president and Smith vice-president. (c) Find the probability that Brown or Jones will be president and Fields vice-president.
2. In a shipment of ten radios, two are defective. If a distributor selects five at random, what is the probability that none are defective?
3. Six companies and four private individuals run boating tours to Buck Island. If three tourists each pick a different tour, what is the probability that exactly two will pick company tours?

SECTION 8. CONDITIONAL PROBABILITY

So far we have found the probability of an event relative to an entire sample space. In this section we are interested in the probability of an event given that another event has occurred. We will see that this may be viewed as finding the probability of an event relative to a subset of the original space, referred to as a *reduced* sample space.

ILLUSTRATIONS OF CONDITIONAL PROBABILITIES

Each illustration given below has two parts. In each, part (a) calls for a probability discussed in preceding sections, while part (b) calls for a conditional probability.

1. (a) The probability that a person chosen at random has a high cholesterol count. (b) The same, given that the person chosen has a high-fat-content diet.

2. (a) The probability that a person chosen at random has sickle-cell anemia. (b) The same, given that the person chosen is an American Black.

3. (a) The probability that a family chosen at random has an annual income over $30,000. (b) The same, with the proviso that the head of the household is under the age of 22.

4. (a) The probability that a letter chosen at random from the text of a volume of the *Encyclopaedia Britannica* is a U. (b) Ditto, provided the letter chosen is preceded by Q.

In illustrations 1, 2, and 4, the probability in (b), which is for a subpopulation, is known to be higher than that for the entire population defined for (a). In 3, however, the conditional probability is lower.

The probability of an event E, given that event F has occurred, is a *conditional probability*. We denote it by $p(E|F)$, which we read "p of E given F."

REDUCED SAMPLE SPACES

In examples 8-1 through 8-4 we show how to find a conditional probability by reducing the sample space to the event whose occurrence constitutes the condition.

Example 8-1

A pair of fair dice is rolled. What is the probability that the sum of the spots showing is seven, given that at least one die shows a three?

The equiprobable sample space for two fair dice has 36 simple events, as shown in the display on page 54. We reduce this to

$$S' = \{(1,3), (2,3), (3,3), (4,3), (5,3), (6,3), (3,1), (3,2), (3,4), (3,5), (3,6)\}$$

each of whose 11 elements is an outcome of the event "at least one die shows a three." There are two simple events in S' for which the sum of the spots is seven: (4,3) and (3,4). Since the original sample space was equiprobable, so is S'. The answer is therefore $\frac{2}{11}$.

Note that the probability of getting the sum seven in the original sample space is $\frac{6}{36} = \frac{1}{6}$.

Example 8-2

On a roll of a pair of fair dice, what is the probability that at least one die shows a three, given that the sum of the spots is seven?

Here the reduced sample space is

$$S'' = \{(1,6), (2,5), (3,4), (4,3), (5,2), (6,1)\}$$

each of whose six elements sums to seven. Of these six simple events, two have at least one die showing a three. So the answer is $\frac{2}{6} = \frac{1}{3}$.

Note that the probability of at least one die showing a three in the original sample space is $\frac{11}{36}$ (see, again, the display on page 54).

NOTE. In examples 8-1 and 8-2, let E = "the sum is seven" and F = "at least one die shows a three." Then in example 8-1 we have that $p(E|F) = \frac{2}{11}$, while in example 8-2 we see that $p(F|E) = \frac{1}{3}$. We conclude that

$$p(E|F) \neq p(F|E)$$

Example 8-3

Find the probability that a family chosen at random from all three-child families has two boys, given that it has at least one girl.

The original sample space is

$$S = \{BBB, BBG, BGB, BGG, GBB, GBG, GGB, GGG\}$$

But for this problem we reduce it to

$$S' = \{BBG, BGB, BGG, GBB, GBG, GGB, GGG\}$$

which is the set of seven types of three-child families with at least one girl. Since there are three simple events in S' with two boys and S' "inherits" equiprobability from S, the desired probability is $\frac{3}{7}$.

Example 8-4

Find the probability that a family chosen at random from among those with three children has at least one girl, given that it has exactly two boys.

We reduce the original sample space of the preceding example to {BBG,BGB,GBB} and note immediately that the answer is 1.

NOTE 1. We can also find conditional probabilities using trees, as we now show for examples 8-3 and 8-4.

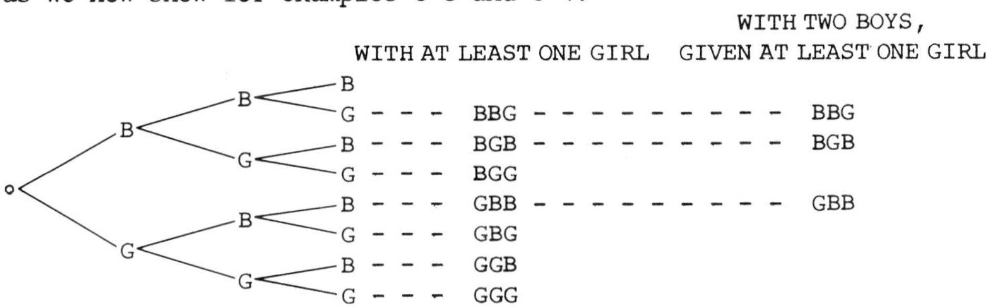

So (example 8-3) the probability that the family has two boys, given that it has at least one girl, is 3 out of 7, or $\frac{3}{7}$.

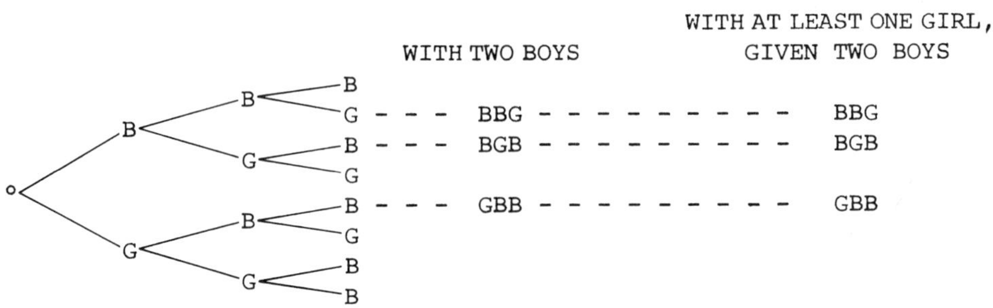

So (example 8-4) the probability that the family has at least one girl, given that it has two boys, is 3 out of 3 or 1.

NOTE 2. If in examples 8-3 and 8-4 we now let E be the event "has two boys" and F be "has at least one girl," we observe again that $p(E|F) \neq p(F|E)$.

In general, these two conditional probabilities are NOT *equal.*

Example 8-5

The table below gives the results of a survey of the drinking and smoking habits of 1200 college students. B = "drinks beer"; B' = "does not drink beer"; C = "smokes cigarettes"; C' = "does not smoke cigarettes."

	B	B'	TOTAL
C	315	165	480
C'	585	135	720
TOTAL	900	300	1200

A student is selected at random from this group. Find these probabilities: (a) $p(C)$; (b) $p(B)$; (c) $p(C \cap B)$; (d) $p(C|B)$; (e) $p(B|C)$.

(a) Of the 1200 students, 480 smoke cigarettes. So $p(C)$ = 480/1200 = 0.40.

(b) Since 900 of the 1200 students drink beer, $p(B)$ = 900/1200 = 0.75.

(c) Note that 315 students both smoke cigarettes and drink beer, out of the total of 1200. Thus $p(C \cap B)$ = 315/1200 = 0.26.

(d) $p(C|B)$ asks for the probability that a student smokes cigarettes, given that he or she drinks beer. Of 900 beer-drinking students, 315 also smoke. So $p(C|B)$ = 315/900 = 0.35.

(e) Finally, $p(B|C)$ = 315/480 = 0.66.

THE FORMULA FOR CONDITIONAL PROBABILITY

Refer again to the table above for Example 8-5 and to the computations for the probabilities in parts (d) and (e). Using $n(X)$ for the number of elements in set X, we have, for (d),

$$p(C|B) = \frac{n(C \cap B)}{n(B)} = \frac{315}{900} = \frac{\frac{315}{1200}}{\frac{900}{1200}} = \frac{p(C \cap B)}{p(B)}$$

and for (e),

$$p(B|C) = \frac{n(B \cap C)}{n(C)} = \frac{315}{480} = \frac{\frac{315}{1200}}{\frac{480}{1200}} = \frac{p(B \cap C)}{p(C)}$$

(Of course, $C \cap B$ and $B \cap C$ are precisely the same set.) In general, the probability that event E occurs, given that F has occurred, is

$$p(E|F) = \frac{p(E \cap F)}{p(F)}$$

The Venn diagram in Figure 8-1 also supports the reasonableness of this formula. The part of E that is within F is $E \cap F$. So asking for the probability that E occurs, given that F has occurred, is equivalent to asking what part of F does $E \cap F$ constitute. If we

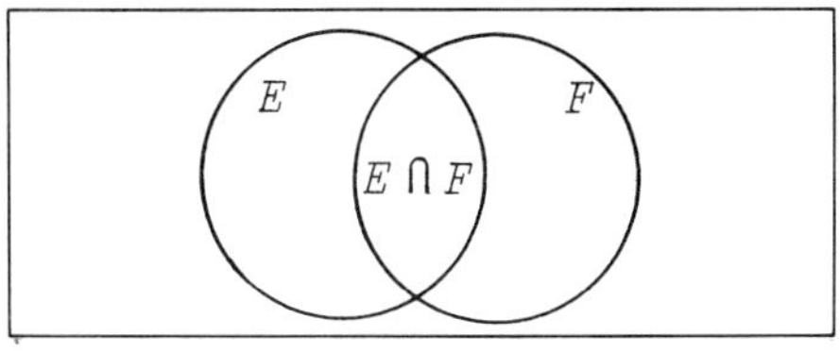

Figure 8.1

let S denote the entire sample space, we have

$$p(E|F) = \frac{n(E \cap F)}{n(F)} = \frac{\frac{n(E \cap F)}{n(S)}}{\frac{n(F)}{n(S)}} = \frac{p(E \cap F)}{p(F)}$$

Here we divided both numerator and denominator by $n(S)$, the number of simple events in the original sample space, to obtain the probabilities in the final fraction.

The centered equation above shows that, given an equiprobable sample space, we can find the conditional probability of E, given F, either by dividing the number of elements in the intersection of E and F by the number of elements in F or by dividing their probabilities. Of course, neither $n(F)$ nor $p(F)$ can equal zero.

Example 8-6

Suppose the probability that someone chosen at random from a population is an alcoholic is 0.05 and the probability that the individual both is an alcoholic and has liver disease is 0.035. What is the probability that a person chosen at random has liver disease, given that he or she is an alcoholic?

Let L = "has liver disease" and A = "is an alcoholic." Then

$$p(\mathrm{L}|\mathrm{A}) = \frac{p(\mathrm{L} \cap \mathrm{A})}{p(\mathrm{A})} = \frac{0.035}{0.05} = 0.70$$

Example 8-7

Suppose that, in addition to the information given in the preceding example, it is also known that 7% of the population has liver disease. Then what is the probability that someone chosen at random and found to have liver disease is also an alcoholic?

With the same symbols, we have

$$p(\mathrm{A}|\mathrm{L}) = \frac{p(\mathrm{L} \cap \mathrm{A})}{p(\mathrm{L})} = \frac{0.035}{0.07} = 0.5$$

Example 8-8

If 90% of the households in a certain region have TV sets and 50% have both TV sets and radios, what is the probability that a household chosen at random and found to have a television set also has a radio?

$$p(\text{radio}|\text{TV}) = \frac{p(\text{radio} \cap \text{TV})}{p(\text{TV})} = \frac{0.50}{0.90} = \frac{5}{9}$$

Example 8-9

If $\frac{5}{6}$ of all people have Rh-positive blood and $\frac{1}{3}$ of all people are both Rh-positive and type A, what is the probability that someone chcsen at random and found to be Rh-positive is also type A?

$$p(\text{A}|\text{Rh+}) = \frac{p(\text{A} \cap \text{Rh+})}{p(\text{Rh+})} = \frac{\frac{1}{3}}{\frac{5}{6}} = \frac{2}{5}$$

Example 8-10

A psychologist interested in right- versus left-handedness ("R" and "L") and in IQ scores collected the following statistics from a random population of 2000:

	R	L	TOTAL
IQ$>$135	190	10	200
IQ$\leqslant$135	1710	90	1800
TOTAL	1900	100	2000

A person is selected at random from this group. Find the following probabilities: (a) that the person is left-handed; (b) that the person's IQ exceeds 135; (c) that the person is both left-handed and has IQ greater than 135; (d) that the person has IQ over 135, given left-handedness; (e) that the person is left-handed, given an IQ in excess of 135.

(a) $p(\text{L}) = 100/2000 = 0.05$.
(b) $p(\text{IQ}>135) = 200/2000 = 0.10$.
(c) $p(\text{L} \cap \text{IQ}>135) = 10/2000 = 0.005$.
(d) $p(\text{IQ}>135 \mid \text{L}) = 10/100 = 0.10$.
(e) $p(\text{L} \mid \text{IQ}>135) = 10/200 = 0.05$.

PROBABILITY OF AN INTERSECTION

From the formulas for conditional probability we obtain, by algebra, formulas for the probability of the intersection of two events, as follows:

$$p(E|F) = \frac{p(E \cap F)}{p(F)} \quad \text{so} \quad p(E \cap F) = p(E|F) \cdot p(F)$$

$$p(F|E) = \frac{p(F \cap E)}{p(E)} \quad \text{so} \quad p(F \cap E) = p(F|E) \cdot p(E)$$

where we note, again, that $F \cap E = E \cap F$.

We now have two formulas for the probability that events E and F

both occur. In any application, we use whichever of the two is appropriate, as the next pair of examples show.

Example 8-11

Suppose the probability of failure of a main life-support system is $p(M) = 0.005$ and the probability of failure of a back-up system, given that the main system has failed, is $p(B|M) = 0.01$. What is the probability that both systems fail?

We want $p(M \cap B)$. Considering the information on hand, the convenient formula here is

$$p(B \cap M) = p(B|M) \cdot p(M)$$

so

$$p(M \cap B) = (0.01) \cdot (0.005) = 0.00005$$

Example 8-12

A vacuum-cleaner salesman estimates that 30% of those who are at home allow him in to demonstrate his product (A) and that 20% of those who let him in buy the cleaner (B). If he finds a family at home, what is the probability that they will both let him in and buy a cleaner?

We have $p(A) = 0.30$ and $p(B|A) = 0.20$. We want $p(A \cap B)$. It is clear that the appropriate formula is

$$\begin{aligned} p(B \cap A) &= p(B|A) \cdot p(A) \\ &= (0.20) \cdot (0.30) = 0.06 \end{aligned}$$

EXERCISES

1. If a pair of fair dice are rolled, what is the probability that (a) the sum of the faces showing equals 9, given that at least one die shows a 5? (b) at least one die shows a 5, given that the sum of the faces showing is 9?
2. Ten Chicanos, 25 Blacks, and 40 Whites apply for a job. (a) If all applicants have equal chances, what is the probability that a Black gets the job? (b) If it becomes known that the successful applicant is not a White, what is the probability that he or she is a Black?
3. Three fair coins are tossed. Find the probability that (a) three heads show, given that at least two heads show; (b) exactly two heads show, given that at least one head shows.
4. Of a group of 20 male patients asking for pain medication, five are given only placebos, ten are given demerol, and the remaining five are given morpheme. If a patient is chosen at random, what is the probability that (a) he received demerol? (b) he received demerol, given that he did not receive a placebo?

5. If $\frac{2}{5}$ of all human beings have type A blood and $\frac{1}{3}$ of all human beings are type A Rh+, what is the probability that someone chosen at random and found to have type A blood is also Rh+?
6. If 30% of all entering college students have high SAT scores and 27% both have high SAT scores and eventually graduate, find the probability that someone chosen at random graduates, if it is known that she entered with high SAT scores.
7. A survey of the sexual preferences of 1127 people yielded the data tabulated below (Ht = heterosexual; Hm = homosexual; M = male; F = female):

	Ht	Hm	TOTAL
M	638	29	667
F	440	20	460
TOTAL	1078	49	1127

A person is selected at random from among those surveyed. Find each of these probabilities: (1) $p(\mathrm{M})$; (b) $p(\mathrm{Hm})$; (c) $p(\mathrm{M} \cap \mathrm{Hm})$; (d) $p(\mathrm{F}|\mathrm{Ht})$; (e) $p(\mathrm{Ht}|\mathrm{M})$.
8. The following data were extracted from *The Commissioner's 1958 Standard Ordinary Table of Mortality*:

AGE	NUMBER SURVIVING
0	10,000,000
20	9,664,994
40	9,241,359
70	5,592,012

Find the probability that (a) a person lives to be 20; (b) a person lives to be 40; (c) a 40-year-old survives to be 70; (d) a 20-year-old dies before reaching age 40.
9. A racing form reports that Attaboy, Ben's-Her, Casanovo, and Do-It-Rihgt have, respectively, a 0.4, 0.3, 0.2, and 0.1 chance of winning a certain race. If Attaboy scratches (that is, is withdrawn before the race), what does Ben's-Her's chance of winning become?
10. The table below shows the results of a survey of the extent of the education of 1000 men and 1000 women ("$\geqslant 4$" means four or more years of college, "< 4" means less than that):

	$\geqslant 4$	< 4	TOTAL
M	211	789	1000
W	130	870	1000
TOTAL	341	1659	2000

A person is chosen at random from among those surveyed. Find the probability that the individual (a) is a woman; (b) had less that four years of college; (c) is a woman with less than four years of college; (d) had four or more years of college, given that the individual is a man; (e) is a woman given that the individual had less than four years of college.

11. Suppose that, in a certain part of the world, in any 50-year period the probability of a major plague is $\frac{1}{3}$, the probability of a major famine is $\frac{3}{4}$, and the probability of both a plague and a famine is $\frac{7}{24}$. What is the probability of (a) a plague, given that there is a famine; (b) of a famine, given that there is a plague?
12. Given a probability of $\frac{2}{3}$ that interest rates will jump this year, and a probability of $\frac{3}{4}$ that if interest rates jump the stock market will decline, what is the probability that interest rates will jump and the stock market will decline?
13. The probability that a particular candidate will win the primary is $\frac{1}{3}$, and the probability that, if he wins the primary, he will also win the election is $\frac{3}{5}$. What is the probability that he will win the election?
14. One takes the bar examination a second time only if one fails on the first try, and one takes it a third time only if one fails on the second. Suppose 80% of those taking the examination for the first time pass, that 90% of those taking it for the second time pass, and that 50% of those trying it a third time are successful. What is the probability that an individual will (a) fail the first try and pass the second? (b) fail the first two tries and pass the third?

TEST A

1. A card is picked at random from a deck of 52. What is the probability that (a) it is a spade, given that it is black? (b) it is a red king, given that it is a face card. (Kings, queens, and jacks are face cards).
2. Suppose the probability that a person picked at random has lung cancer is 0.02 and the probability that the person will both have lung cancer and be a heavy smoker is 0.015. Given that someone picked at random has lung cancer, what is the probability that the person is a heavy smoker?
3. Suppose the probability that a company chosen at random obtains a government contract is $\frac{1}{3}$ and that the probability of obtaining a subsidiary one if the major one is obtained is $\frac{1}{2}$. What is the probability that a randomly selected company gets both contracts?

TEST B

1. A pair of dice are rolled, one red and one green. If the red die shows a six, what is the probability that the sum is ten? If at least one die shows a six, what is the probability that

the sum is ten?

2. Suppose that the probability that the income of a family chosen at random is under \$5000 is 0.2 and that the probability that its income is under \$5000 and it receives food stamps is 0.1. Given that the income of a randomly chosen family is less than \$5000, what is the probability that it receives food stamps?
3. Suppose the probability that the Yankees win the pennant is 0.4 and the probability of their winning both the pennant and the World Series is 0.3. What is the probability that the Yankees win the Series, given that they have won the pennant?

SECTION 9. INDEPENDENT EVENTS

In §8 we discussed $p(E|F)$, the conditional probability of E given F. Sometimes $p(E|F)$ is equal to $p(E)$; that is, the probability of the occurrence of E does not depend on the occurrence of F. In this case, we say that events E and F are *independent*.

Example 9-1

Two fair dice are rolled. Find the probability that (a) the sum is seven; (b) the sum is seven, given that a three shows on the first die.

(a) Refer if necessary to the display of sample space S_2 on page 54. S_2 has 36 elements and the event "sum is seven" has six elements: (1,6), (2,5), (3,4), (4,3), (5,2), (6,1). So the answer is $\frac{1}{6}$.

(b) The reduced sample space for this part is

$$\{(3,1),\ (3,2),\ (3,3),\ (3,4),\ (3,5),\ (3,6)\}$$

Since the sum is seven only in the one simple event (3,4) of this sample space, the answer is again $\frac{1}{6}$.

If we let the event that the sum is seven be E and the event that a three shows on the first die be F, then we have

$$p(E) = p(E|F) = \frac{1}{6}$$

The probability of event E is thus not affected by the occurrence of F. The events "the sum is seven" and "a three shows on the first die" are therefore independent.

Example 9-2

A card is selected at random from a deck of 52 cards. Find the probability that (a) the card is a spade; (b) the card is a spade provided it is an ace; (c) the card is a spade provided it is black.

The probability that (a) the card is a spade is $\frac{13}{52} = \frac{1}{4}$; (b) the card is a spade provided it is an ace is $\frac{1}{4}$; and (c) the card is a spade provided it is black is $\frac{13}{26} = \frac{1}{2}$. If E, F, and G are the events that the card is a spade, is an ace, and is black, then we see that

$$p(E) = p(E|F) \quad \text{but} \quad p(E) \neq p(E|G)$$

Therefore drawing a spade and drawing an ace are independent events, but drawing a spade and drawing a black card are not. We say that events E and G in this example are *dependent*.

FORMULAS FOR INDEPENDENT EVENTS

Events E and F are independent if and only if $p(E) = p(E|F)$. Using the formula for $p(E|F)$ from §8 (page 66), we have

$$p(E) = p(E|F) = \frac{p(E \cap F)}{p(F)} \qquad \text{provided } p(F) \neq 0$$

from which it follows that

$$p(E \cap F) = p(E) \cdot p(F)$$

This is sometimes called the *multiplication theorem*. Note especially that this formula holds IF AND ONLY IF EVENTS E AND F ARE INDEPENDENT.

Example 9-3

For a particular two-engine plane, the probability that either engine fails is 0.05; the failure of either is independent of that of the other. What is the probability that (a) both engines fail: (b) at least one engine fails?

Letting E and F be the events that the port and starboard engines fail, respectively, we have, for (a), using the formula just given,

$$p(E \cap F) = p(E) \cdot p(F) = (0.05) \cdot (0.05) = 0.0025$$

and for (b),

$$p(E \cup F) = p(E) + p(F) - p(E \cap F) = 0.05 + 0.05 - 0.0025 = 0.0975$$

Example 9-4

Suppose that the probability that a company's bid on contract A is successful is 0.3, that its bid on contract B is successful is 0.2, and that both bids are successful is 0.05. Is the success of one bid independent of that of the other?

The product $p(\text{A}) \cdot p(\text{B})$ is $(0.2) \cdot (0.3) = 0.06$. But this is not the same as $p(\text{A} \cap \text{B})$, which is given as 0.05. So the two events are not independent.

Example 9-5

Refer to the table in Example 8-5 (page 64). Are the events that the person smokes and that the person is a beer-drinker independent?

Since from part (a) (of Example 8-5) we have $p(C) = 0.40$, and from part (b) that $p(B) = 0.75$), it follows that $p(C) \cdot p(B) = 0.30$. But this does not equal $p(C \cap B)$, which is 0.26, from part (c). The events are therefore not independent.

Example 9-6

Refer to the table in Example 8-10 (page 66). Are the events that the person is left-handed and that the person has IQ greater than 135 independent?

Since $p(\text{L})\cdot p(\text{IQ}>135) = (0.05)\cdot(0.1) = 0.005$ and $p(\text{L} \cap \text{IQ}>135)$ also equals 0.005, the events are independent. This implies that for the given population the probability of a high IQ is not affected by handedness.

INDEPENDENCE OF THREE OR MORE EVENTS

Events E_1, E_2, ..., E_n are independent if and only if

$$p(E_1 \cap E_2 \cap \ldots \cap E_n) = p(E_1)\cdot p(E_2)\cdots p(E_n)$$

AND this equation holds also for EVERY SUBSET of the set of events.

Example 9-7

Suppose a reputed psychic in an ESP experiment has called heads or tails correctly on ten successive tosses of a coin. What is the probability that guessing would yield this perfect score?

Let E_1, E_2, ..., E_{10} be the events of correct guesses on the respective tosses. On the assumption that these events are all independent, we have

$$\begin{aligned} p(E_1 \cap E_2 \cap \ldots \cap E_{10}) &= p(E_1)\cdot p(E_2)\cdots p(E_{10}) \\ &= \frac{1}{2}\cdot\frac{1}{2}\cdots\frac{1}{2} = \frac{1}{1024} \end{aligned}$$

It is interesting to note that we can get the same answer by reasoning with a sample space each of whose simple events is a possible sequence of ten heads and tails. There are 2^{10} simple events in this sample space, of which only one is the favorable event corresponding to the series of calls by the supposed psychic. So the probability of occurrence of that one is $(1/2)^{10} = \frac{1}{1024}$.

NOTE: INDEPENDENT VERSUS MUTUALLY EXCLUSIVE

It is important not to confuse the concepts of independent events and of mutually exclusive events. Two events E and F are mutually exclusive if $p(E \cap F) = 0$. They are independent, as we have seen above, if $p(E \cap F) = p(E)\cdot p(F)$. The only way E and F can be both independent and mutually exclusive is if either $p(E) = 0$ or $p(F) = 0$. (Why?)

Example 9-8

Suppose $p(E) = 0.3$ and $p(F) = 0.2$. Find $p(E \cup F)$ if (a) E and F are independent; if (b) E and F are mutually exclusive.

(a) If E and F are independent then $p(E \cap F) = p(E) \cdot p(F)$, so $p(E \cup F) = p(E) + p(F) - p(E \cap F) = 0.3 + 0.2 - (0.3) \cdot (0.2) = 0.44$

(b) If E and F are mutually exclusive then $p(E \cap F) = 0$; so

$$p(E \cup F) = p(E) + p(F) - p(E \cap F) = 0.3 + 0.2 - 0 = 0.5$$

EXERCISES

1. Two fair dice are rolled. Which of the following pairs of events are independent?

	EVENT E	EVENT F
(a)	the sum is seven	six shows on second die
(b)	the sum is seven	six shows on at least one die
(c)	four shows on first die	three shows on second die
(d)	the sum is seven	first die shows even number
(e)	the sum is even	first die shows even number

2. A card is chosen at random from a deck of 52. Which of the following pairs of events are independent? That the card is

	EVENT E	EVENT F
(a)	a seven	a heart
(b)	a face card	red
(c)	a red ace	the ace of hearts
(d)	between five and ten inclusive	a spade or a heart
(e)	between five and eight inclusive	between four and seven inclusive

3. Suppose the probabilities that a student scores above 600 on the mathematics SAT, above 600 on the verbal SAT, and above 600 on both, are respectively 0.1, 0.15, and 0.025. Are the two events independent?

4. Suppose the probabilities that an individual chosen at random has had chicken pox, mumps, and both chicken pox and mumps are, respectively, 0.8, 0.7, and 0.56. Are the two events independent?

5. (a) Two fair dice are rolled twice in succession. Find the probability of obtaining the sum seven on the first roll followed by the sum eleven on the second. (b) Two fair dice are rolled three times in succession. Find the probability of obtaining, in order, the sums two, three, and 12.

6. Suppose $p(E) = 0.5$ and $p(F) = 0.4$. Find $p(E \cup F)$ if (a) E and F are independent; (b) E and F are mutually exclusive; (c) F is a subset of E.

7. Suppose $p(E) = 0.3$ and $p(E \cup F) = 0.4$. Find $p(F)$ if (a) E and F are independent; (b) E and F are mutually exclusive; (c) E is a subset of F.

8. Suppose $p(E) = p(E|F)$. Show that $p(F) = p(F|E)$.
9. Consult the table for the group surveyed in Exercise 7 of §8 (page 68). Determine if the sexual preference of an individual is independent of the individual's sex.
10. Consult the statistics tabulated in Exercise 10 of §8 (page 68). Do they indicate that the education of an individual and the individual's sex are independent?

TEST A

1. Two fair dice are rolled. Let E, F, and G be respectively the events "six on the first die," "six on the second die," and "the sum is 12." (a) Are E and F independent? (b) Are E and G independent?
2. Suppose that on any given day the probabilities that the value of the dollar drops, that the price of gold rises, and that both occur are respectively 0.6, 0.65, and 0.55. Are the two events independent?
3. Suppose $p(E) = 0.2$ and $p(F) = 0.1$. What is $p(E \cup F)$ if E and F are (a) independent? (b) mutually exclusive?

TEST B

1. A card is chosen at random from a deck of 52. Let E, F, and G be respectively the events "the card is a spade," "the card is red," and "the card is an ace." Are E and F independent? Are F and G independent?
2. Suppose that, for any given year, the probabilities that the stock market declines, that women's hemlines lower, and that both happen are respectively 0.4, 0.35, and 0.3. Are the two events independent?
3. Suppose $p(E) = 0.4$ and $p(E \cup F) = 0.6$. Find $p(F)$ if E and F are (a) independent; (b) mutually exclusive.

SECTION 10. BAYES' LAW

Suppose an automobile manufacturer makes a certain part at two different factories, F1 and F2. Suppose, further, that he knows the probabilities that a part made at F1 or at F2 will be defective. He chooses a part at random and finds it to be defective. If he now asks what the probability is that the part was made at factory F1, he is asking a question calling for the use of a theorem named for an early eighteenth-century British mathematician, the Reverend Thomas Bayes (1702-1761).

In earlier sections dealing with two-stage experiments, with Q an event occurring at stage one and R an event at stage two, we have asked for the conditional probability $p(R|Q)$ (§8, page 61). In this section we ask for the *reverse* conditional probability, $p(Q|R)$. We must be careful to distinguish between these probabilities, as the following examples show.

Example 10-1

Suppose, in the case described above, that 60% of the parts are made at factory F1, the rest at F2. Suppose also that 2% of F1's output are defective, and that 1% of F2's parts are defective. If a part chosen at random is found to be defective, what is the probability that it came from (a) factory F1? (b) factory F2?

Figure 10.1 is a probability tree for this two-stage "experiment." The outcomes at stage one (involving the factories) are labeled F1 and F2; at stage two, we label the outcomes "D" for a defective

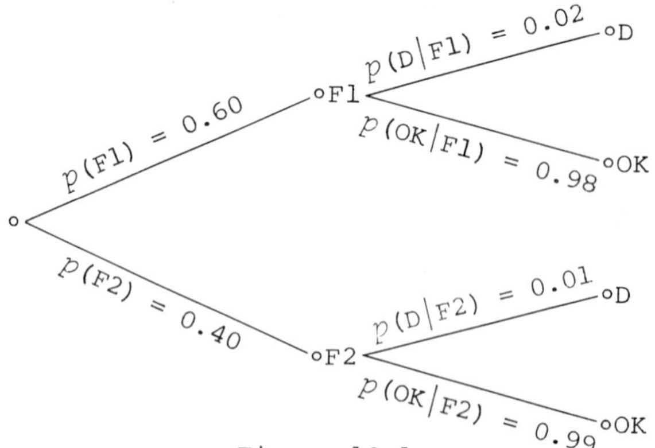

Figure 10.1

part, "OK" for a nondefective one. We then assign probabilities to the various events, based on the percentages given in the example. At stage one, the probabilities are just the percentages of parts made by the two factories. At stage two, each probability is a conditional one: for the occurrence of the event, given that the preceding (stage-one) event has occurred. For example, the uppermost probability at stage two, $p(D|F1)$, is the conditional probability that a part will be defective, given that it was made at factory F1. It equals 0.02 because 2% of F1's parts are defective. Since a part from F1 is either defective or nondefective (OK), we compute $p(OK|F1)$ by subtracting $p(D|F1)$ from one; so $p(OK|F1) = 1 - 0.02 = 0.98$. The conditional probabilities for the other two second-stage events are obtained similarly.

We now find the reverse probabilities asked for. Part (a) calls for $p(F1|D)$, part (b) for $p(F2|D)$. From §8 (page 64) we have

$$p(F1|D) = \frac{p(F1 \cap D)}{p(D)} \quad \text{and} \quad p(F2|D) = \frac{p(F2 \cap D)}{p(D)} \qquad (*)$$

The theorem for the probability of an intersection (page 66) yields

$$p(F1 \cap D) = p(F1) \cdot p(D|F1) \quad \text{and} \quad p(F2 \cap D) = p(F2) \cdot p(D|F2)$$

Substituting from the tree, we get

$$p(F1 \cap D) = (0.60) \cdot (0.02) = 0.012 \quad \text{and} \quad p(F2 \cap D) = (0.40) \cdot (0.01) = 0.004$$

We also have to find $p(D)$. For this, we note that event D is the union of two mutually exclusive events, as shown both in the tree (Figure 10.1) and in the Venn diagram (Figure 10.2). Therefore

$$p(D) = p(F1 \cap D) + p(F2 \cap D) = 0.012 + 0.004 = 0.016$$

We can now complete the answers to (a) and (b) by substituting the proper values in the equations above marked (*):

$$\text{(a)} \quad p(F1|D) = \frac{0.012}{0.016} = 0.75 \qquad \text{(b)} \quad p(F2|D) = \frac{0.004}{0.016} = 0.25$$

Thus, given that a randomly selected part is defective, the probability that it came (a) from factory F1 is 0.75, and (b) from factory F2 is 0.25. The sum of these two probabilities is one. (Why

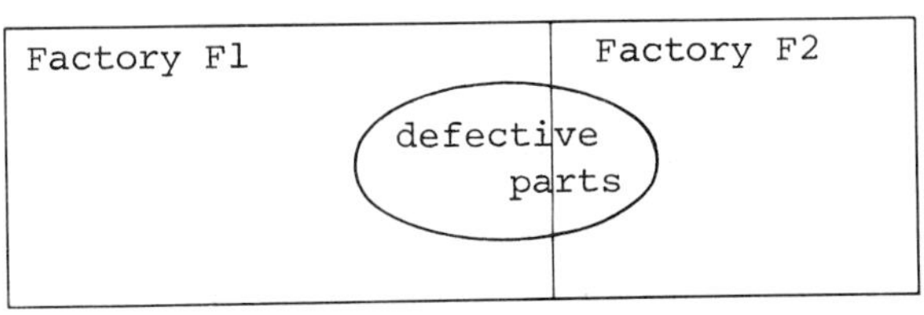

$D = (F1 \cap D) \cup (F2 \cap D)$

Figure 10.2

should it be?)

Note especially, by comparing probabilities on the tree (Figure 10.1) with those just above, that

$$p(\mathrm{D}|\mathrm{F1}) \neq p(\mathrm{F1}|\mathrm{D}) \qquad\qquad p(\mathrm{D}|\mathrm{F2}) \neq p(\mathrm{F2}|\mathrm{D})$$

In the Note on page 62 we pointed out that, in general, $p(E|F)$ is different from $p(F|E)$. Here, once again, we see that the order in which variables appear in a conditional probability is important.

In Figure 10.3 we add the probabilities of the intersections to the tree. As it shows, there are four simple events in this two-stage experiment: $\mathrm{F1} \cap \mathrm{D}$, $\mathrm{F1} \cap \mathrm{OK}$, $\mathrm{F2} \cap \mathrm{D}$, $\mathrm{F2} \cap \mathrm{OK}$. The sum of the probabilities of these events is one. (Why?) Also, the probability of each intersection is just the product of the probabilities that appear along the appropriate branch of the tree.

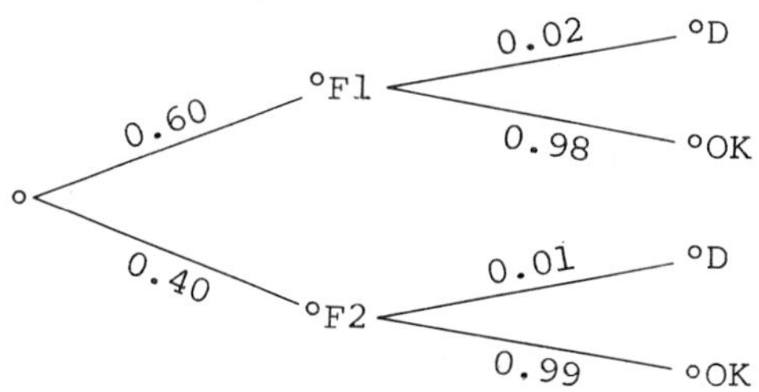

$$p(\mathrm{F1} \cap \mathrm{D}) = (0.60)\cdot(0.02) = 0.012$$

$$p(\mathrm{F1} \cap \mathrm{OK}) = (0.60)\cdot(0.98) = 0.588$$

$$p(\mathrm{F2} \cap \mathrm{D}) = (0.40)\cdot(0.01) = 0.004$$

$$p(\mathrm{F2} \cap \mathrm{OK}) = (0.40)\cdot(0.99) = 0.396$$

$$\text{sum} = 1.000$$

Figure 10.3

FORMULATION OF THE LAW

The computations for $p(\mathrm{F1}|\mathrm{D})$ and $p(\mathrm{F2}|\mathrm{D})$ in the preceding example are applications of Bayes' law: Given the mutually exclusive events F1 and F2, whose union is the sample space, we have

$$p(\mathrm{F1}|\mathrm{D}) = \frac{p(\mathrm{F1} \cap \mathrm{D})}{p(\mathrm{D})} = \frac{p(\mathrm{F1})\cdot p(\mathrm{D}|\mathrm{F1})}{p(\mathrm{F1} \cap \mathrm{D}) + p(\mathrm{F2} \cap \mathrm{D})}$$

$$= \frac{p(\mathrm{F1})\cdot p(\mathrm{D}|\mathrm{F1})}{p(\mathrm{F1})\cdot p(\mathrm{D}|\mathrm{F1}) + p(\mathrm{F2})\cdot p(\mathrm{D}|\mathrm{F2})}$$

For $p(\mathrm{F2}|\mathrm{D})$ one replaces F1 by F2 in each numerator of the equation.

Or, in general: If the two events A and B form a partition of a sample space S (that is, A and B are mutually exclusive and $A \cup B = S$) then for any event E of S we have

$$p(A|E) = \frac{p(A \cap E)}{p(E)} = \frac{p(A \cap E)}{p(A \cap E) + p(B \cap E)}$$

$$= \frac{p(A)\cdot p(E|A)}{p(A)\cdot p(E|A) + p(B)\cdot p(E|B)}$$

And for $p(B|E)$ we replace A by B in the numerator of each fraction.

Example 10-2

Suppose that 1% of a clinic's outpatients are known to have cancer. A blood test is developed which is positive in 95% of patients with cancer but is also positive in 2% of patients who do not have cancer. If a person chosen at random from the clinic's outpatients is given the test and it comes out positive, what is the probability that the person actually has cancer?

Figure 10.4 shows the probability tree for this example, and the probabilities for the intersections: C = has cancer; C′ = does not have cancer; Ps = blood test positive; Ng = test negative.

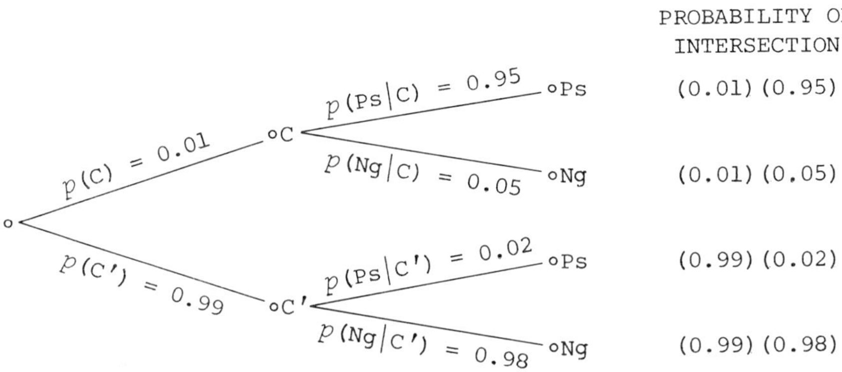

Figure 10.4

To find the probability that a person actually has cancer if the blood test is positive, we evaluate $p(\mathrm{C}|\mathrm{Ps})$, using Bayes' law:

$$p(\mathrm{C}|\mathrm{Ps}) = \frac{p(\mathrm{C} \cap \mathrm{Ps})}{p(\mathrm{Ps})} = \frac{p(\mathrm{C} \cap \mathrm{Ps})}{p(\mathrm{C} \cap \mathrm{Ps}) + p(\mathrm{C}' \cap \mathrm{Ps})}$$

$$= \frac{(0.01)(0.95)}{(0.01)(0.95) + (0.99)(0.02)} = \frac{0.0095}{0.0095 + 0.0198}$$

$$= \frac{0.0095}{0.0293} \approx 0.324$$

[The student should draw a suitable Venn diagram to show that $p(\mathrm{Ps}) = p(\mathrm{C} \cap \mathrm{Ps}) + p(\mathrm{C}' \cap \mathrm{Ps})$.]

We conclude that the probability that a person actually has cancer, given that the blood test is positive, is about 0.324.

Example 10-3

Suppose that four branches, B1, B2, B3, and B4, of a chain of supermarkets average respectively 900, 900, 1200, and 600 customers per day. Suppose, further, that the probabilities that a customer spends more than $50 in the stores are, respectively, $\frac{1}{2}$, $\frac{1}{3}$, $\frac{1}{4}$, and $\frac{1}{6}$. A customer is chosen at random. Find the probability that the customer (a) spends more than \$50; (b) shops at branch B2, given that she spends over \$50.

We note that the size of the population here under consideration is $900 + 900 + 1200 + 600 = 3600$, and that $900/3600 = \frac{1}{4}$, $1200/3600 = \frac{1}{3}$, and $600/3600 = \frac{1}{6}$. We are now ready to construct a probability tree (Figure 10.5: M = spends more than \$50; L = spends \$50 or less.)

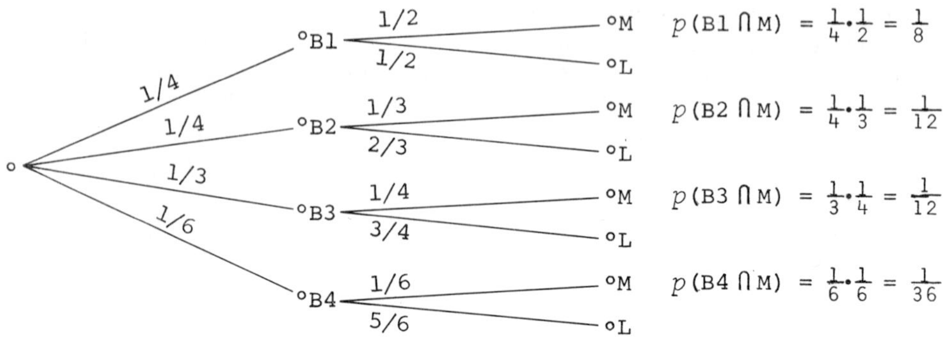

Figure 10.5

For completeness we draw the entire tree, including the subbranches for the complementary event "\$50 or less." Note that for the specific probabilities sought in the example these additional subbranches are not needed. It is instructive, however, to notice that the sum of the probabilities of *all eight* intersections (all the simple events of this sample space) is one.

In Figure 10.6 we show the relevant Venn diagram.

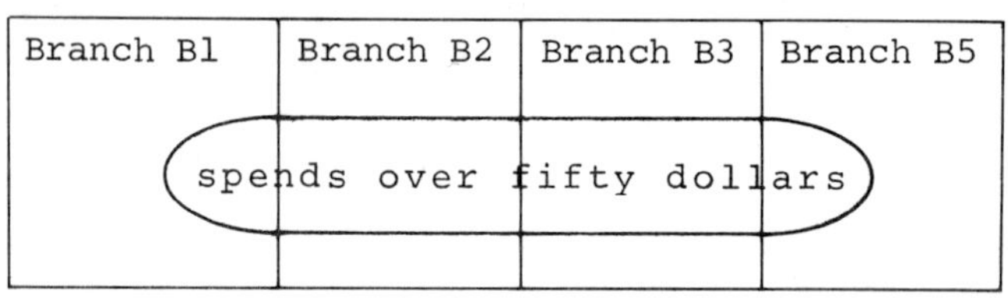

Figure 10.6

For the answer to (a), we note that $p(\mathrm{M})$ is the sum of the four probabilities listed on the right in Figure 10.5:

$$p(\mathrm{M}) = \frac{1}{8} + \frac{1}{12} + \frac{1}{12} + \frac{1}{36} = \frac{23}{72}$$

For part (b) we use Bayes' law. The law was stated above (page 78) in terms of a sample space of only two events; here we must extend it to a sample space of four events:

$$p(\mathrm{B2}|\mathrm{M}) = \frac{p(\mathrm{B2} \cap \mathrm{M})}{p(\mathrm{M})} = \frac{p(\mathrm{B2} \cap \mathrm{M})}{p(\mathrm{B1} \cap \mathrm{M}) + p(\mathrm{B2} \cap \mathrm{M}) + p(\mathrm{B3} \cap \mathrm{M}) + p(\mathrm{B4} \cap \mathrm{M})}$$

$$= \frac{\frac{1}{12}}{\frac{1}{8} + \frac{1}{12} + \frac{1}{12} + \frac{1}{36}} = \frac{\frac{1}{12}}{\frac{23}{72}} = \frac{6}{23}$$

The computations show that the probability that a randomly selected customer (a) will have spent over \$50 is $\frac{23}{72}$; (b) will have spent it at branch B2 is $\frac{6}{23}$.

Verify that the probabilities that a customer who spends over \$50 spends it in branch B1 is

$$p(\mathrm{B1}|\mathrm{M}) = \frac{\frac{1}{8}}{\frac{23}{72}} = \frac{9}{23}$$

in branch B3 is

$$p(\mathrm{B3}|\mathrm{M}) = \frac{\frac{1}{12}}{\frac{23}{72}} = \frac{6}{23}$$

and in branch B4 is

$$p(\mathrm{B4}|\mathrm{M}) = \frac{\frac{1}{36}}{\frac{23}{72}} = \frac{2}{23}$$

Also, if a customer selected at random is found not to spend as much as \$50, then the probability that the customer shopped in, say, branch B3 is given by

$$p(\mathrm{B3}|\mathrm{L}) = \frac{p(\mathrm{B3} \cap \mathrm{L})}{p(\mathrm{L})} = \frac{\frac{1}{3}\cdot\frac{3}{4}}{\frac{1}{4}\cdot\frac{1}{2} + \frac{1}{4}\cdot\frac{2}{3} + \frac{1}{3}\cdot\frac{3}{4} + \frac{1}{6}\cdot\frac{5}{6}} = \frac{\frac{1}{4}}{\frac{49}{72}} = \frac{18}{49}$$

Note that $p(\mathrm{M}) + p(\mathrm{L}) = 1$; also that

$$p(\mathrm{B1}|\mathrm{M}) + p(\mathrm{B2}|\mathrm{M}) + p(\mathrm{B3}|\mathrm{M}) + p(\mathrm{B4}|\mathrm{M}) = 1$$

Justify these two equations.

APPLYING BAYES' LAW

We now generalize from the preceding examples.

Suppose we are considering a two-stage experiment with simple events B_1, B_2, and B_3 comprising the sample space at stage one and with A one of the events at stage two. We are interested in evaluating $p(B_1|A)$, $p(B_2|A)$, and $p(B_3|A)$.

STEP 1. We draw a probability tree (see Figure 10.7), assign appropriate probabilities, and then compute the probabilities of the

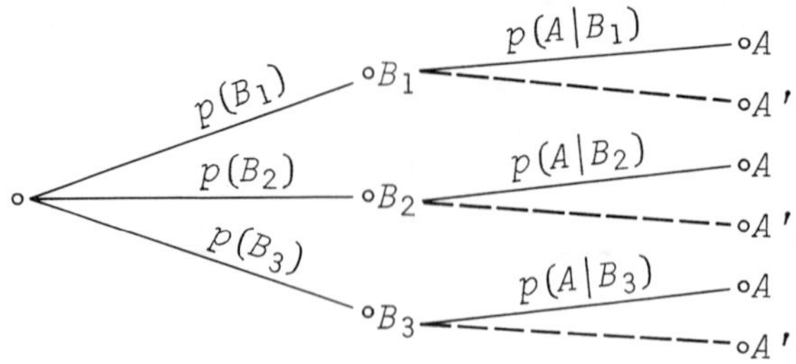

$$p(B_1 \cap A) = p(B_1) \cdot p(A|B_1)$$

$$p(B_2 \cap A) = p(B_2) \cdot p(A|B_2)$$

$$p(B_3 \cap A) = p(B_3) \cdot p(A|B_3)$$

Figure 10.7

relevant intersections. Note the following points:

(1) We have used dotted lines for A', the event complementary to A. This is for completeness; only event A interests us.

(2) The sum $p(B_1) + p(B_2) + p(B_3)$ must equal one.

(3) The probability of any intersection is the product of the probabilities that appear along the appropriate branch of the tree.

STEP 2. We draw a suitable Venn diagram (Figure 10.8) to find $p(A)$. As the diagram shows,

$$p(A) = p(B_1 \cap A) + p(B_2 \cap A) + p(B_3 \cap A)$$

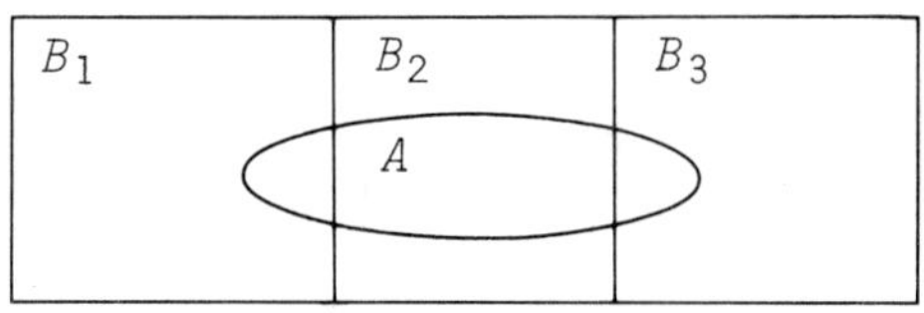

Figure 10.8

STEP 3. We now compute the conditional probabilities, using Bayes' law:

$$p(B_1|A) = \frac{p(B_1 \cap A)}{p(A)}$$

$$p(B_2|A) = \frac{p(B_2 \cap A)}{p(A)} \qquad p(B_3|A) = \frac{p(B_3 \cap A)}{(A)}$$

where the value for $p(A)$ is obtained from the formula in Step (2).

When only a few events are involved, as in our examples so far, constructing a tree is very helpful in obtaining the desired conditional probabilities. However, when there are many events, often the application of Bayes' law is itself more useful than a tree. We therefore state the general formula here:

If the events $B_1, B_2, \ldots, B_n$ partition a sample space, then for any event A, the probability of B_i (where $i = 1,2,\ldots,n$), given that A has occurred, is

$$p(B_i|A) = \frac{p(B_i \cap A)}{p(B_1 \cap A) + p(B_2 \cap A) + \ldots + p(B_n \cap A)}$$

$$= \frac{p(B_i) \cdot p(A|B_i)}{p(B_1) \cdot p(A|B_1) + p(B_2) \cdot p(A|B_2) + \ldots + p(B_n) \cdot p(A|B_n)}$$

Verify the application of Bayes' law to the set of four events in Example 10.3.

EXERCISES

1. A home-builder obtains 75% of the electrical parts he needs from a company whose defect rate is 0.006 and the rest from a company whose defect rate is 0.01. (a) What proportion of the electrical parts can be expected to be defective? If a part is defective, what is the probability that it came (b) from the first company? (c) from the second company?
2. In a certain country 5% of the teenagers are married, 20% of the married teenagers have children, and 10% of the unmarried teenagers have children. (a) Find the probability that a teenager has a child. (b) If a teenager has a child, what is the probability that the child's parents are not married?
3. The probability that a person will show a certain gene-transmitted trait is one if both parents show the trait, 0.6 if one parent shows the trait, and 0.05 if neither parent shows the trait. Suppose the children at a certain school come from families in 10% of which both parents show the trait and in 25% of which one parent shows the trait. (a) What is the probability that a given child chosen at random from the school shows the trait? Given that a child chosen at random shows the trait, what is the probability that (b) both parents show the trait? (c) one parent shows the trait? (c) neither parent shows the trait?
4. A poll shows that the percentages of Democrats, Republicans, and Independents in a certain city are respectively 50, 30, and 20. In a recent election 80% of the Democrats, 10% of the Republicans, and 60% of the Independents voted for Candidate A. (a) What percentage voted for Candidate A? Find the probability that a person chosen at random who turns out to have voted for Candidate A is (b) a Democrat; (c) a Republican; (d) an Independent.
5. Suppose that the percentages of employees working on each of three different shifts at a certain plant are respectively 50, 30, and 20. The probabilities that an employee is absent on the shifts are respectively 0.01, 0.02, and 0.03. (a) Find the probability that a randomly chosen employee is absent. If an employee is absent, find the probability that he works on (b) the first shift; (c) the second shift; (d) the third shift.
6. At a college located on an island near Puerto Rico, 75% of the students are native islanders, 20% are Puerto Rican, and the rest are mainlanders. Suppose 30% of the native islanders speak Spanish, all the Puerto Ricans speak Spanish, and 5% of the mainlanders speak Spanish. (a) What is the probability that a student chosen at random speaks Spanish? If a student does speak Spanish, what is the probability that he or she is (b) a native islander? (c) a Puerto Rican? (d) a mainlander?
7. A television dealer receives $\frac{1}{2}$ his sets from plant P1, $\frac{1}{3}$ from

plant P2, and the rest from plant P3. Suppose that 1% of the sets from P1, 2% from P2, and 5% from P3 are defective. (a) What is the probability that a set chosen at random will be defective? If a set is found defective, what is the probability that it came from (b) plant P1? (c) plant P2? (d) plant P3?

8. Suppose that with regard to a particular gene the percentages of genotypes AA, Aa, and aa in a particular population are respectively 64, 32, and 4. Suppose, further, that the percentages of these genotypes that contract a certain disease are respectively 1, 10, and 50. (a) What is the probability that a person selected at random from the population will contract the disease? If a person does contract the disease, what is the probability that the person is of (b) genotype AA? (c) genotype Aa? (d) genotype aa?

9. At a certain college 10% of the women students in business administration take one mathematics course, 20% take two, 30% take three, and the rest take four. Of those taking just one mathematics course, 10% successfully complete all their business courses on the first attempt, while this success rate rises to 25%, 50%, and 80% respectively for those taking two, three, and four mathematics courses. (a) Find the probability that a woman business student chosen at random at this college will complete all her business courses on the first attempt. If a student does complete them all on the first attempt, what is the probability that she took (b) four mathematics courses? (c) three? (d) two? (e) just one? (f) at least three?

10. Jar I has three black and four white marbles. Jar II has two black, three white, and four red marbles. We select a jar at random and draw a marble from it. Find the probability that (a) we obtain a white marble; that, if the marble drawn is white, it came (b) from jar I; (c) from jar II.

11. A fair die is rolled. If a five or six shows, a marble is drawn from box B1; if one, two, three, or four shows, a marble is drawn from box B2. Box B1 holds four red and three white marbles; box B2 holds two red, three white, and four black marbles. Draw a probability tree. Find the probability that (a) a red marble is drawn; (b) that if the marble drawn is white, it came from box B2; (c) that if a black marble is drawn, it came from box B2.

12. A consolidated school district has regularly administered a test to all first-graders, on the basis of which a child is labeled "exceptional" (E) or "not exceptional" (E'). On the basis of information other than the scores on this test, some of the children have been placed in a special class for the gifted. To evaluate the test, statistics have been collected and are being analyzed. They show that 75% of those who have succeeded in the special class had been labeled exceptional, that 20% of those who have not succeeded in the class had also been

labeled exceptional, and that 80% of the children in the special class have been successful. What is the probability that a child labeled exceptional on the basis of the test will succeed in a special class?

13. Do exercise 14 of §8 (page 69) by drawing a probability tree.

TEST A

1. A travel agent books passages on three different tours, with half his customers choosing tour T1, one-third T2, and the rest T3. The agent has noted that three-quarters of those who take tour T1 return to book passage again, two-thirds of those on T2 come back for more, and one-half of those who take T3 come back. (a) What is the probability that a customer chosen at random returns for another tour? If someone does come back, what is the probability that the person first went on (b) tour T1? (c) tour T2? (d) tour T3?

TEST B

1. Suppose the Air Force receives 25% of its parachutes from company C1 and the rest from company C2. The probability that a parachute from C1 will fail to open is 0.001, while for C2 the probability of failure is 0.0008. (a) What is the probability that a randomly chosen parachute fails? If one does fail what is the probability that it came from (b) company C1? (c) company C2?

SECTION 11. FINITE STOCHASTIC PROCESSES

In this section we solve problems involving a sequence of experiments, often referred to as a *finite stochastic process*. A probability tree is most useful in analyzing these multi-stage experiments, as illustrated in the examples.

Example 11-1

Suppose there are three boxes; box I has two white balls and one black one; box II has three white and two black balls; box III contains five white and two black balls. Choose a box at random, then draw a ball from it at random. What is the probability that the ball is (a) white? (b) black?

In the probability tree for this problem, I, II, and III are the three boxes, W = white, and B = black. The outcomes at stage two

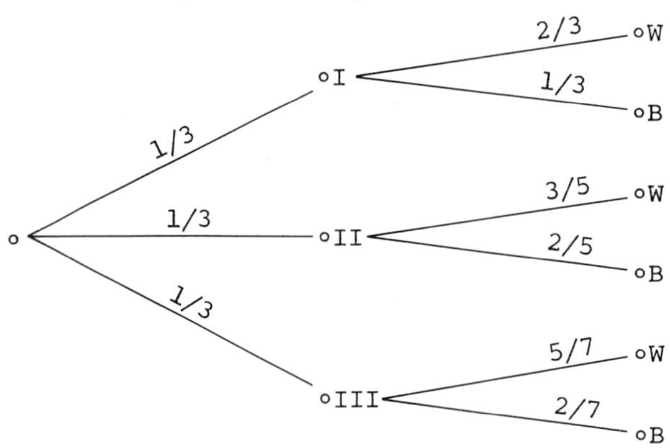

$$p(\mathrm{I} \cap \mathrm{W}) = \frac{1}{3} \times \frac{2}{3} = \frac{2}{9}$$

$$p(\mathrm{I} \cap \mathrm{B}) = \frac{1}{3} \times \frac{1}{3} = \frac{1}{9}$$

$$p(\mathrm{II} \cap \mathrm{W}) = \frac{1}{3} \times \frac{3}{5} = \frac{1}{5}$$

$$p(\mathrm{II} \cap \mathrm{B}) = \frac{1}{3} \times \frac{2}{5} = \frac{2}{15}$$

$$p(\mathrm{III} \cap \mathrm{W}) = \frac{1}{3} \times \frac{5}{7} = \frac{5}{21}$$

$$p(\mathrm{III} \cap \mathrm{B}) = \frac{1}{3} \times \frac{2}{7} = \frac{2}{21}$$

depend on the outcomes at stage one; we have indicated this on the tree by writing conditional probabilities on the subbranches for stage two. The calculations to the right of the diagram show that the probability of any intersection is, as in §§9 and 10, just the product of the probabilities that appear along the corresponding branches of the tree. To find the required probabilities we add together the probabilities of those *mutually exclusive events* (a) whose union is "a white ball is selected" and those (b) whose union is "a black ball is selected," obtaining

$$\text{(a)}\quad p(\mathrm{W}) = \frac{2}{9} + \frac{1}{5} + \frac{5}{21} = \frac{208}{315} \qquad \text{(b)}\quad p(\mathrm{B}) = \frac{1}{9} + \frac{2}{15} + \frac{2}{21} = \frac{107}{315}$$

The probabilities in (a) and (b) sum to one. Why?

We note in this example that there are ten white and five black balls altogether. It may be tempting to conclude that the answer to (a) is $\frac{10}{15} = \frac{2}{3}$, and, to (b), $\frac{5}{15} = \frac{1}{3}$. Why are these answers wrong?

Example 11-2

A blindfold player on a TV game show is invited to draw a card from a hat containing three cards. One card yields the player a prize of $1; a second allows the participant to toss a coin with heads netting a $100 prize, tails $500; the third card allows the player to roll a die. If the die comes up one, two, or three, the contestant wins $1000; a four or five wins $5000; and a six earns $10,000. What is the probability of winning each of these amounts, from $1 through $10,000?

The figure shows the stages and possible outcomes of this experiment together with the respective probabilities of the outcomes. We

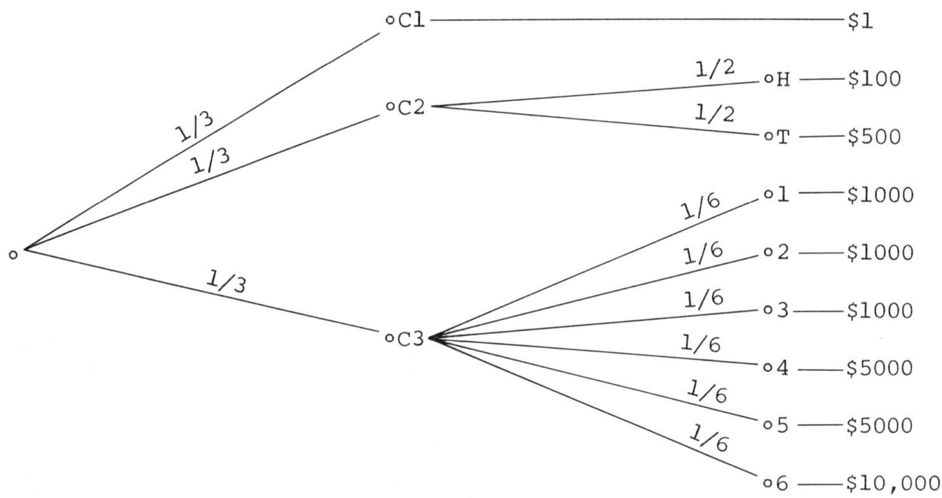

note that each path through the tree consists of independent events. It therefore follows from §9 that the probability of a particular path or branch is the product of the probabilities along that path:

$$p(\$1) = \frac{1}{3} \qquad p(\$100) = \frac{1}{3}\times\frac{1}{2} = \frac{1}{6}$$

$$p(\$500) = \frac{1}{3}\times\frac{1}{2} = \frac{1}{6} \qquad p(\$1000) = \frac{1}{3}\times\frac{3}{6} = \frac{1}{6}$$

$$p(\$5000) = \frac{1}{3}\times\frac{2}{6} = \frac{1}{9} \qquad p(\$10{,}000) = \frac{1}{3}\times\frac{1}{6} = \frac{1}{18}$$

Example 11-3

A college student is trying to decide whether to go to law school or to medical school. Suppose that the probability that the student will opt for, and be admitted to, law school is $\frac{2}{3}$, medical school $\frac{1}{3}$. If three quarters of the students admitted to law school eventually become lawyers and four fifths of those admitted to medical school ultimately become doctors, what is the probability that the student will become (a) a lawyer? (b) a doctor? (c) neither?

The figure shows the probability tree. Note that the probabilities at stage two are conditional ones, for outcomes contingent on outcomes at step one.

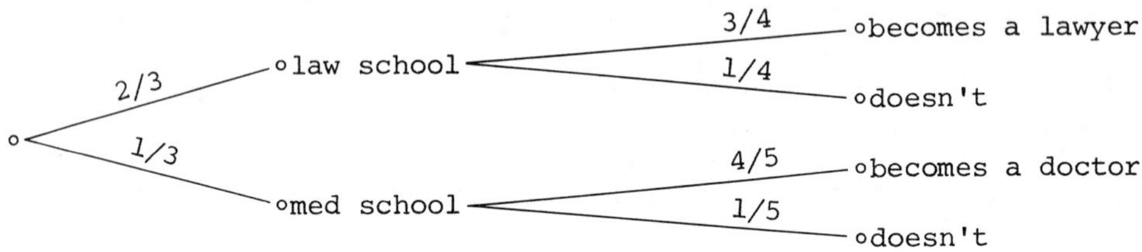

(a) The probability that the student will become a lawyer is

$$\frac{2}{3} \times \frac{3}{4} = \frac{1}{2}$$

(b) The probability of becoming a doctor is

$$\frac{1}{3} \times \frac{4}{5} = \frac{4}{15}$$

(c) The probability of becoming neither can be figured in either of two ways: by adding up the appropriate products from the tree,

$$\frac{2}{3} \times \frac{1}{4} + \frac{1}{3} \times \frac{1}{5} = \frac{7}{30}$$

or by subtracting from one the probability of becoming either a doctor or a lawyer:

$$1 - \left(\frac{1}{2} + \frac{4}{15}\right) = \frac{7}{30}$$

Example 11-4

With the boxes and balls as in Example 11-1, suppose that the ball selected is set aside and a second ball is then drawn from the same box. Find the probability of obtaining two white balls.

The probability tree is shown at the top of the facing page. Let $p(\text{WW})$ represent the probability asked for. Then we have

$$\begin{aligned} p(\text{WW}) &= p(\text{I} \cap \text{W} \cap \text{W}) + p(\text{II} \cap \text{W} \cap \text{W}) + p(\text{III} \cap \text{W} \cap \text{W}) \\ &= \left(\frac{1}{3} \times \frac{2}{3} \times \frac{1}{2}\right) + \left(\frac{1}{3} \times \frac{3}{5} \times \frac{2}{4}\right) + \left(\frac{1}{3} \times \frac{5}{7} \times \frac{4}{6}\right) = \frac{233}{630} \end{aligned}$$

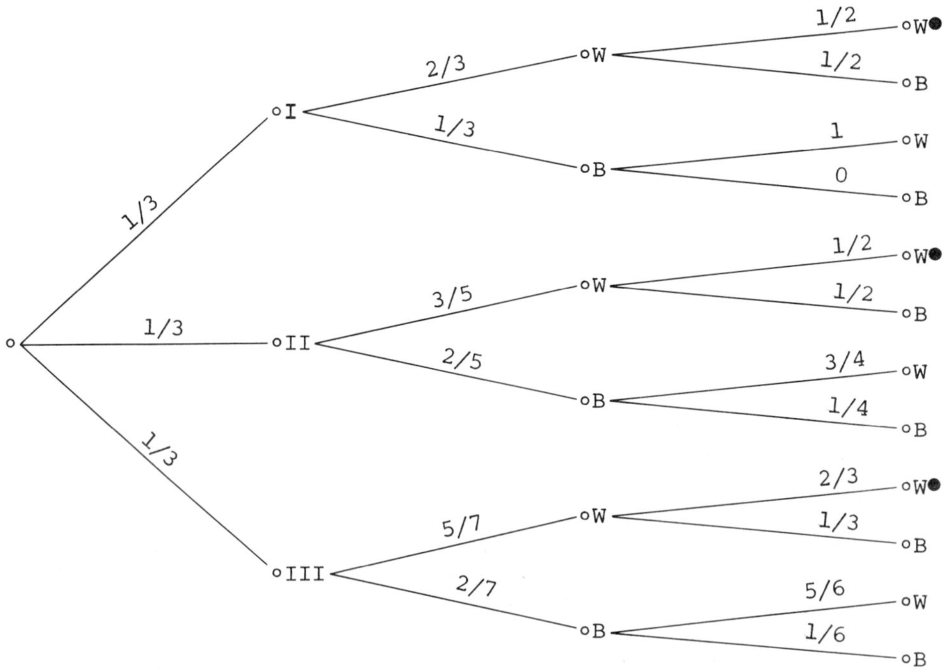

Example 11-5

A box contains two white and three red marbles. Two marbles are drawn one after the other without replacement. Find the probability that (a) both marbles are red; (b) the two marbles are of different colors; (c) the second marble is white.

Here is the probability tree:

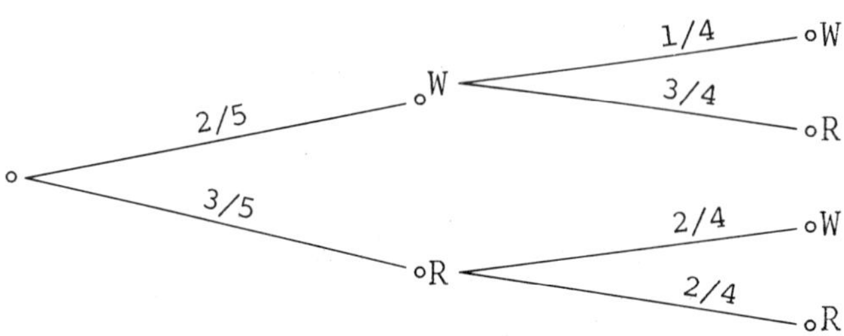

$$p(\text{WW}) = \frac{2}{5} \times \frac{1}{4} = \frac{1}{10}$$

$$p(\text{WR}) = \frac{2}{5} \times \frac{3}{4} = \frac{3}{10}$$

$$p(\text{RW}) = \frac{3}{5} \times \frac{2}{4} = \frac{3}{10}$$

$$p(\text{RR}) = \frac{3}{5} \times \frac{2}{4} = \frac{3}{10}$$

Note that there are four outcomes in the sample space: {WW,WR,RW,RR}. At the right in the figure we obtain the probability of each outcome (along a path), as in the preceding examples, by multiplying the probabilities on its branches. The sum of the probabilities of the outcomes is (of course!) one.

The answer to (a) is

$$p(\text{RR}) = \frac{3}{5}\times\frac{2}{4} = \frac{3}{10}$$

The answer to (b) is

$$p(\text{WR}) + p(\text{RW}) = \frac{2}{5}\times\frac{3}{4}+\frac{3}{4}\times\frac{2}{5} = \frac{3}{5}$$

And the answer to (c) is

$$p(\text{WW}) + p(\text{RW}) = \frac{2}{5}\times\frac{1}{4}+\frac{3}{4}\times\frac{2}{5} = \frac{2}{5}$$

EXERCISES

1. A hero must choose randomly between two doors. Behind one is a tiger, behind the other a beautiful maiden. If the tiger's door is opened there is a probability of $\frac{2}{3}$ that it will kill the hero. If the maiden's door is opened there is a probability of $\frac{3}{4}$ that she will marry the hero. What is the probability that (a) the hero will be killed? (b) he will marry the maiden?
2. Suppose that on an evening selected at random $\frac{1}{2}$ of those watching commercial TV networks are watching CBS, $\frac{1}{3}$ ABC, and $\frac{1}{6}$ NBC. If on any evening the probability that CBS will preempt regular programming with a news special is $\frac{2}{3}$, that ABC will preempt with a sports special is $\frac{3}{4}$, and that NBC will preempt with a movie special is $\frac{1}{2}$, what is the probability that a viewer selected at random (who stays tuned to the initially dialed channel despite any changes in programming) sees (a) a news special? (b) a sports special? (c) a movie special?
3. Among voters declaring a political preference in each of three cities, the percentages were found to be

	CITY X	CITY Y	CITY Z
REPUBLICAN	40	50	30
DEMOCRAT	60	50	70

   Suppose one of these cities is picked at random and a voter is then chosen at random from among those who have declared a political preference. What is the probability that the voter chosen is (a) a Republican? (b) a Democrat?
4. Given the conditions stated for Exercise 3, suppose a city is chosen at random, then a voter is selected, and then a second voter is chosen from the same city. What is the probability of choosing (a) two Republicans? (b) two Democrats? (c) one of each? (Assume that the populations of the cities are large enough that the probabilities can be treated as unchanged by the selection of the first voter.)

5. Suppose there are three boxes: box 1 has two white balls and two black ones; box 2 has one white and two black; box 3 has three white and no black. A box is chosen at random and a ball is selected at random from the box. What is the probability that the ball is (a) white? (b) black?
6. With the initial conditions of Exercise 5, suppose that after the first ball is drawn and its color is noted, it is set aside and a second ball is then chosen from the same box. What is the probability that (a) both balls are white? (b) both balls are black? (c) one ball is white and one black?
7. Suppose a box contains six white and four red marbles. Two marbles are drawn in succession without replacement. Find the probability that (a) both marbles are white; (b) the two marbles are of different colors.
8. With the initial conditions of Exercise 7, a marble is drawn, its color noted, and it is replaced; then again a marble is drawn. Find the probability in this case that (a) each drawing is of a white marble; (b) one drawing yields a white marble and one a red.
9. City A has twice as many apartment complexes as single-family houses, while city B has equal numbers of the two types. A salesman has a 50% chance of making a sale in an apartment complex but only a 25% chance in a single-family house. If the salesman is equally likely to be sent to either city, what is the probability he will make a sale in the first building he randomly chooses to enter?
10. A businessman is competing for three contracts. He has a 0.5 probability of winning the first contract. If successful in obtaining any one contract, the probability of winning the next increases by 0.1; if unsuccessful, the probability of winning the next one decreases by 0.1. What is the probability of winning (a) all three contracts? (b) exactly two contracts? (c) only one contract? (d) no contract?
11. Two players toss a coin for $1 per toss; they agree to quit if either is ever $2 ahead. What is the probability that the game will end after (a) exactly two tosses? (b) exactly three tosses? (c) exactly four tosses? (d) an odd number of tosses? (e) $2n$ tosses, for n a fixed integer?
12. A mouse is put in a maze with three different paths, only one of which leads to food. The mouse tries paths at random until it reaches the food, but never chooses the same incorrect path a second time. Draw a tree for this situation, then find the probability that the mouse gets to the food (a) on the first try; (b) on the second try; (c) not until the third try.

TEST A

1. The probability is $\frac{7}{8}$ that a certain patient will receive the

proper medication. If he receives the proper medication, there is a change of $\frac{5}{7}$ that he will show marked improvement, a chance of $\frac{2}{7}$ that he will show only slight improvement. However, if he does not receive the proper medication, there is a $\frac{2}{5}$ chance that he will show no change and a $\frac{3}{5}$ chance that his condition will deteriorate. What is the probability for (a) marked improvement? (b) slight improvement? (c) no change? (d) deterioration?

2. Suppose that in a certain year the probability that a Democrat will be elected President is $\frac{2}{3}$ and that a Republican will be elected is $\frac{1}{3}$. If the probability that a Democrat will successfully combat inflation is $\frac{1}{3}$ and that a Republican will is $\frac{1}{2}$, what is the probability that inflation will be successfully combatted?

TEST B

1. Suppose that $\frac{2}{3}$ of a certain population can be hypnotized. Of those that can be hypnotized, $\frac{1}{10}$ can be put into a very deep trance, $\frac{1}{2}$ into a mild trance, and the rest into only a slight trance. What is the probability that an individual chosen at random from this population can be put into (a) a deep trance? (b) a mild trance? (c) only a slight trance?
2. The probability that a particular company will have a successful year is $\frac{2}{3}$. If it has a successful year, there is a $\frac{7}{8}$ chance the current president will be kept on; if the year is unsuccessful, there is only a $\frac{1}{4}$ chance of keeping the same president. What is the probability that the president will be kept on?

SECTION 12. BINOMIAL PROBABILITIES

An important special case of a stochastic process is one in which the successive experiments or trials of a sequence are all identical and independent, each experiment having the same two possible outcomes. The outcomes are called *success* and *failure*; the independent experiments are referred to as *binomial* or *Bernoulli* trials*, and these same descriptive terms apply to the resulting probabilities. To compute these probabilities we will use the techniques of §11. Since we will be concerned with the number of successes (or of failures) that might occur, rather than with the order of their occurrence, we will also use the results obtained in §5 on combinations. The examples that follow illustrate Bernoulli trials and yield a formula for the probability of obtaining k successes in a sequence of n independent trials.

Example 12-1

What is the probability of obtaining exactly two heads when a fair coin is tossed (a) three times? (b) ten times?

In the case of three tosses the sample space is

$$\{HHH, HHT, HTH, HTT, THH, THT, TTH, TTT\}$$

Each simple event here has probability $\frac{1}{8}$. We note also that three of the simple events have exactly two heads; namely, HHT, HTH, and THH. The answer to (a) is therefore $3 \times \frac{1}{8} = \frac{3}{8}$.

In the case of ten tosses, the sample space has $2^{10} = 1024$ simple events, so a new approach--something other than listing--is in order.

The probability of the particular ordering TTHTTTTTHT, with exactly two heads, is found by multiplying the probabilities of the independent events, yielding

$$\frac{1}{2}\times\frac{1}{2}\times\frac{1}{2}\times\frac{1}{2}\times\frac{1}{2}\times\frac{1}{2}\times\frac{1}{2}\times\frac{1}{2}\times\frac{1}{2}\times\frac{1}{2} = \left(\frac{1}{2}\right)^{10}$$

But note that the probability of the ordering HHTTTTTTTT, or of the ordering THTHTTTTTT, each of which likewise has exactly two heads, is also $(\frac{1}{2})^{10}$. In how many different ways can the result be two

*James Bernoulli (1654-1705), a member of a Swiss family of famous mathematicians, was a pioneer in the theory of probability.

heads and eight tails? This is equivalent to asking for the number of ways in which exactly two heads can occur in ten tosses of a coin (the remaining eight necessarily coming up tails). We considered just such a question in Example 5-9 (page 33). The number of ways of selecting two positions out of ten is

$$C(10,2) = \frac{10!}{2!8!} = 45$$

The answer to part (b) is therefore

$$45\left(\frac{1}{2}\right)^{10} = \frac{45}{1024}$$

Example 12-2

Suppose a coin is weighted so that the probability of heads is $\frac{2}{3}$, the probability of tails $\frac{1}{3}$. What is the probability of obtaining exactly three heads in seven tosses of the coin?

The probability that any one sequence will contain three heads and four tails is

$$\frac{2}{3}\times\frac{2}{3}\times\frac{2}{3}\times\frac{1}{3}\times\frac{1}{3}\times\frac{1}{3}\times\frac{1}{3} = \left(\frac{2}{3}\right)^3\times\left(\frac{1}{3}\right)^4$$

The number of different sequences with three heads and four tails is just the number of ways in which three of seven tosses of a coin can show heads; that is,

$$C(7,3) = \frac{7!}{4!3!} = 35$$

So the probability in question is

$$35\left(\frac{2}{3}\right)^3\left(\frac{1}{3}\right)^4 = \frac{280}{2187}$$

Example 12-3

Suppose a baseball player's batting average is 0.300; that is, the probability of his getting a hit any time he comes to bat is 0.3. What is the probability that in five at-bats he will get exactly two hits?

Since the probability of a hit is 0.3, the probability of not getting a hit is $1 - 0.3 = 0.7$. The probability that any sequence will consist of two hits and three misses is

$$0.3\times 0.3\times 0.7\times 0.7\times 0.7 = (0.3)^2\times(0.7)^3$$

The number of different sequences of five at-bats, each with exactly two hits, is

$$C(5,2) = \frac{5!}{3!2!} = 10$$

The final answer is therefore $10(0.3)^2(0.7)^3 = 0.3087$.

Formula for Binomial Probability

The examples above share the following characteristics:

(1) Each involves a sequence of identical experiments or trials.
(2) Each experiment has only two outcomes, success and failure.
(3) The experiments are all independent; that is, the outcome of any experiment does not affect that of any other.

If we let p denote the probability of success and q the probability of failure, where $q = 1 - p$, then the (binomial) probability of exactly k successes in n trials (which implies exactly $n - k$ failures) is given by the formula*

$$C(n,k)p^k q^{n-k} = \frac{n!}{k!(n-k)!}p^k q^{n-k}$$

Example 12-4

Suppose the probability that a tire chosen at random is defective is 0.1. In a sample of five tires, find the probability that (a) none are defective; (b) exactly one is defective; (c) more than one are defective.

We have, for (a), (b), and (c) respectively,

$$p(0 \text{ defective}) = C(5,0)(0.1)^0(0.9)^5 = (0.9)^5 = 0.59049$$

$$p(1 \text{ defective}) = C(5,1)(0.1)^1(0.9)^4 = 5(0.1)(0.9)^4 = 0.32805$$

$$p(>1 \text{ defective}) = C(5,2)(0.1)^2(0.9)^3 + C(5,3)(0.1)^3(0.9)^2 + C(5,4)(0.1)^4(0.9) + (0.1)^5 = 0.08146$$

The last result can be obtained much more quickly by using the complementary event, as follows:

$$p(>1 \text{ defective}) = 1 - p(\leqslant 1 \text{ defective}) = 1 - (0.59049 + 0.32805) = 0.08146$$

Example 12-5

A salesman makes a sale in one-third of the houses he visits. If he visits five houses what is the probability that he will make (a) exactly two sales? (b) fewer than three sales? (c) at least four sales?

---

*These probabilities are called "binomial" because they are the terms in the expansion of the binomial $p + q$ to the nth power; that is, of $(p + q)^n$.

We have, respectively for (a), (b), and (c),

$$p(2 \text{ sales}) = C(5,2)\left(\frac{1}{3}\right)^2\left(\frac{2}{3}\right)^3 = \frac{80}{243}$$

$$p(<3 \text{ sales}) = p(0 \text{ sales}) + p(1 \text{ sale}) + p(2 \text{ sales}) =$$

$$\left(\frac{2}{3}\right)^5 + 5\left(\frac{1}{3}\right)\left(\frac{2}{3}\right)^4 + \frac{80}{243} = \frac{192}{243} = \frac{64}{81}$$

$$p(4 \text{ or } 5 \text{ sales}) = p(4 \text{ sales}) + p(5 \text{ sales}) = 5\left(\frac{1}{3}\right)^4\left(\frac{2}{3}\right) + \left(\frac{1}{3}\right)^5 = \frac{11}{243}$$

USING BINOMIAL PROBABILITIES FOR LARGE POPULATIONS

When a population is large and a sample taken from it is relatively small, the techniques of finding probabilities described in §7 often involve formidable computations. Fortunately, in many such cases binomial probabilities, which are much easier to calculate, provide excellent approximations. This is so because the probabilities for a large population are hardly altered when only a few members are removed. In effect, we may assume that the sampling process in such problems involves replacement. The next two examples illustrate this technique.

Example 12-6

On an examination administered nationally to 100,000 students, 70,000 scored higher than 85. If a sample of five students is chosen at random from this group, what is the probability that exactly three of them scored higher than 85?

The exact answer, using the procedures of §7 on equally likely outcomes (see the rule on page 58), is

$$\frac{C(70\,000,\ 3) \times C(30\,000,\ 2)}{C(100\,000,\ 5)} = 0.3087074\ldots$$

Using the formula for binomial probabilities, we get

$$C(5,3)(0.7)^3(0.3)^2 = 0.3087$$

This is an extremely good approximation to the exact answer, and assumes only that removing a very few members of the large population does not alter the fact that if you choose a student at random you have a 0.7 chance of getting one who did 85 or better on the examination.

Example 12-7

For an opinion poll, a sample of ten people is selected at random from a list consisting of 100,000 names of Republicans and 50,000 names of Democrats. Find the probability that exactly seven people in the sample are Republicans.

The exact answer is

$$\frac{C(100\,000,\,7)\times C(50\,000,\,3)}{C(150\,000,\,10)} = 0.26013\ldots$$

Now consider this parallel question: The probability is $\frac{2}{3}$ that a person chosen at random (from a large population) is Republican; it is $\frac{1}{3}$ that the person selected is a Democrat. What is the probability that there will be exactly seven Republicans in a sample of size ten?

The answer to this question is

$$C(10,7)\left(\frac{2}{3}\right)^7\left(\frac{1}{3}\right)^3 = 0.260123$$

This is very close indeed to the answer to the original problem.

EXERCISES

1. A coin is weighted so that the probability of heads is $\frac{3}{4}$. What is the probability of obtaining (a) exactly four heads in six tosses? (b) five heads in five tosses? (c) at least one head in seven tosses?
2. A baseball player has a 0.200 batting average. What is the probability of (a) exactly one hit in four at-bats? (b) at least two hits in seven at-bats? (c) fewer than five hits in six at-bats?
3. The probability that a participant in a rifle competition will hit the bull's-eye is 0.8. Find the probability of her hitting at least three bull's-eyes in four shots.
4. Approximately 50% of those who take the medication imuran suffer a side-effect known as alopecia; that is, loss of hair. In a group of five people on imuran, what is the probability that at least two will suffer from alopecia?
5. Sixty percent of the residents in a village favor a bond issue. If a sample of five residents is surveyed, find the probability that a majority will be in favor of the issue.
6. Medical statistics show that the probability of recovery from melanoma is 0.04. The implication is that 96% of those diagnosed as having melanoma will be dead within five years. In a hospital with ten such patients, what is the probability that at least one will survive at least five years?
7. A pair of fair dice is thrown six times. If the sum 7 appears exactly once, player A wins; otherwise player B wins. What is the probability that A will win?
8. A judge chosen at random will reach a just decision roughly 75% of the time. What is the probability that on randomly chosen cases at least two out of three judges will reach just decisions?

9. A golf course provides motorized carts for those who want them. If there is a probability of 0.5 that a party selected at random will want a cart, and if eight parties arrive at a time when only three carts are available, what is the probability that everyone wanting a cart will get one?
10. Johnny boasts that he can tell the difference between two brands of beer 75% of the time. Frankie agrees to pay Johnny \$5 if he distinguishes between them correctly at least four out of five trials, but will collect \$5 if Johnny fails. What is the probability that Frankie will lose \$5 if (a) Johnny's estimate of his ability is accurate? (b) Johnny is in fact merely guessing each time?
11. A true-false test has ten questions. What is the probability that you will get nine or ten right if you guess randomly on each question?
12. Each of the five questions on a multiple-choice examination offers four alternatives. What is the probability that pure guessing can yield a score of at least 80, if each correct answer earns 20 points?
13. How many times must a fair coin be tossed so that the probability of getting at least one tail exceeds 0.95? (Hint: use trial and error.)

In problems 14, 15, and 16 use binomial probabilities to obtain approximate answers.

14. At a small northeastern college, 1000 out of 3000 students came down with influenza during the winter. If five students at this college are interviewed, what is the probability that exactly two will have had the flu?
15. An insurance company owns 100 stocks which increased in value during the past year and 25 stocks which decreased. If an auditor selects three of the company's stocks at random, what is the probability that a majority will have increased in value during the past year?
16. A rental agency has 100 cars, 30 of which have air-conditioning. If a company rents four of the cars, selected at random, what is the probability that at least two have air-conditioning?

TEST A

1. A coin is weighted so that the probability of heads is $\frac{3}{5}$. If the coin is tossed five times, what is the probability of getting (a) exactly three heads? (b) at least three heads? (c) at least one tails?
2. Suppose 3% of the television sets manufactured by a company are defective. If a retailer purchases five TV sets from this man-

ufacturer, what is the probability that at least one is defective?
3. Among members of a State Bar Association, 300 lawyers oppose capital punishment while 100 favor it. If a sample of five members is surveyed, what is the probability that a majority favor capital punishment? (Approximate the answer with binomial probabilities.)

TEST B

1. A baseball player has a batting average of 0.250. In four at-bats what is the probability of (a) exactly one hit? (b) at least two hits? (c) at most two hits?
2. Suppose that 60% of the residents of a certain town are Tories and the rest are Whigs. If a sample of five people is surveyed, what is the probability that a majority will be Tories?
3. Fifty of 125 employees are members of a union. If a committee of four employees is chosen by lot, what is the probability that at least two are union members? (Approximate the answer with binomial probabilities.)

SECTION 13. EXPECTED VALUE

A student in a mathematics class can find his average quiz grade from a set of grades consisting of three 10's, two 8's, and one 5 as follows:

$$\frac{10 + 10 + 10 + 8 + 8 + 5}{6} = \frac{51}{6} = 8\frac{1}{2}$$

This is equivalent to the following computation:

$$10\left(\frac{3}{6}\right) + 8\left(\frac{2}{6}\right) + 5\left(\frac{1}{6}\right)$$

which we interpret as a *weighted average*: one in which each grade is weighted by the probability of its occurrence, the probability being based on the frequency in a sample of grades.

The *expected value* in an experiment is essentially a weighted average. A gambler is interested in the number of heads he can expect when a coin is tossed ten times, or in the expected sum of spots in a roll of a pair of dice; a TV picture-tube manufacturer cares about the expected number of defective items in a given sample; a medical researcher who has statistics on the effect of a drug on a large population may want to know how many patients can be expected to improve in a randomly chosen group. In this section we define expected value and related terms in the context of several applications. We start with examples.

Example 13-1

Attendance at a local college football game during the winter depends on the weather: on a clear day the game usually will attract 50,000 fans; a light snow will bring the attendance down to 40,000; during a heavy snow, only 15,000 rooters are present. According to a long-range forecast, the probability that a winter day chosen at random is clear is 0.8, that there is light snow is 0.15, and that the snow is heavy is 0.05. What is the expected attendance at a game?

The following table shows the attendance and probability for each outcome:

SNOW:	NONE	LIGHT	HEAVY
PROBABILITY:	0.8	0.15	0.05
ATTENDANCE:	50,000	40,000	15,000

The expected attendance at a winter game is thus

$$50{,}000(0.8) + 40{,}000(0.15) + 15{,}000(0.05) = 46{,}750$$

We note from the information given in the problem that the actual attendance at any particular winter game will be 50,000, 40,000, or 15,000, never 46,750. But 46,750 is the average attendance expected over the course of many games.

Example 13-2

In a weekly lottery 10,000 tickets are sold for $1 each. The winner gets $7500. What is the expected value of this lottery for a person who buys one ticket?

OUTCOME:	WINS	LOSES
PROBABILITY:	$\frac{1}{10{,}000}$	$\frac{9999}{10{,}000}$
PAYOFF($):	+7499	-1

The term "payoff" here denotes the amount the player wins (marked with a plus sign) or loses (a minus sign). In this example the payoff to the winner takes into account that his "investment" of $1 is not returned. The expected value of this lottery for the purchaser of one ticket is, in dollars,

$$(7499)\left(\frac{1}{10{,}000}\right) - (1)\left(\frac{9999}{10{,}000}\right) = -0.25$$

Of course, in any particular lottery a person who buys a ticket *either* wins $7499 *or* loses $1; one cannot lose 25¢ in a single lottery. The expected value in this case is interpreted as follows: a player who buys one lottery ticket each week or so can in the long run expect an average loss of 25¢ per lottery.

DEFINITION OF EXPECTED VALUE

If the n outcomes of an experiment are assigned values x_1, x_2, ..., x_n and these outcomes have probabilities p_1, p_2, ..., p_n respectively, then the expected value is defined to be

$$E = x_1p_1 + x_2p_2 + \ldots + x_np_n$$

In the introductory example on grades, the values of the outcomes were 10, 8, and 5 (the student's quiz grades), which occurred with relative frequencies, and hence with assignable probabilities, $\frac{3}{6}$, $\frac{2}{6}$, and $\frac{1}{6}$ respectively.

In example 13-1, the outcomes were different weather conditions, the values of the outcomes were the corresponding attendance figures (which depended on the weather), and the probabilities of the outcomes were those provided in a long-range weather forecast.

In example 13-2, on the lottery, the two outcomes were a win and a loss, the values assigned to these outcomes were the payoffs, and the probabilities for a win and for a loss were determined by the description given of the lottery.

Example 13-3

A committee of two members is selected at random from a group of five men and ten women. What is the expected number of women on the committee?

To evaluate the relevant probabilities here, we apply the methods of §7. Note that the value of an outcome in this example is just the number of women on the committee.

OUTCOME:	NO WOMEN	ONE WOMAN	TWO WOMEN
PROBABILITY:	$\frac{C(5,2)}{C(15,2)} = \frac{2}{21}$	$\frac{C(5,1)C(10,1)}{C(15,2)} = \frac{10}{21}$	$\frac{C(10,2)}{C(15,2)} = \frac{9}{21}$
VALUE:	0	1	2

The expected number of women on the two-person committee is thus

$$0\left(\frac{2}{21}\right) + 1\left(\frac{10}{21}\right) + 2\left(\frac{9}{21}\right) = 1\frac{1}{3}$$

This answer makes vivid a point touched on earlier: that the expected value is not, in general, one assigned to any outcome in the experiment. In this example the answer is not even a possible one for the number of women on any particular two-person committee; it is, rather, as noted, the *average* number of women per committee if many such committees are selected. This also indicates why mathematicians and others may prefer to use a more technical phrasing, asking, in this example, "What is the expected value for the number of women?" or "What is the expected number of women on a two-person committee?" instead of asking the seemingly more natural "How many women can one expect on a two-person committee?"

EXPECTED VALUE IN GAMES OF CHANCE

In gambling, the expected value E of a game of chance is interpreted as the value of the game to the player (as opposed to the dealer or to the "house"). If E is positive, the game is *favorable* to the player; if E is negative the game is said to be *unfavorable* to the player. The lottery of example 13-1 is unfavorable to the player, since the expected value is -25¢. A fair game is one in which $E = 0$.

Example 13-4

A coin is weighted so that the probability of heads is $\frac{3}{4}$. In a game the coin is tossed three times in succession, the player winning \$12, \$10, \$5, and \$1 for three, two, one, and no heads respectively. (a) What is the expected value of this game for the player? (b) How much should the player be willing to pay per game for it to be fair?

Each outcome has a binomial probability, which is computed according to the methods of §12. The payoffs listed below are based on the assumption that it costs the player nothing to play:

OUTCOME:	3 HEADS	2 HEADS	1 HEAD	0 HEADS
PROB.:	$\left(\frac{3}{4}\right)^3 = \frac{27}{64}$	$3\left(\frac{3}{4}\right)^2\left(\frac{1}{4}\right) = \frac{27}{64}$	$3\left(\frac{3}{4}\right)\left(\frac{1}{4}\right)^2 = \frac{9}{64}$	$\left(\frac{1}{4}\right)^3 = \frac{1}{64}$
PAYOFF($):	12	10	5	1

(a) The expected value of this game to the player is, in dollars,

$$12\left(\frac{27}{64}\right) + 10\left(\frac{27}{64}\right) + 5\left(\frac{9}{64}\right) + 1\left(\frac{1}{64}\right) = 10$$

(b) It seems reasonable for a player to pay \$10 per game for it to be fair. This inference can be justified mathematically as follows. Each payoff in the game is reduced by \$10 if a player hands over \$10 to play. Thus the player wins (net) \$2 for three heads, plays for free if exactly two heads show, but loses \$5 if only one head shows and \$9 if no heads show. His expected value, then, paying \$10 per game, is (in dollars)

$$2\left(\frac{27}{64}\right) + 0\left(\frac{27}{64}\right) - 5\left(\frac{9}{64}\right) - 9\left(\frac{1}{64}\right) = \frac{54 - 54}{64} = 0$$

Example 13-5

How much should you be willing to pay per ticket in the lottery of example 13-2 for it to be a fair game?

Does the answer "75¢" seem obvious? If we let c be the price in dollars per ticket for a fair lottery, then we can find c as follows. First construct a payoff table:

OUTCOME:	WINS	LOSES
PROBABILITY:	$\frac{1}{10,000}$	$\frac{9999}{10,000}$
PAYOFF($):	$7500 - c$	$-c$

Then, since the expected value E must equal zero, we conclude that c must satisfy the equation

$$(7500 - c)\left(\frac{1}{10,000}\right) - c\left(\frac{9999}{10,000}\right) = 0$$

or

$$\begin{aligned} 7500 - c - 9999c &= 0 \\ 10,000c &= 7500 \\ c &= 0.75 \end{aligned}$$

This answer makes sense, since an average loss of 25¢ per game resulted when each ticket cost \$1. With a reduction of the ticket price to 75¢, a player can (theoretically) expect to break even in the long run. (Or think of it this way: a player who buys all ten thousand tickets, at 75¢ apiece, pays out \$7500 but is bound to win the prize of \$7500, thus breaking even.)

Example 13-6

The probability is 0.7 that a contestant will hit the bull's-eye in a rifle competition. If he has five shots, how many bull's-eyes can he be expected to get?

We can calculate the binomial probabilities, then compute the expected value here as in example 13-4. We note that

$$\begin{aligned} (0.7)^5 &= 0.16807 & \qquad 10(0.7)^2(0.3)^3 &= 0.1323 \\ 5(0.7)^4(0.3)^1 &= 0.36015 & 5(0.7)^1(0.3)^4 &= 0.02835 \\ 10(0.7)^3(0.3)^2 &= 0.3087 & (0.3)^5 &= 0.00243 \end{aligned}$$

These probabilities and the corresponding payoffs appear in the following table:

OUTCOME:	5 HITS	4 HITS	3 HITS	2 HITS	1 HIT	0 HITS
PROBABILITY:	0.16807	0.36015	0.3087	0.1323	0.02835	0.00243
PAYOFF:	5	4	3	2	1	0

The expected value is therefore

$$5(0.16807) + 4(0.36015) + 3(0.3087) + 2(0.1323) + 1(0.02835) + 0(0.00243) = 3.5$$

It is no coincidence that the expected value is equal precisely to the probability of a bullseye (0.7) times the number of shots (5). This is true in general for binomial trials.

Formula for Expected Value in Experiments involving Binomial Trials

If the probability of success on each binomial trial is p, then in n trials the expected value is given by

$$E = np$$

If a fair coin is tossed ten times, then the expected number of heads, by this formula, is $10 \times \frac{1}{2} = 5$. This result can also be obtained by the method used in the preceding example (13-6), but that procedure is in this case almost prohibitively long.

Example 13-7

In a west-coast city, 60% of the voters are known to favor a much debated bond issue. If a random sample of 20 votes is surveyed, how many can be expected to be in favor of the bond issue?

We need only recognize that this "experiment" involves binomial trials. Then, for 20 "trials," each with probability 0.6 of "success," the formula just given yields $20 \times 0.6 = 12$. So 12 voters in the sample can be expected to favor the bond issue.

EXERCISES

1. The average annual incomes of high-school and college graduates in a midwestern town are \$12,000 and \$15,000 respectively. If a company hires personnel only with at least a high-school diploma, and 20% of its employees have been through college, what is the expected income of an employee selected at random?
2. Attendance at a team's home baseball games this season depends on the records of its visiting opponents last season. Excellent, average, and poor standings by the visitors bring in respectively 40,000, 25,000, and 15,000 fans. If two of ten visiting teams had excellent records last year, three did poorly, and the rest have so-so records, what attendance can be expected on the average at home games this season?
3. At a warehouse sale 100 customers are invited to choose one of 100 identical boxes. Five boxes contain \$500 color-television sets; 25 boxes contain \$60 cameras; and the remaining boxes contain \$10 radios. (a) What should a customer be willing to pay to participate in the sale? (b) If each customer must pay \$60, how much does the warehouse-owner make?
4. An insurance company charges \$300 annually for car insurance. The policy specifies that the company will pay \$500 for a minor accident and \$5000 for a major accident. If the probability of a motorist having a minor accident during the year is 0.2 and of a major accident is 0.03, how much can the insurance company expect to make on a policy?
5. In a chess tournament a player gets one point for each win, one-half point for each draw, and no points for each loss. Based on his previous tournament record, a player figures that he will score 0, $\frac{1}{2}$, 1, $1\frac{1}{2}$, 2, $2\frac{1}{2}$, or 3 points in a three-game tournament with probabilities 0.1, 0.1, 0.1, 0.2, 0.25, 0.2, and 0.05 respectively. How many points can the player expect to make in the next three-game tournament?
6. In a set of ten radio tubes, four are defective. If a sample of three tubes is chosen at random, how many can be expected to be defective?
7. There are five black and eight white mice in a cage. If a random sample of four mice is taken from the cage, how many black mice can you expect in the sample?
8. A coin is weighted so that the probability of heads is one-third. How many heads can you expect in six tosses?
9. A baseball player has a 0.250 batting average. How many hits can he expect in ten at-bats?

10. A pair of fair dice are rolled four times in succession and the sum of the spots is noted. The player wins \$1, \$5, \$10, or \$100 if a seven is rolled one, two, three, or four times respectively, but loses \$2 if no seven is thrown. (a) What is the expected value of this game to the player? (b) How much should the player pay per game for it to be fair?
11. In a box of 12 hand calculators, two are defective. You select and test them one by one until you find one that is okay. How many should you expect to have to test?
12. In a game of chance a player tosses three fair coins. If three heads show, he wins \$5; if exactly two heads show, he wins \$2; if only one head shows, \$1. How much should he be willing to pay to play the game?

TEST A

1. At the grand opening of a new store, various door prizes are distributed at random, no more than one to a customer. One prize is worth \$1000, ten are worth \$100 each, and 100 are valued at \$1 each. What is the expected value for a customer of this giveaway if 10,000 are expected to participate?
2. A two-member committee is chosen at random from a group of county legislators of whom three are Populists and four are Conservatives. (a) How many Populists can be expected on the committee? (b) Suppose a friend agrees to pay you \$5 if both members of the committee are Conservatives. How much should you be willing to pay your friend if at least one member is a Populist?
3. In a southern city, statistics show that the chance of rain on a day chosen at random is 0.1. How many rainy days can be expected in a 30-day month?

TEST B

1. A highway engineer's crew can lay ten miles of new road on a sunny day, five miles on a rainy day. If the chance of rain is 0.2, what is the expected number of miles of new road per day?
2. There are two defective switches in a set of ten. A random sample of two of the ten is examined. If both are defective the manufacturer will be fined \$1000; if one is defective the fine is \$100. What is the expected value of the fine?
3. If one out of three marriages in this country is successful, how many successful marriages can be expected in 1,500,000 marriages?

SUMMARY OF §§6-13

You should now be able to

(1) describe a suitable sample space for an experiment and determine a valid probability assignment for it;
(2) find the probability of an event, a complementary event, or a union of events;
(3) use Venn diagrams to calculate probabilities;
(4) apply the addition formula for $p(X \cup Y)$;
(5) compute probabilities in equally likely sample spaces;
(6) calculate probabilities that involve combinations;
(7) evaluate conditional probabilities;
(8) distinguish between the conditional probabilities $p(E|F)$ and $p(F|E)$;
(9) find the probability of the intersection of events;
(10) find the probability of two or more independent events;
(11) distinguish between independent and mutually exclusive events;
(12) construct a probability tree;
(13) use Bayes' law to compute conditional probabilities resulting from multistage experiments;
(14) draw probability trees for and find the probabilities resulting from multistage experiments;
(15) compute binomial probabilities;
(16) use binomial probabilities for large populations;
(17) find the expected value of an outcome in a game of chance or in other experiments;
(18) find the expected value of an outcome in experiments involving binomial trials.

REVIEW EXERCISES

1. Consider the experiment of tossing, in order, a penny, a nickel, and a dime. Describe two sample spaces: (a) with reference only to how many tails appear; (b) with reference to the order in which heads and tails appear.
2. A sample space consists of four simple events. Find the probability of each if $p(e_2) = 3p(e_1)$, $p(e_3) = 2p(e_2)$, and $p(e_4) = 2p(e_1)$.
3. A fair die is tossed. What is the probability of getting (a) a two? (b) an odd number? (c) a number greater than one?
4. Find $p[(X \cup Y)']$ if $p(X) = \frac{1}{3}$, $p(Y) = \frac{2}{5}$, and events X and Y are mutually exclusive.
5. The probability that a sophomore in a business school passes accounting is 0.8, that she passes finite mathematics is 0.85, and that she passes both is 0.75. What is the probability that (a) she passes at least one of the two courses? (b) she fails both courses? (c) she fails exactly one course? (d) she fails at least one course?
6. The numerals 1, 2, 3, 4, 5, and 6 are written on slips of paper. Two of these are drawn at random without replacement, and their sum is noted. (a) Describe a sample space for which the simple events are equiprobable. How many simple events are there in this sample space? (b) Find the probability that the sum is 4. (c) Find the probability that the sum is greater than 3.
7. In a shipment of 20 calculators, five are known to be defective. If a dealer randomly selects three from this shipment, what is the probability that (a) none is defective? (b) at most one is defective?
8. A drawer contains six identical blue socks and four identical black ones. If you reach in and choose two socks blindly, what is the probability that you draw a matching pair?
9. A card is drawn at random from a regular deck of 52. If

   H is the event that the card drawn is a heart;
   R is the event that the card drawn is red;
   J is the event that the card drawn is a black jack;
   F is the event that the card drawn is a face card;

   find (a) $p(H|R)$; (b) $p(R|H)$; (c) $p(J|F)$; (d) $p(J|H)$; (e) $p(F|R)$.

10. A survey of 183 senior citizens on their use of hearing aids yields the following data, where J = the person is under 70, S = the person is 70 or older, Y = an aid is used, and N = no aid is used:

	Y	N	TOTAL
J	35	60	95
S	48	40	88
TOTAL	83	100	183

A senior citizen is selected at random from this group. Find (a) $p(J)$; (b) $p(N)$; (c) $p(J \cap N)$; (d) $p(S|Y)$; (e) $p(Y|S)$.

11. It is determined in a given population that the probability that a businessman is overweight is $\frac{1}{3}$ and that the probability that a businessman has high blood pressure, given that he is overweight, is $\frac{1}{2}$. If a businessman is chosen at random from this group what is the probability that he both is overweight and has high blood pressure?

12. In exercise 9 four events were specified (H, R, J, and F). Which pairs of those four are independent? Justify your conclusions.

13. Suppose that on a given day the probabilities that the prime interest rate will rise, that the closing Dow-Jones index will be down, and that both of these events will occur, are respectively 0.55, 0.5, and 0.35. Are the two events independent? Why or why not?

14. Suppose $p(E) = 0.5$ and $p(E \cup F) = 0.8$. Find $p(F)$ if (a) events E and F are independent; (b) E and F are mutually exclusive.

15. In the freshman class at a college, 10% have mathematics SAT scores above 700, 50% have between 500 and 700 inclusive, and the rest have under 500. During their first semester, 90% of those with scores above 700 elect a mathematics course, 75% of the group with scores between 500 and 700 choose such a course, but only 30% of those who scored under 500 do. Draw a tree for this situation; then answer the following questions: (a) What is the probability that a freshman chosen at random is taking a mathematics course? (b) If a freshman selected at random is in a mathematics course, what is the probability that the freshman's score on the mathematics SAT was above 700? (c) If one selected at random is in a mathematics course, what is the probability that his or her mathematics SAT score was below 500?

16. The number of nickels, dimes, and quarters in each of three boxes is as follows:

BOX	NICKELS	DIMES	QUARTERS
I	2	3	1
II	3	1	2
III	4	2	4

A box is chosen at random, and from it a coin is drawn at random. What is the probability that it is (a) a nickel? (b) a dime? (c) a quarter?

17. An examination has ten true-false questions. If you guess blindly, what is the probability that you will get at most one wrong?
18. A baseball player has a batting average of 0.300. What is the probability (a) exactly two hits in five at-bats? (b) at least one hit in three at-bats?
19. In a straw vote of 3000 students, 1200 voted in favor of drafting women for military service. If four students are chosen at random from this group, what is the probability that exactly three will have voted against drafting women?
20. In a game of chance a player tosses two fair coins. If two heads show, he wins \$2; if one head shows, he wins \$1; if no head shows, he wins \$4. How much should the player be willing to pay per game for it to be fair?
21. There are three defective radios in a shipment of 12. If a sample of two is chosen, how many can be expected to be defective?

CHAPTER TEST

1. A pair of fair dice are rolled and the sum is noted. Describe an equiprobable sample space for this experiment; then find the probability that the sum is (a) six; (b) three or four; (c) less than or equal to ten.
2. Find $p[(X \cap Y)']$ if $p(X) = 0.4$, $p(Y) = 0.5$, and $p(X \cup Y) = 0.7$.
3. Find $p(E \cup F)$ if $p(E) = 0.55$, $p(F) = 0.40$, and if E and F are (a) independent; (b) mutually exclusive.
4. A card is picked at random from an ordinary bridge deck of 52 cards. What is the probability that it is (a) a spade? (b) a club or a face card? (c) not a diamond and not an ace? (d) a face card, given that the card is black?
5. Ten high-school seniors, seven girls and three boys, are eligible for three prizes. If the prize-winners are chosen from this group at random, what is the probability that (a) all three winners are boys? (b) at least one winner is a boy? (c) all three winners are boys, given that at least one is a boy?
6. Suppose that 7% of a particular population is color-blind and that color-blind males comprise 6% of the population. What is the probability that a person chosen at random and found to be color-blind will be male?
7. The probability that a human being has Rh-positive blood is 0.82. If it is known that a person is Rh-positive, then the probability that he or she is type A is 0.40. What is the probability that someone chosen at random has type A Rh-positive blood?
8. A fair coin is tossed in succession three times. Consider the events

   E: the coin shows heads at least twice in succession;
   F: a head never appears;
   G: a tail comes up on the first toss;
   H: a head shows on the second toss.

   (a) Prove that events E and F are dependent. (b) Prove that events G and H are independent. (c) Prove that events G and H are not mutually exclusive. (d) Find $p(H|G)$.
9. Calculators are manufactured at only two plants, with 70% of the output coming from plant I. It has been found that 3% of I's output is defective but only 1% of II's are imperfect. (a) What is the probability that a calculator chosen at random is defective? (b) Suppose a calculator selected at random is found to be defective; what is the probability that it was made at plant II?

10. The number of balls of various colors in two urns is given below:

URN NO.	RED	WHITE	BLUE
I	5	2	3
II	3	1	4

An urn is selected at random and one ball is selected from it at random. Find the probability that the ball (a) is blue; (b) is blue and came from urn II; (c) came from urn II, given that it is blue; (d) is white, given that it came from urn I.

11. Find the probability in a family of five children that (a) all five will be girls; (b) at least two will be boys.
12. Slips of paper numbered 1 through 11 inclusive are placed face down on the floor. A blindfolded player chooses a slip. If he guesses the number on it, he wins $5. How much should he be willing to pay to play this game if he insists on its being fair?
13. A multiple-choice test has 20 questions, each with four choices. If each correct answer is worth five points, write an expression for (but do not evaluate) the probability that a student who guesses blindly will get at least 85 points.
14. With the data of question 13, what is the expected value on the test for a student?
15. In a survey of 2000 adults, 60% indicated a preference for prohibiting smoking on airplanes during flights. If five members of this population are interviewed, what is the probability that at least one does not favor prohibiting smoking during flights?

Chapter 3

Linear functions, equations, and systems

One of the major topics of this book is linear programming, to be taken up in a later chapter. At that point we will consider how to set up and solve problems of an important class, for which the material of the present chapter is essential. We begin with the concept of function, then focus specifically on linear functions and their applications.

SECTION 14. FUNCTIONS AND FUNCTIONAL NOTATION

One of the most fundamental concepts in mathematics is that of function. The demand for hand calculators, for example, is a function of--that is, depends (in part) on--their price; the amount of a monthly mortgage payment on a house is a function of the prevailing interest rate; the size of your life-insurance premium is a function of your age. Indeed, all the applications in this book, whether to business, physical and biological science, or social science, involve functions.

When you complete this section you should be able to (1) recognize when a relationship is a function, (2) specify the domain and range of a given function, and (3) evaluate a function at an element of its domain.

DEFINITION AND EXAMPLES

A *function* is a correspondence between two sets that associates with each element of one set, the *domain*, one and only one element of the other set, the *range*.

Example 14-1

The distance covered by a jet flying at an average speed of 500 miles per hour (mph) is a function of the time the jet is in transit. If s represents the distance in miles and t the time in hours, then the function is

$$s = 500t$$

If the total time is $7\frac{1}{2}$ hours (h) then the domain is $0 \leqslant s \leqslant 7\frac{1}{2}$; the range is $0 \leqslant s \leqslant 3750$. This function associates a *unique* distance s with each time t in the domain.

Example 14-2

The association between a particular set of companies on the New York Stock Exchange and their closing prices on a specific day defines a function. The domain is the particular set of companies; the range is the set of prices at which they closed on a given day. Each company closed at only *one* price.

Example 14-3

The cost of producing Presto hand calculators is given by the equation

$$C = 500 + 2x$$

where x is the number of calculators and C is the cost in dollars. If Presto manufactures no more than 10,000 calculators, then the domain of this function is the set of positive integers between 0 and 10,000. (Limiting the domain to positive integers says that Presto manufactures only whole calculators!)

Example 14-4

The circumference of a circle is a function of its radius. This is usually expressed by the equation

$$C = 2\pi r$$

where the domain is $r \geqslant 0$ (with $r = 0$ only for a "point" circle).

Example 14-5

If you earn simple interest on an investment of $4000 at the rate of $8\frac{1}{2}\%$ per year (yr), then the function

$$I = 0.08(4000)t$$

gives the interest I earned in dollars after t yr.

Example 14-6

If A is the set of people in a certain city and B is the set of their mothers, then there is a function from A into B since each person has a unique mother. However, since many mothers have more than one child, the correspondence between the sets B and A with B as domain is not a function.

This example shows that we can think of a function as mapping its domain into its range. (A map of the United States is a function from the set of cities, towns, villages, and other actual topographical features to the points on the paper that identify their relative locations.)

Example 14-7

A function may be defined by a table, such as this one:

HIGH TEMPERATURE (°C) IN SELECTED U.S. CITIES ON 24 NOVEMBER 1976

City	°C	City	°C
Albany	4	Honolulu	27
Albuquerque	16	Juneau	4
Boston	7	Miami Beach	21
Chicago	- 1	Oklahoma City	16
Denver	16	San Francisco	22
El Paso	19		

The domain of this function is the set of cities in the left-hand column; the range includes the set of numbers {-1,4,7,16,19,21,22,27}. This correspondence is a function since each city listed maps into a single temperature. Note that two or more cities may (and sometimes do!) map into the same temperature.

Example 14-8

A function may be defined by a graph:

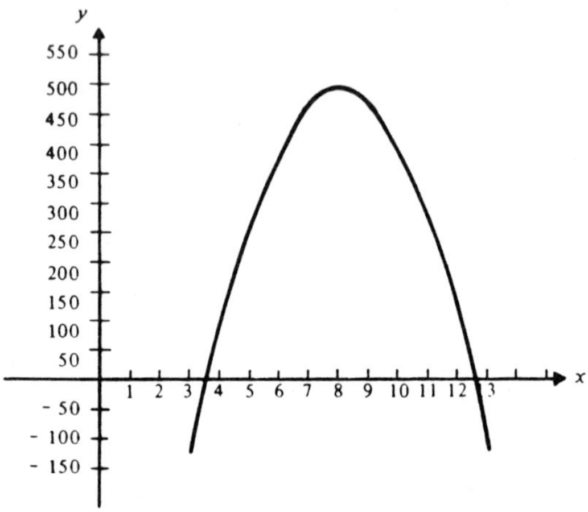

Figure 14.1

Here the domain is $3 \leqslant x \leqslant 13$ and the range is $-125 \leqslant y \leqslant 500$. For each x in the domain of the function there is a unique y. But different x's may (and sometimes do!) produce the same y: for example, when $x = 6$, $y = 400$, and when $x = 10$, again $y = 400$.

FINDING THE DOMAIN OF A GIVEN FUNCTION

The functions of most interest to us are those whose domains are sets of numbers. *If the domain is not specified, we take it to be the largest subset of the reals for which the function is defined.* (The *reals* are the numbers corresponding to all the points on a number line. We will use the symbol R for the entire set of reals.)

Example 14-9

Find the domain of each function:

(a) $y = \dfrac{1}{x - 2}$ (b) $y = \dfrac{x + 1}{x^2 - x}$ (c) $y = \sqrt{x - 9}$

(a) Since we cannot divide by zero, no denominator may equal zero. The domain is therefore the set of all reals except 2. This is often written "$x \neq 2$."

(b) The denominator can be factored: $x^2 - x = x(x - 1)$. Since this denominator equals zero if $x = 0$ or 1, the domain is all reals except these numbers. We can write "$x \neq 0,1$."

(c) The square root (or any other even root) of a negative number is not a real number. Since $x - 9$ is negative, the domain of $y = \sqrt{x - 9}$ is the set of numbers greater than or equal to 9. Note that $\sqrt{0} = 0$. Remember that if b is positive, then $\sqrt{b}$ is *defined* to be the *positive* number c such that $c^2 = b$. Thus, for example, $\sqrt{9} = 3$, $\sqrt{(-4)^2} = 4$, $\sqrt{8} = 2\sqrt{2}$.

Generally, then, a function is not defined at any number that leads to division by zero or to an even root of a negative number. Remember that zero divided by any number except zero is well defined!--it equals zero.

NOTATION

The notation used to represent a function f of a variable x is "$f(x)$," read "f of x." Also, this same notation denotes the element in the range associated by the function with x. If $f(x) = 3x - 1$, then

$$\begin{aligned} f(-1) &= 3(-1) - 1 = -4 \\ f(0) &= 3(0) - 1 = -1 \\ f(5) &= 3(5) - 1 = 14 \\ f(a) &= 3a - 1 \\ f(-x) &= 3(-x) - 1 = -3x - 1 \end{aligned}$$

Note that in the expression "$f(x) = 3x - 1$" the symbol "$f(x)$" simply replaces the symbol "y" in "$y = 3x - 1$." Also, "$f(5) = 14$" is a precise and brief way of saying "the value of the function f when $x = 5$ is 14."

When y is a function of x we often write "$y = f(x)$" and refer to x as the *independent variable* and to y as the *dependent variable*. Of course, letters other than f are also used to represent functions.

Example 14-10

If $g(x) = x^2 + 1$, find the domain and range of g. Evaluate $g(0)$, $g(-2)$, $g(\sqrt{2})$, $g(1 + h)$.

The domain of g is the set R of real numbers; its range consists of all the reals that are greater than or equal to 1.

We have

$$\begin{aligned} g(0) &= 0^2 + 1 = 1 \\ g(-2) &= (-2)^2 + 1 = 4 + 1 = 5 \\ g(\sqrt{2}) &= (\sqrt{2})^2 + 1 = 2 + 1 = 3 \\ g(1+h) &= (1+h)^2 + 1 = 1 + 2h + h^2 + 1 = 2 + 2h + h^2 \end{aligned}$$

EXERCISES

1. Which of the descriptions in (a) through (c) below define functions (i) with domain S and range T; (ii) with domain T and range S?
   (a) S is the set of 40,000 people who live on St. Croix and T is the set of twelve months during which one may be born.
   (b) S is the set of all the people alive in the world today and T is their right-thumb prints.
   (c) S is the set of 4000 students at a certain college and T is the set of their weights rounded to the nearest pound.
2. Find the domain of each function in (a) through (j):

   (a) $f(x) = x^2 - x + 3$ (b) $y = \dfrac{3}{x}$

   (c) $g(x) = \dfrac{2x}{x - 3}$ (d) $f(x) = \sqrt{x}$

   (e) $h(x) = \dfrac{x + 3}{x^2 - 4}$ (f) $y = \sqrt{x + 3}$

   (g) $g(x) = \dfrac{5}{2x^2 + 1}$ (h) $f(x) = x^{100} - 25x^2$

   (i) $h(x) = \dfrac{1 - x^2}{4x - x^2}$ (j) $y = \sqrt{2 - x}$

3. The cost of producing x calculators in Example 14-3 is given by the function $C(x) = 500 + 2x$. How much does it cost to produce ten calculators? 250? 1000? What happens to this cost function as increases? Find $C(0)$ and suggest an economic interpretation for it.
4. If $f(x) = 8$, find $f(-2)$, $f(0)$, $f(33)$, $f(b)$. What is the domain of this function? What is its range?
5. If $f(x) = 2x^2$, find $f(-1)$, $f(0)$, $f(1)$, $f(5)$, $f(a)$.
6. Let $f(x) = x^2 + 2$. Find $f(-3)$, $f(0)$, $f(4)$.
7. If $g(x) = (x + 5)^2$, evaluate $g(2)$, $g(-2)$, $g(0)$. For what x does $g(x)$ equal zero?
8. If $g(x) = \dfrac{2}{x - 3}$, find $g(1)$, $g(-1)$, $g(0)$, $g(1 + h)$.
9. In (a) through (e), tell whether each statement is true or false:
   (a) "$f(x)$" means "f times x."
   (b) If $f(2) = 2x$, then $f(x) = 2f(x)$.
   (c) If $f(x) = x^2$, then $f(-x) = f(x)$.
   (d) If $f(x) = 3x + 1$, then $f(1) + f(2) = f(1 + 2)$.
   (e) If $f(x) = 4x$, then $f(-x) = -f(x)$.

TEST A

1. Find the domain of each function:

(a) $f(x) = 4x + 3$ (b) $f(x) = \dfrac{x - 1}{x + 2}$

(c) $f(x) = \sqrt{x + 3}$ (d) $f(x) = \dfrac{1}{4x - x^2}$

2. Find $f(1)$, $f(-1)$, $f(0)$ if

(a) $f(x) = 4$ (b) $f(x) = x^2 - 1$

(c) $f(x) = \dfrac{1}{x - 2}$ (d) $f(x) = \sqrt{2 - x}$

TEST B

1. Find the domain of each function:

(a) $f(x) = x^2 - x - 1$ (b) $f(x) = \dfrac{1 - x^2}{x - 2}$

(c) $f(x) = \sqrt{5 - x}$ (d) $f(x) = \dfrac{x - 1}{x^2 - 1}$

2. Find $f(2)$, $f(-2)$, $f(0)$ if

(a) $f(x) = 2x + 500$ (b) $f(x) = (x - 2)^2 + 2$

(c) $f(x) = \dfrac{1}{x^2 - 1}$ (d) $f(x) = \sqrt{4 - x}$

SECTION 15. EQUATIONS OF LINES

Many applications of mathematics involve linear functions, whose graphs are straight lines. Moreover, even if a function is nonlinear, we often use a linear function to approximate its behavior. In this chapter and in the next two we will use both linear equations and linear inequalities to solve many practical problems.

In this section we aim to find and graph equations of lines and to note properties of linear functions and of lines.

The function

$$f(x) = mx + b \qquad \text{or} \qquad y = mx + b$$

where m and b are fixed real numbers, is called the *linear function*. Since its graph is a nonvertical straight line, we often call the function the line and vice versa.

The *y-intercept* of any graph is the point where the graph intersects the y-axis. We can therefore find it by letting $x = 0$ in the equation for y. So the y-intercept of the line $y = mx + b$ equals $m \cdot 0 + b$ or b (see Figure 15.1).

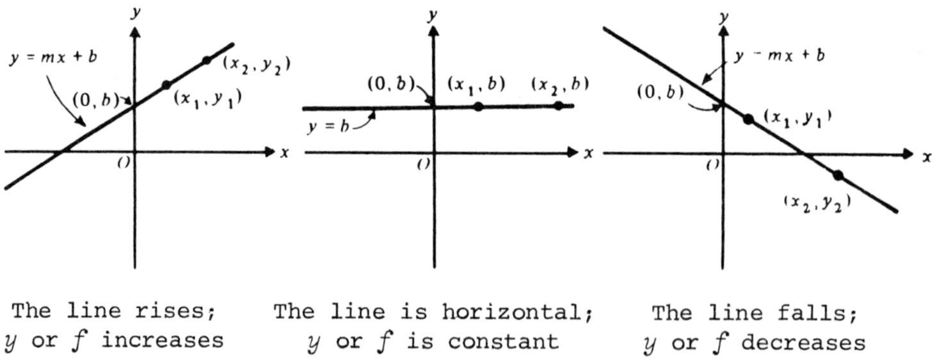

The line rises; y or f increases | The line is horizontal; y or f is constant | The line falls; y or f decreases

Figure 15.1

If two different points, say (x_1,y_1) and (x_2,y_2), lie on the line $f(x) = mx + b$, then

$$f(x_1) = mx_1 + b = y_1 \tag{1}$$
$$f(x_2) = mx_2 + b = y_2 \tag{2}$$

Subtracting (1) from (2) yields

$$m(x_2 - x_1) = y_2 - y_1$$

and

$$m = \frac{y_2 - y_1}{x_2 - x_1}$$

The number m is the ratio of the change in y to the change in x. It measures the steepness of the line and is called the *slope* of the line.

In Figure 15.1 we see the graph of $y = mx + b$ when $m > 0$, $m = 0$, and $m < 0$. In each sketch we choose $x_2 > x_1$; that is, $x_2 - x_1 > 0$; this is equivalent to "moving" along the line from left to right.

When $m > 0$, then $y_2 > y_1$. We say the function is *increasing*.

When $m = 0$, we have the function $f(x) = b$ (or $y = b$), whose graph is a horizontal line. This is call the *constant function*; the value of y is the same for every x and equals b.

When $m < 0$, then $y_2 < y_1$, and the function *decreases*.

Example 15-1

Graph the equation $x = a$, where a is a constant.

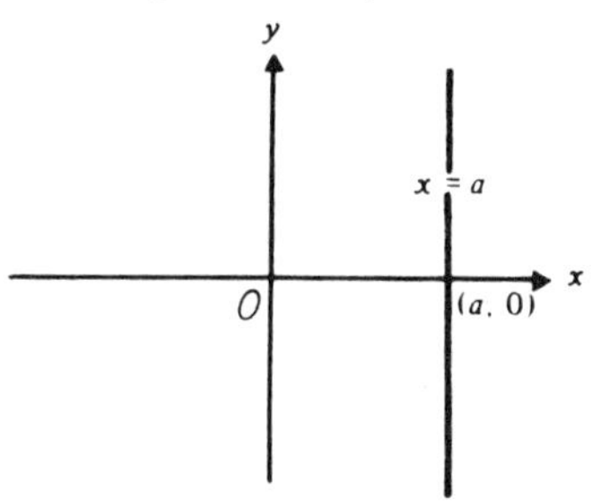

Figure 15.2

The graph is a vertical line through the point $(a,0)$. But note that this graph does *not* represent a function. (Why?)

Example 15-2

Graph $y = -3x + 2$.

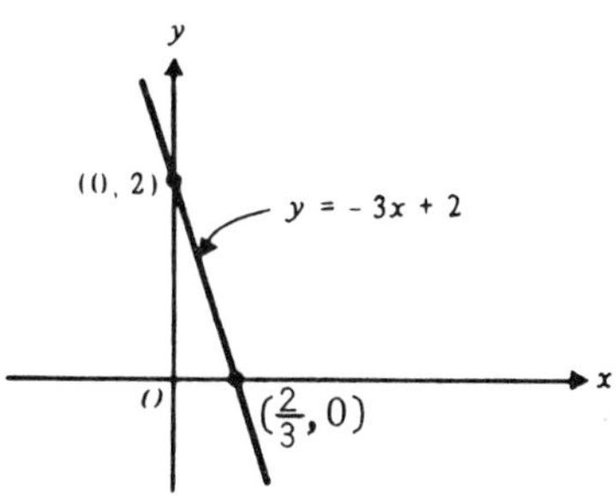

Figure 15.3

The easiest way to graph a linear function is to plot both intercepts. When $x = 0$, $y = 2$; when $y = 0$, $0 = -3x + 2$, $3x = 2$, and the x-intercept is $\frac{2}{3}$. Plot these points, then draw the line through them.

Example 15-3

Graph the line through the points $(-1,2)$ and $(3,4)$ and find its equation.

The slope of the line is

$$m = \frac{4-2}{3-(-1)} = \frac{2}{4} = \frac{1}{2}$$

So we have

$$y = \tfrac{1}{2}x + b$$

We find b by using the fact that (-1,2) lies on the line:

$$2 = \tfrac{1}{2}(-1) + b$$

$$\tfrac{5}{2} = b$$

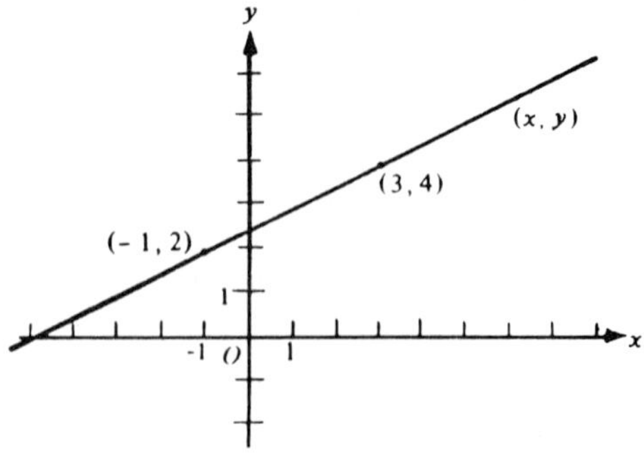

Figure 15.4

So the desired equation is

$$y = \frac{1}{2}x + \frac{5}{2}$$

We could also have used the point (3,4) to find b; instead, we can verify that (3,4) satisfies the equation:

$$4 \stackrel{?}{=} \frac{1}{2}(3) + \frac{5}{2} \qquad 4 \stackrel{?}{=} \frac{3}{2} + \frac{5}{2} \qquad 4 = 4$$

Although we found the equation of the line above by determining b, this is not obligatory. Since the coordinates of any point on the line are denoted by (x,y), and since we already know that the slope of the line through (-1,2) and (3,4) is $\frac{1}{2}$, it follows that

$$\frac{y-4}{x-3} = \frac{1}{2} \tag{1}$$

which yields

$$y - 4 = \frac{1}{2}(x-3) \tag{2}$$

or

$$y = \frac{1}{2}x + \frac{5}{2}$$

as before. Note that Equation (1) holds for all points other than (3,4) but that Equation (2) is satisfied also by (3,4).

Forms of Linear Equations

If a line with slope m contains the point (x_1,y_1), then its equation is

$$y - y_1 = m(x - x_1)$$

This is called the *point-slope form* of the equation for a line, while

$$y = mx + b$$

is called the *slope-intercept* form.

Example 15-4

It costs the Presto Company $(500 + 2x)$ dollars to produce x calculators (see Example 14-3). Draw the graph of the function $C(x) = 500 + 2x$.

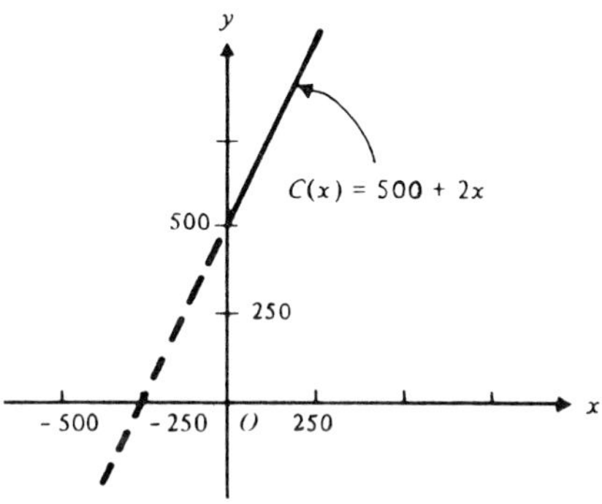

Figure 15.5

Note that the y-intercept is 500. This is the fixed cost that does not depend on x. The x-intercept is -250. With these points plotted, we can draw the line quickly. We dot the part of the line corresponding to $x < 0$ because the function is meaningless there. By drawing the graph for $x \geqslant 0$ continuously (without any breaks or holes), we're implying that x may be *any* nonnegative number, not just an integer. Although this does not make sense practically, we make such assumptions in most problems of this sort for ease of mathematical handling.

We see immediately in this example that the cost increases as x does, either from the graph or from the equation itself. The equation tells us that the slope is 2.

Additional Important Facts about Lines

(1) A vertical line has no slope; that is, slope is not defined for such a line. (As has already been noted, the slope of a horizontal line is zero.)

(2) Nonvertical parallel lines have equal slopes.

(3) If two nonvertical lines are perpendicular, then the product of their slopes is -1; or, the slope of either line is the negative reciprocal of the slope of the other.

(4) The line $y = mx$ goes through the origin. When $m > 0$, then y is said to *vary directly* as x, or to be *directly proportional* to x. We call m the *factor of proportionality*.

(5) The equation of any line can be written in the form

$$Ax + By = C$$

where A and B are not both zero.

Example 15-5

Find the equation of the line parallel to $3x + 4y = 5$ and containing the point $(6,-1)$.

Since the given line equation can be written in the form

$$y = -\frac{3}{4}x + \frac{5}{4}$$

its slope is $-\frac{3}{4}$. Any line parallel to it is therefore of the form

$$y = -\frac{3}{4}x + b$$

because parallel lines have equal slopes. Since the point (6,-1) lies on the required line, we have

$$-1 = -\frac{3}{4}(6) + b$$

so that

$$b = \frac{7}{2}$$

The equation is therefore

$$y = -\frac{3}{4}x + \frac{7}{2} \qquad \text{or} \qquad 3x + 4y = 14$$

Example 15-6

Find the equation of the line passing through (-4,-3) and perpendicular to $2x - 5y = 10$.

We can rewrite the given equation in the form

$$y = \frac{2}{5}x - 2$$

Any line perpendicular to this one has slope $-\frac{5}{2}$ and is of the form

$$y = -\frac{5}{2}x + b$$

Since (-4,-3) lies on the line, we have

$$-3 = -\frac{5}{2}(-4) + b \qquad -3 = 10 + b \qquad b = -13$$

The required equation is therefore

$$y = -\frac{5}{2}x - 13 \qquad \text{or} \qquad 5x + 2y = -26$$

Example 15-7

An airline charter service charges \$25 plus 50¢ a mile. Draw a graph of their rate schedule.

See Figure 15.6. Converting cents to dollars we have that the rate in dollars for a flight of x miles is $R(x) = 25 + 0.5x$. Note that R is defined only for nonnegative x.

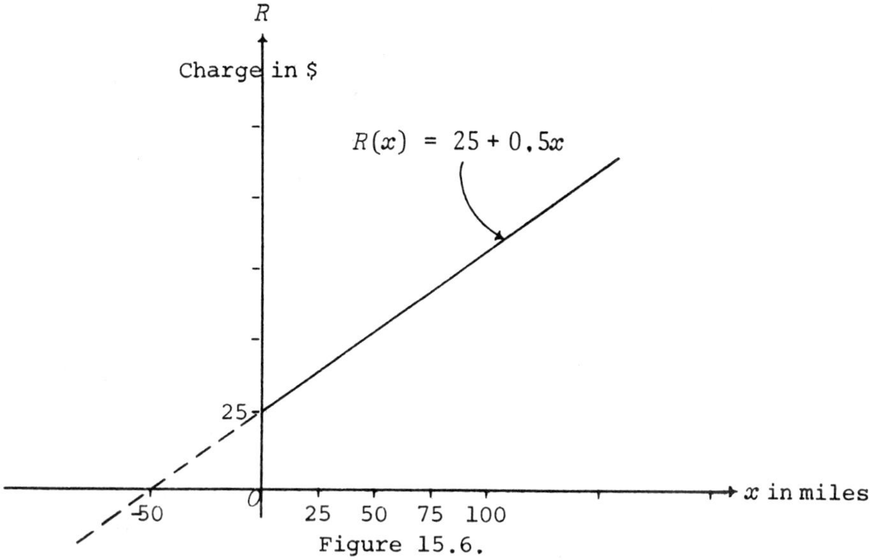

Figure 15.6.

STRAIGHT-LINE DEPRECIATION

One acceptable technique for depreciating certain property for tax purposes is called *linear* or *straight-line depreciation*. With this method, it is assumed that the loss in value of the property is the same each year during its useful life. If, for example, a \$7000 car is sold for \$1000 after three years, then its annual depreciation (for each of the three years) is (7000 - 1000)/3 = \$2000. The net cost C is equal to the original cost minus any scrap or salvage value. To evaluate the remaining undepreciated value v, after t years, we multiply the allowable annual depreciation by t and subtract from C. So, letting d be the annual depreciation, we have

$$v = C - dt = 6000 - 2000t$$

for the car. Since C and d are both constants, we see that the equation defines the undepreciated value v as a linear function of t.

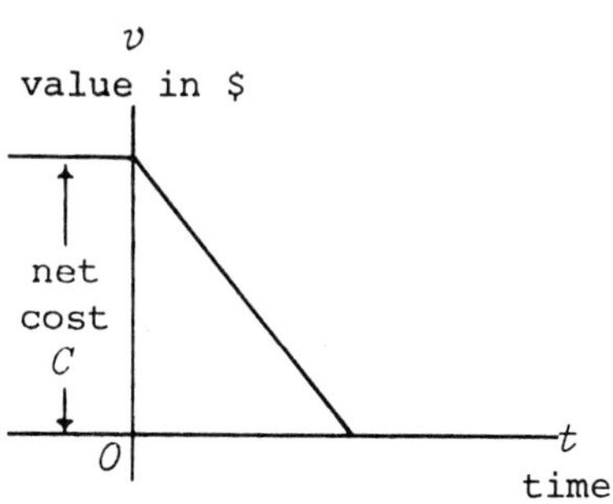

The slope of the line is $-d$, where d is the annual depreciation or loss in value each year. The y-intercept is C, the net cost.

Figure 15.7.

Example 15-8

A piece of farm equipment costing \$20,000 has a useful life expectancy of 10 years, after which its scrap value will be \$3000. Find its value after t years.

Since the net cost, in dollars, is $20{,}000 - 3000 = 17{,}000$ and its useful life is 10 years, the annual depreciation is $17{,}000/10 = 1700$. Its value v, after t years, is therefore (in dollars)

$$v = 17{,}000 - 1700t$$

A Note on Depreciation Methods

As of 1972 the depreciation rules of the U.S. Internal Revenue Service have been radically modified, mostly to allow more realistically for the fact that the actual value of business facilities and equipment diminishes more rapidly when it is relatively new than it does later on. As a result of these changes, the straight-line method described and illustrated above can be used only in a minority of cases. (Even the definition of the term "straight-line" has been modified!) Familiarity with the traditional straight-line method nevertheless continues to be useful.

A POPULATION PROBLEM

The methods of this section can also be used to express a population as a function of time, when the population is assumed to vary linearly with time.

Example 15-9

Suppose there are 25,000 bacteria on a Petri dish and that a chemical is introduced which kills all the bacteria gradually over a five-hour interval. Express the population P as a function of time t, where $0 \leqslant t \leqslant 5$, assuming that the relation is linear.

We know two points on the graph of $P(t)$, namely $(0, 25\,000)$ and $(5,0)$. The slope of the graph is therefore

$$(25\,000 - 0)/(0 - 5)$$

or -5000. Since the y-intercept is 25,000, the equation is

$$P(t) = -5000t + 25{,}000$$

where $P(t)$ is the number of bacteria alive after t hours.

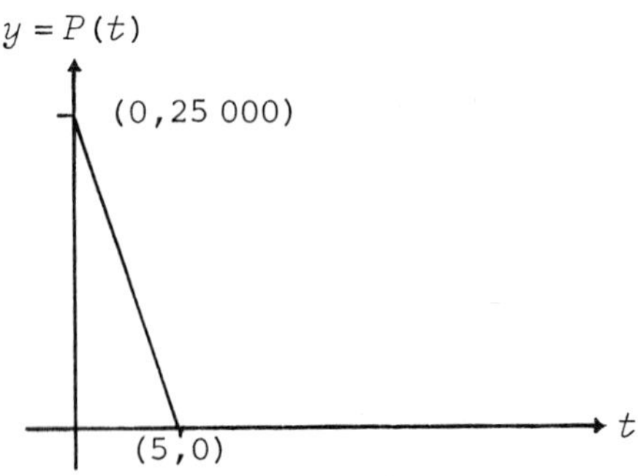

Figure 15.8

EXERCISES

1. Find the slope, if it exists, of the line through

(a) (2,3) and (4,9) (b) (-1,3) and (4,-7)
(c) (2,-1) and (2,3) (d) (0,3) and (-2,0)
(e) (-1,0) and (-1,4) (f) (3,-2) and (-4,-2)
(g) (2,4) and $(2+h, 2(2+h))$
(h) $(x, 3x-2)$ and $(x+h, 3(x+h)-2))$

2. Find the equation of the line

(a) whose slope is 4 and whose y-intercept is 1.
(b) with y-intercept -2 and slope $\frac{1}{2}$.
(c) with slope 3 and containing the point (1,2).
(d) with slope $-\frac{1}{2}$ and whose x-intercept is -2.
(e) with y-intercept 5 and slope 0.
(f) passing through (1,-3) and with slope 0.

3. Find the equation of the line through each pair of points in Problem 1 above.
4. Graph:

(a) $x - 3y = 3$ (b) $2x + y + 4 = 0$ (c) $f(x) = x - 2$
(d) $2x + 5y = 3$ (e) $y = 3x$ (f) $g(x) = -3$
(g) $x = 4$ (h) $x = 0$ (i) $y - x = 0$
(j) $y = 0$

5. Graph each of the following pairs of lines on a single set of axes; then estimate the coordinates of any point of intersection.

(a) $\begin{cases} 3x - y = 0 \\ 2x + 3y = 11 \end{cases}$ (b) $\begin{cases} 2x - 3y = 6 \\ 4x - 6y = 12 \end{cases}$

(c) $\begin{cases} x - 3y = -2 \\ 2x + y = -4 \end{cases}$ (d) $\begin{cases} y = 3x - 2 \\ 3x - y + 2 = 0 \end{cases}$

6. Find the equation of a line that

(a) is horizontal and passes through (2,-4).
(b) is parallel to the line $y = 2x + 1$ and goes through (-1,2).
(c) is perpendicular to the line $2x + y - 3 = 0$ and has y-intercept 2.
(d) is parallel to $2x - 3y = 2$ and has y-intercept -2.
(e) is perpendicular to $x - 3y = 2$ and goes through (2,1).

7. Which of the following linear functions are increasing? decreasing? neither?

(a) $f(x) = x - 2$ (b) $g(x) = 2$ (c) $h(x) = -x + 1$
(d) $y = 0.01x + 500$ (e) $f(x) = -5$ (f) $g(x) = 250x - 251$

8. The Wick Candle Company has fixed costs of \$750 per month and it costs them 20¢ to produce a single candle. (a) Find the total

monthly cost of producing x candles. (b) How much does it cost to produce 10,000 candles a month? (c) What is the domain of the cost function?

9. Each of the following represents the initial cost, scrap value, and useful life of a piece of machinery. Find a function representing the value v of the machinery after t years, assuming straight-line depreciation.

   (a) \$5000, \$1000, 10 years (b) \$8500, \$2000, 25 years
   (c) \$15,000, \$0, 50 years (d) \$125, \$5, 6 years

10. A pharmaceutical company claims that its mouthwash can kill 240,000 bacteria in one minute. If we assume that the death rate is linear, find a function expressing the number of bacteria killed in t seconds, where $0 \leqslant t \leqslant 60$.
11. A function $n(p)$ for the number of items that can be sold at p dollars each is called a *demand* function. If 75 seaside efficiency units can be rented at \$50 a night, but only 50 units at \$60 each, find the demand function, assuming it is linear. How many units can be rented at \$30 a night?
12. Solve the equation of Problem 11 for p in terms of n. What daily price should be charged to rent 100 units?
13. A function $S(p)$ for the number S of items a seller will supply at unit price p is called a *supply* function. Suppose an electronics firm is willing to supply 10,000 calculators at \$20 each but only 8,000 at \$19. Find the supply function, if it is assumed to be linear. How many calculators will the firm supply at \$22 each?
14. Suppose the interest earned on savings is directly proportional to the amount invested. If \$650 earns \$39 in one year, then (a) how much will \$900 earn in one year? (b) how much must be invested to earn \$300 interest in one year?
15. Temperatures in degrees Fahrenheit (F) and in degrees Celcius (C) are related as follows:

$$C = \frac{5}{9}(F - 32)$$

   (a) Find C when F = 32, 98.6, 212. (b) Express F as a function of C. (c) Find F when C = 40, 50, 100.
16. Fund the particular value of k for which

   (a) $kx - 2y + 7 = 0$ has slope -3.
   (b) $4x + y = k$ has x-intercept 6.
   (c) $y = kx + 3$ passes through (-2,5).

TEST A

1. Find the slope of the line containing the points

(a) (1,4) and (3,8) (b) (0,-2) and (-1,0)
(c) (-3,4) and (5,0)

2. Find the equation of each line and express it in the form $y = mx + b$:

(a) The line containing the points (0,2) and (-1,2).
(b) The line whose slope is -2 and that passes through (3,-1).
(c) The line parallel to $2x - y = 4$ and with y-intercept 5.

3. Graph:

(a) $f(x) = -x + 1$ (b) $3y + x = 6$
(c) $y = 5$ (d) $f(x) = 0.5x - 1$

TEST B

1. Find the slope of the line containing the points

(a) (25, 100) and (35, 150) (b) (0,0) and (-1,1)
(c) (2,-5) and (0,-2)

2. Find the equation of the line

(a) containing the points (2,3) and (3,2).
(b) whose slope is 0 and that passes through (-1,-2).
(c) perpendicular to $x + 3y = 5$ and with y-intercept -2.

3. Graph:

(a) $y = 2x - 3$ (b) $g(x) = 0$
(c) $f(x) = -x - 1$ (d) $x - y = 1$

SECTION 16. SYSTEMS OF LINEAR EQUATIONS

In §15 we noted that the equation of any line can be written in the form $Ax + By = C$, where A, B, C are constants, A and B not both zero. This linear equation which involves the two variables or unknowns x and y is the general equation of a line in the plane. In this section we will also consider linear equations involving more than two variables. Since the variables arising from applications often satisfy more than one linear equation, we shall be particularly interested in the set of such equations, called a *system of linear equations*. We will see how different systems of linear equations can be derived from a variety of applications.

DEFINITION OF LINEAR EQUATION

A linear equation in three unknowns (x, y, and z) can be written

$$Ax + By + Cz = 0 \qquad (A, B, C \text{ not all zero})$$

Its graph, which is in three-space, is a plane.

When more than three variables are involved, it is often simpler to use lower-case letters with subscripts. Thus, in general, the *linear equation in n unknowns* (x_1, x_2, ..., x_n) is written

$$a_1x_1 + a_2x_2 + \ldots + a_nx_n = b$$

Here the constant coefficients a_1, a_2, ..., a_n cannot all equal zero. When $n \geqslant 4$, the graph cannot be drawn; indeed, in our three-dimensional world it cannot even be visualized.

Note that the linear equation in the two variables x and y is a special case of the general equation, with $n = 2$, $a_1 = A$, $a_2 = B$, $x_1 = x$, $x_2 = y$. A similar remark holds for the linear equation in three variables.

Some examples of linear equations are

$$3x + 2y = 5 \qquad x - y + 2z = 3 \qquad 4x = y + z + w$$

None of the following, however, is a linear equation (Why?):

$$x^2 + y = 2 \qquad \frac{1}{y} = x + 1 \qquad xy = 3$$

SYSTEMS OF LINEAR EQUATIONS

A system of linear equations is also called simply a *linear sys-*

tem. Some examples are

$$\begin{cases} x + 2y = 3 \\ 4x - y = 2 \end{cases} \qquad \begin{cases} x + y + z = 5 \\ 2x - y - z = 2 \end{cases} \qquad \begin{cases} x + 2y - z = 3 \\ 2x - y + 3z = 1 \\ x - 4y - z = 0 \end{cases}$$

The leftmost of these three is a linear system of two equations in two unknowns; the central one is a linear system of two equations in three unknowns; the one on the right is a linear system of three equations in three unknowns.

To *solve* a linear system is to find the values of the variables that satisfy all the equations.

Geometric Interpretation of Solutions

Although we postpone to §17 procedures for obtaining solutions of systems of equations, we note here that there are only three possibilities: (i) a unique solution; (ii) no solution; and (iii) an infinite number of solutions.

For a set of two linear equations in two unknowns, these possibilities can be interpreted geometrically as follows:

(i) The two lines intersect in a single point (that satisfies both equations and is therefore the *unique solution* of the system).

(ii) The lines are parallel (and so do not intersect at all, corresponding to *no solution* for the system).

(iii) The lines are identical (so all the points on the line satisfy both equations, yielding *infinitely many solutions* for the system).

Note that in (i) the slopes of the lines are different, whereas in (ii) and (iii) the slopes are the same.

Example 16-1

Discuss the solutions of each system:

(a) $\begin{aligned} x + 3y &= 2 \\ 2x + y &= 3 \end{aligned}$ (b) $\begin{aligned} -x + 2y &= 1 \\ -x + 2y &= 3 \end{aligned}$ (c) $\begin{aligned} 2x - y &= 2 \\ 4x - 2y &= 4 \end{aligned}$

(a) We can rewrite these equations as

$$y = -\frac{1}{3}x + \frac{2}{3}$$

$$y = -2x + 3$$

Since the slopes, $-\frac{1}{3}$ and -2, are different, the lines intersect in a single point.

(b) Rewriting here yields

$$y = \frac{1}{2}x + \frac{1}{2}$$

$$y = \frac{1}{2}x + \frac{3}{2}$$

Since the slopes are the same but the lines are different (they have different x-intercepts), the lines are parallel. They do not intersect; there is no solution.

(c) For the third system we have

$$y = 2x - 2$$

$$y = 2x - 2$$

Here the lines are identical: they "intersect" everywhere--along the entire line.

APPLICATIONS

We now illustrate how linear systems arise in a variety of situations.

Example 16-2

The Sunshine Resort Hotel has 50 rooms for rent. At \$40 a night per room, the revenue $R(x)$ from renting x rooms is given, in dollars, by $R(x) = 40x$. If the fixed overhead is \$450 and it costs \$10 a night in upkeep for each room that is rented, then the cost $C(x)$ of renting x rooms is given, in dollars, by $C(x) = 450 + 10x$. The point at which the revenue and cost "curves" (lines here, of course) intersect is called the *break-even point*. At this point the revenue equals the cost and the resort owner neither makes nor loses money. The break-even point in this example is the solution of the system

$$y = 40x$$
$$y = 450 + 10x$$

where $R(x)$ and $C(x)$ have both been replaced by y to reinforce the geometric interpretation of drawing both lines on the same coordinate system.

Example 16-3

At a selling price of p dollars per unit, a clothing manufacturer is willing to produce (that is, "will supply") $25p$ velvet evening gowns. Consumers are willing to buy ("will demand") $6000 - 5p$ such gowns at p dollars each. If we replace, by y, both the number of units the manufacturer will supply and the number of units consumers will demand, we obtain the system

$$y = 25p$$
$$y = 6000 - 5p$$

The solution of this system, the point of intersection of the lines, is called the *market* or *equilibrium price*. At this price, supply equals demand.

Example 16-4

There are at the moment 50 tons of waste in a river. Every week factories dump two more tons of waste into the river. However, a purification plant just starting up will remove three tons of waste per week. To determine how many weeks it will take to rid the river of (all!) pollution we would solve the linear system

$$y = 50 + 2w$$
$$y = 3w$$

where we use y to represent both the amount of waste remaining after w weeks and the amount of waste removed after w weeks.

NOTE. This is an interesting illustration of how mathematics simplifies or "idealizes" the real world. Mathematically we can easily determine, from the model, precisely when the river will be waste-free, but it is highly unlikely that that state of affairs would ever really be reached. Actually, the rate at which waste can be removed is probably *not* independent of the rate at which it is dumped in. Our extremely simple mathematical model obscures that reality, thus affording us only a "first approximation." More complicated mathematical models, of course, can yield closer approximations.

Example 16-5

A dealer sells three TV models for \$100, \$400, and \$550 respectively. During one week he sells 35 sets altogether for a total of \$12,500. What system of equations represents the week's sales?

First we introduce variables for the numbers of each model sold during the week: let these be x, y, and z respectively. Since the dealer sold a total of 35 sets, we have $x + y + z = 35$. The total revenue, in dollars, from sales of x sets at \$100, y at \$400, and z at \$550 is given by $100x + 400y + 550z = 12,500$. The system is thus

$$x + y + z = 35$$
$$100x + 400y + 550z = 12,500$$

Example 16-6

A pharmacist is to prepare 100 milligrams of a medication that is a mixture of demerol, phenergan, and compazine. The prescription specifies that one tenth be compazine and that there be twice as much demerol and phenergan. Find a system of equations for the mixture.

We let x, y, and z be the number of milligrams respectively of demerol, phenergan, and compazine in the mixture. Noting that one tenth of 100 is 10, we have the system

$$\begin{aligned} x + y + z &= 100 \\ z &= 10 \\ x &= 2y \end{aligned}$$

Example 16-7

For a public-opinion survey a researcher has lists of names of Democrats, Republicans, and independents from which to choose for a poll. He wishes to interview 300 people, using three times as many Democrats and Republicans and with the total number of Democrats and Republicans to be twice the number of independents. If we let x, y, and z be the numbers respectively of Democrats, Republicans, and independents, we get the linear system

$$\begin{aligned} x + y + z &= 300 \\ x &= 3y \\ x + y &= 2z \end{aligned}$$

SUMMARY

The best way to develop skill in handling word problems is to try lots of them. In finding functions described in words, it will often help to consider the following suggestions:

(1) Read the problem carefully (often more than once); make a list of what information is given and what is sought. Introduce variables, but only where necessary.
(2) Try to decompose the problem into smaller (and usually easier!) problems. For example: profit equals revenue minus total cost; revenue equals number sold times unit cost; total cost equals fixed cost plus variable cost; and so on.
(3) Draw a sketch if relevant, labeling the parts with variables or with the given data.
(4) Express algebraically any relationships that are given; write down relevant formulas, using any information given.
(5) It may be useful to find the value of the function for one or two particular values of the variable, so you can generalize your procedure.
(6) If your choice of variable leads to an equation that is too complicated, consider an alternative.
(7) Be careful about units, of either money or measurement!
(8) Simplify your results arithmetically and algebraically.

EXERCISES

1. Which of the following are linear equations?

(a) $x + 3y = 2$ (b) $xy - z = 0$
(c) $x + y^2 + z = 3$ (d) $5x + 3 = 0$
(e) $x + y + 2x = w + 5$ (f) $3x_1 - 2x_2 + 4x_3 = 1$

2. In each of the following systems determine whether the lines are identical, parallel, or intersect in a single point.

(a) $3x - 2y = 5$
$4x + y = 2$

(b) $y = 2x - 1$
$4x - 2y = 2$

(c) $x = 3y - 1$
$3y - x = 2$

(d) $y = x + 1$
$x = y + 1$

(e) $2x - y = 3$
$y = 3x - 2$

(f) $-x + 2y - 1 = 0$
$3 + 3x = 6y$

3. The Salty Dog Company owns a beach shack from which they rent snorkel equipment to tourists at \$9 per day. Their overhead for building and salary is \$80 per day, and wear and tear on rented equipment is estimated to cost \$2 per rented set per day. Write a system of equations whose solution is the break-even point.

4. Prestige Press publishes a weekly shopping guide which is distributed free to the public. The press charges each advertiser \$6; overhead costs amount to \$200 weekly; and the cost of printing is \$1 per ad. Find a system of equations whose solution is the break-even point.

5. At a selling price of p dollars per unit a ship-builder is willing to produce $0.01p$ luxury power boats. Consumers are willing to purchase $1000 - 0.03p$ such boats at this price. Represent this situation by a system of equations whose solution is the market price.

6. A 12-ounce can of mixed nuts contains peanuts, cashews, and walnuts. By weight there are twice as many walnuts as cashews and three times as many peanuts as cashews. Represent this mixture by a system of equations.

7. An automobile dealer sells a stripped-down model for \$6500, a standard model for \$8500, and a luxury version for \$11,000. In a certain month he sells 20 cars for a total of \$157,500. What system of equations describes the month's sales?

8. After surgery a doctor orders folbesyn and tetracycline in normal saline solution. A pharmacist plans to prepare 1500 cm^3 of the mixture. He knows there should be 100 cm^3 of saline solution for every 1 cm^3 of tetracycline, and 250 cm^3 of saline solution for every 1 cm^3 of folbesyn. Find a system of equations

for the mixture.

9. A pharmacological firm is preparing a 10,000-gram concentrate from which to make vitamin-B pills. The concentrate is to contain equal amounts of vitamins B-1 and B-2, ten times as much B-6 as B-1, and 400 times as much B-6 as B-12. Describe the concentrate by a system of equations.

TEST A

1. Which of the following are linear equations?

   (a) $3x + 4y = z^2$ (b) $2x - y = 0$
   (c) $x_1 + 2x_2 = x_3$ (d) $4xy = 1$

2. In each system determine whether the lines are identical, parallel, or intersect in a single point.

   (a) $2x = y - 5$, $3y - 6x = 15$ (b) $x - y = 3$, $y = 2x - 3$ (c) $2y = x - 1$, $x = 2y - 1$

3. At a selling price of p dollars per unit, a machine shop is willing to turn out $0.05p$ units of a particular piece of heavy equipment, while consumers are apparently willing to purchase $200 - 0.15p$ units. Find a system of equations whose solution is the market price.
4. A nursery sells apple, pear, and cherry trees for \$10, \$12, and \$15 apiece respectively. One day the nursery sold 40 trees for a total of \$470. Write a system of equations for the day's sales.

TEST B

1. Which of the following are linear equations?

   (a) $4x_1 = x_2x_3$ (b) $5x = y + z$
   (c) $(x + 3y + z)^2 = 4$ (d) $2x + w = z + 5$

2. In each system determine whether the lines are identical, parallel, or intersect in a single point.

   (a) $3x - y = 5$, $5y = 3x - 1$ (b) $x + y = 1$, $5 - 5y = 5x$ (c) $x = y - 3$, $2y = 2x + 3$

3. It costs a publisher \$3 per copy to manufacture a certain book and his overhead is \$10,000. If the book sells for \$15 a copy, find the system of equations whose solution is the break-even point.
4. In a survey of 750 college students it was found that the same number smoked heavily as didn't smoke at all, while eight times as many smoked lightly as the number who smoked heavily. Represent these data by a system of equations.

SECTION 17. SOLUTIONS OF SYSTEMS OF LINEAR EQUATIONS

When we have found a system of linear equations as a mathematical model for some practical problem, we are usually interested in the (simultaneous) solution of the system: the values of the variables that satisfy all the equations of the system. In this section we review two methods of solution. Other methods, involving matrices, will be presented in the next chapter.

We first consider two linear equations in two unknowns.

SOLUTION BY SUBSTITUTION

Let the variables be x and y. We solve one of the equations for one of the variables, say x, and substitute this result into the other equation to obtain the numerical value for y.

Example 17-1

Solve the system

$$x + 2y = 7 \qquad \text{(E1)}$$
$$2x + 3y = 11 \qquad \text{(E2)}$$

First solve equation (E1) for x:

$$x = 7 - 2y \qquad \text{(E3)}$$

Now substitute this result for x in (E2):

$$2(7 - 2y) + 3y = 11$$

Simplify and solve for y:

$$14 - 4y + 3y = 11$$
$$y = 3$$

Now replace y by 3 in equation (E3):

$$x = 7 - 2(3) = 1$$

Check $x = 1$, $y = 3$ in the original equations, (E1) and (E2):

$$1 + 2(3) \stackrel{\checkmark}{=} 7$$
$$2(1) + 3(3) \stackrel{\checkmark}{=} 11$$

Example 17-2

Find the simultaneous solution of the system

$$4x - 3y = 6 \qquad \text{(E1)}$$
$$5x - y = 13 \qquad \text{(E2)}$$

It is easier here to solve (E2) for y:

$$y = 5x - 13 \qquad \text{(E3)}$$

Substitute this value of y in (E1):

$$4x - 3(5x - 13) = 6$$

Simplify and solve for x:

$$\begin{aligned} 4x - 15x + 39 &= 6 \\ -11x &= -33 \\ x &= 3 \end{aligned}$$

Replace x by 3 in (E3):

$$y = 5(3) - 13 = 2$$

Check the solution (3,2) in the original equations, (E1) and (E2):

$$4(3) - 3(2) \stackrel{\checkmark}{=} 6$$
$$5(3) - 2 \stackrel{\checkmark}{=} 13$$

Example 17-3

Suppose it costs $500 + 20x$ dollars to produce x units and that the revenue from this output is $30x$ dollars. What is the break-even point?

Since we seek x, the production level, for which the cost is the same as the revenue, the linear system to be solved can be written

$$\begin{aligned} y &= 500 + 20x \\ y &= 30x \end{aligned}$$

Setting these expressions for y equal to each other yields

$$\begin{aligned} 500 + 20x &= 30x \\ x &= 50 \end{aligned}$$

The break-even point is thus 50 units. The cost (and revenue) at this production level is \$1500.

Example 17-4

Solve the system

$$\begin{aligned} 2x + y &= 5 \\ 2x &= 6 - y \end{aligned}$$

Solving the first equation for y yields

$$y = 5 - 2x$$

When we replace y by this expression in the second equation, we get

$$2x = 6 - (5 - 2x)$$

which reduces to $0 = 1$. This is impossible, showing that the given system has no solution. More careful inspection reveals that the two lines are parallel; therefore they do not intersect.

Example 17-5

To solve the pair of equations

$$\begin{aligned} x + 3y &= 4 \\ 2x &= 8 - 6y \end{aligned}$$

we solve the first for x in terms of y, getting

$$x = 4 - 3y$$

then substitute in the second equation,

$$2(4 - 3y) = 8 - 6y$$

which yields $0 = 0$. A closer look at this system shows that the lines are identical. Thus every point on the common line is a solution of the system.

SOLUTION BY ADDITION AND SUBTRACTION

Two Equations in Two Unknowns

This method requires first that we operate on the equations in such a way as to get the same numerical coefficient for one of the variables, say for x. We then add (or subtract) the equations to eliminate the term in that variable.

Example 17-6

Solve the system

$$x + 3y = 12 \qquad \text{(E1)}$$
$$2x - 5y = 2 \qquad \text{(E2)}$$

Multiplying equation (E1) by two yields the equivalent system*

$$2x + 6y = 24 \qquad \text{(E1')}$$
$$2x - 5y = 2 \qquad \text{(E2)}$$

Now subtract (E2) from (E1'):

$$11y = 22$$

and solve for y:

$$y = 2$$

*Two systems of equations are equivalent if their solutions are identical.

Substitute this value for y in either of the original equations--we here choose (E1)--and solve for x:

$$\begin{aligned} x + 3(2) &= 12 \\ x &= 6 \end{aligned}$$

Now check $x = 6$, $y = 2$ in the original equations:

$$6 + 3(2) \overset{\checkmark}{=} 12$$

$$2(6) - 5(2) \overset{\checkmark}{=} 2$$

Example 17-7

To solve the system

$$\begin{aligned} 3x - 2y &= 12 \qquad \text{(E1)} \\ 4x + 3y &= -1 \qquad \text{(E2)} \end{aligned}$$

multiply (E1) by 3 and (E2) by 2:

$$\begin{aligned} 9x - 6y &= 36 \qquad \text{(E1}'\text{)} \\ 8x + 6y &= -2 \qquad \text{(E2}'\text{)} \end{aligned}$$

add (E1′) and (E2′): $17x = 34$

solve for x: $x = 2$

substitute 2 for x in (E2): $4(2) + 3y = -1$

solve for y: $y = -3$

Note that in the last two steps we could just as well have substituted 2 for x in (E1).

Always check in the original equations:

$$3(2) - 2(-3) \overset{\checkmark}{=} 12$$

$$4(2) + 3(-3) \overset{\checkmark}{=} -1$$

Example 17-8

Suppose that at a selling price of p dollars per unit, manufacturers are willing to produce $50p$ units and that consumers are willing to buy $300 - 25p$ units. What is the market price?

An appropriate model for this supply-demand problem is the linear system

$$\begin{aligned} y &= 50p \\ y &= 300 - 25p \end{aligned}$$

Since the coefficients of y are the same in the two equations, we can subtract immediately, to get

$$0 = -300 + 75p$$

Solving yields $p = 4$. We obtain the same result, of course, by solving

$$50p = 300 - 25p$$

The market price is therefore \$4. At this price, the supply is 200 units and equals the demand.

Three Equations in Three Unknowns

The strategy here is to use the method of addition and subtraction to eliminate one of the three variables from two different pairs of the three equations. This yields a system of two equations in two unknowns, which we can solve by the methods already described above.

Example 17-9

Find the simultaneous solution of

$$2x + y + z = 6 \qquad \text{(E1)}$$
$$x - y + 2z = 9 \qquad \text{(E2)}$$
$$3x + 2y - z = 1 \qquad \text{(E3)}$$

FIRST: retain (E1):	$2x + y + z = 6$	
multiply (E2) by 2:	$2x - 2y + 4z = 18$	
subtract:	$3y - 3z = -12$	
simplify:	$y - z = -4$	(E4)
NEXT: multiply (E2) by 3:	$3x - 3y + 6z = 27$	
retain (E3):	$3x + 2y - z = 1$	
subtract:	$-5y + 7z = 26$	(E5)

Now we have a system of two equations, (E4) and (E5), in two unknowns.

NEXT: multiply (E4) by 5:	$5y - 5z = -20$
retain (E5):	$-5y + 7z = 26$
add:	$2z = 6$
solve for z:	$z = 3$
NEXT: replace z by 3 in (E4) or (E5) (we choose (E4)):	$y - 3 = -4$
solve for y:	$y = -1$
FINALLY: substitute $y = -1$ and $z = 3$ in (E1), (E2), or (E3) (we choose (E1)):	$2x + (-1) + 3 = 6$
solve for x:	$x = 2$

But now, as always, CHECK the solution (2,-1,3) in the original equations:

$$2(2) + (-1) + (3) \stackrel{\checkmark}{=} 6$$
$$(2) - (-1) + 2(3) \stackrel{\checkmark}{=} 9$$
$$3(2) + 2(-1) - (3) \stackrel{\checkmark}{=} 1$$

The graph of a linear equation in three unknowns is, as mentioned earlier, a plane in three-dimensional space. Two planes may intersect in a line, may be parallel, or may be identical. If they intersect in a line, a third plane may contain that line, may be parallel to it, or may intersect it in a single point. Most of the systems of three equations in three unknowns considered in this text have a unique solution; that is, the planes (the graphs of the three equations) share a single point. However, bear in mind that the planes may not intersect at all, or that their intersection may be more involved. We will say more about this in chapter 4.

EXERCISES

In problems 1 through 10 find the simultaneous solution of each system of linear equations.

1. $x + 3y = 17$
   $4x - y = 3$

2. $-2x + y = 4$
   $3x - 2y = -7$

3. $3x - 4y = 19$
   $2x + 5y = 5$

4. $8x + 2y = 3$
   $-5x + 4y = 6$

5. $x - 3y = 0$
   $-4x + 6y = -2$

6. $3x = 4 - 2y$
   $4y - 9x - 4$

7. $3x + 2z = 0$
   $x + y + z = 2$
   $2x - y + 2z = 1$

8. $2x - 3y + 4z = 3$
   $3x + 6y + 2z = 0$
   $-x + 6y + 6z = -1$

9. $x + y = 4$
   $2x - z = 9$
   $3x + 2z = -1$

10. $w + x + y + z = 0$
    $w - x + 2y - z = -3$
    $2w - y + 3z = 4$
    $4x - 5y + z = 6$

11. Suppose that the cost of producing x units is $600 + 5x$ dollars while the revenue from x units is $10x$ dollars. What is the break-even point?
12. Suppose that at a selling price of p dollars per unit, manufacturers are willing to produce $40p$ units, while consumers are willing to purchase $900 - 20p$ units. What is the market price?
13. An automobile agency rents cars for \$10 per day. If the agency has fixed daily expenses of \$100, and in addition the upkeep for each rented car is \$5, how many cars must be rented to break even?
14. A dealer sells metal desks at \$125 each and fine wooden desks at \$450 each. One week he sells 20 desks and collects \$5100. How many of each kind were sold that week?
15. A pharmacist is preparing 100 mg of a medication that is to be a mixture of demerol and phenergan. If the medication is to have three times as much demerol as phenergan, how much should it contain of each?
16. An 18-ounce can of mixed nuts is to contain peanuts, cashews, and walnuts. By weight there are to be twice as many walnuts as cashews and six times as many peanuts as cashews. How many ounces of each nut should the can contain?
17. An investor wishes to put half of her money in a bank, and

split the rest between stocks and bonds with twice as much in stocks as in bonds. If she has \$24,000, how should her total investment be split?

18. A mixture of three chemicals is to be made, subject to the following conditions: the mixture is to weigh 240 g; there is to be five times as much of the second chemical as of the first; the weight of the third chemical is to equal the combined weights of the first and second. How many grams of each chemical should the mixture have?

TEST A

1. Find the simultaneous solution of each linear system:

(a) $\begin{aligned} -2x + 3y &= 0 \\ 3x - 4y &= 1 \end{aligned}$ (b) $\begin{aligned} -x &= 2y - 3 \\ 6 &= 4y + 2x \end{aligned}$ (c) $\begin{aligned} x - y - z &= -3 \\ 2x + y + 2z &= 3 \\ 3x - 2y + z &= 0 \end{aligned}$

2. If the cost of producing tires amounts to \$15 per tire plus \$10,000 in overhead expenses, and the tires sell for \$40 each, what is the break-even point?
3. A company sells two kinds of crystal glasses. It requires three man-hours of labor to produce a set of type A, which sells for \$40; a set of type B sells for \$30 and requires only two man-hours of labor. One month, 675 man-hours of labor were used to produce sets of both kinds that sold altogether for \$9500. How many sets of each kind were produced?

TEST B

1. Find the simultaneous solution of each linear system:

(a) $\begin{aligned} x &= 2y - 1 \\ 2y &= x - 1 \end{aligned}$ (b) $\begin{aligned} 3x &= 2y + 3 \\ x - 1 &= 4y \end{aligned}$ (c) $\begin{aligned} 2x + y + z &= 0 \\ x - y - 3z &= 2 \\ 4x + 3y + 2z &= 1 \end{aligned}$

2. At a selling price of p dollars per unit, a book publisher is willing to print $100p$ books; at this price consumers are willing to buy $1800 - 50p$ books. What is the market price?
3. A total of 300 supporters of a charitable organization made contributions of either \$5 or \$10 each. If \$2075 was collected, how many \$5 contributions were made?

SECTION 18. SYSTEMS OF LINEAR INEQUALITIES

A *linear inequality* is like a linear equation except that instead of an equals sign between the two sides there is a sign of inequality, thus:

EQUATIONS:	$x = 0$	$y = 3x$	$2x + 3y = 7$
INEQUALITIES:	$x < 0$	$y < 3x$	$2x + 3y < 7$
	$x > 0$	$y > 3x$	$2x + 3y > 7$
	$x \leqslant 0$	$y \leqslant 3x$	$2x + 3y \leqslant 7$
	$x \geqslant 0$	$y \geqslant 3x$	$2x + 3y \geqslant 7$

Many important real-world problems can be dealt with mathematically if we represent them by the kind of mathematical model called a *linear program.* A linear program consists of a set of linear inequalities, together with a linear function called the *objective* function which is to be maximized or minimized subject to the conditions (or *constraints*) imposed by the inequalities. The technique of linear programming will be dealt with in detail below in chapter 5. Those techniques require the procedures presented in this section.

One approach in solving a linear program is geometric. This approach requires that we graph a system of linear inequalities. In this section we first graph single linear inequalities in two variables; then we set up systems of inequalities to meet the specifications given in word problems, and graph those systems.

THE GRAPH OF A LINEAR INEQUALITY

The graph of an inequality in two variables is the set of points in the plane whose coordinates satisfy the inequality. The graph is obtained by using the related equality (that is, equation), whose graph is a straight line, as illustrated below.

Example 18-1

Graph $y \geqslant 3x$.

We first graph the line whose equation is $y = 3x$. The line passes through the origin and has slope 3 (see figure 18.1(a)). If (a,b) satisfies the equation $y = 3x$--that is, if the point (a,b) is *on* the line--then, for any $b' > b$, (a,b') satisfies the inequality

$y > 3x$. The graph of $y \geqslant 3x$ therefore consists of all the points on the line together with those in the half-plane above it. In figure 18.1(b) the shaded region is the graph of $y \geqslant 3x$.

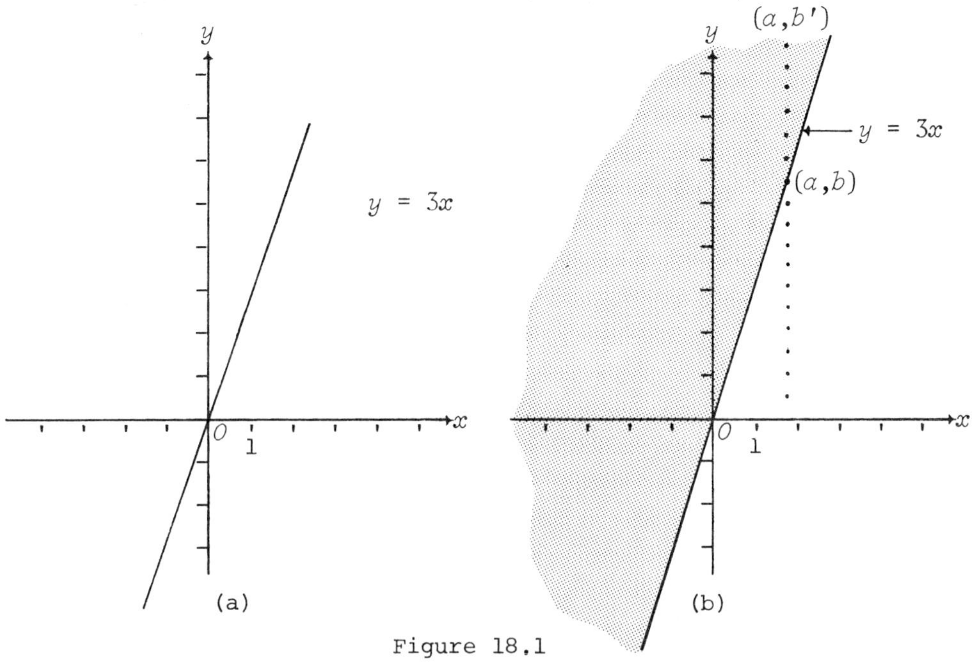

Figure 18.1

NOTE. Heed carefully the difference between $y > 3x$ (a "strict" inequality, which precludes the possibility that $y = 3x$) and $y \geqslant 3x$. The graph of the former would *not* include the points on the line. There are various ways of showing this on a graph; for example, by drawing a dotted line instead of a solid one. However, in this text we avoid the difficulty since no problems call for the graphing of strict inequalities.

TESTING AND GRAPHING LINEAR INEQUALITIES

To determine which side of a line--that is, which half-plane--is in the graph of a linear inequality, first graph the line. Then

(1) Choose any point not on the line and substitute its coordinates in the inequality.

(2) If the point satisfies the inequality then its side of the line is the desired half-plane; if not, the other side is.

(3) Shade the half-plane that satisfies the inequality.

Example 18-2

Graph $2x + y \leqslant 6$.

The x- and y-intercepts of the line are 3 and 6 respectively; the line is sketched in figure 18.2(a). Note that the origin, (0,0), is not on the line. We therefore test this point as indicated above:

(1) Clearly, $2(0) + (0) = 0 \leqslant 6$.

(2) Since the origin satisfies the linear inequality, the half-plane containing the origin is the desired graph.

(3) In figure 18.2(b) we shade the side of the line containing the point (0,0). The graph of $2x + y \leqslant 6$ is the shaded half-plane and its boundary.

Example 18-3

Graph (a) $x \geqslant 2$; (b) $y \geqslant 0$.

The shaded areas in Figure 18.3, respectively (a) and (b), together with their boundaries, are the desired graphs.

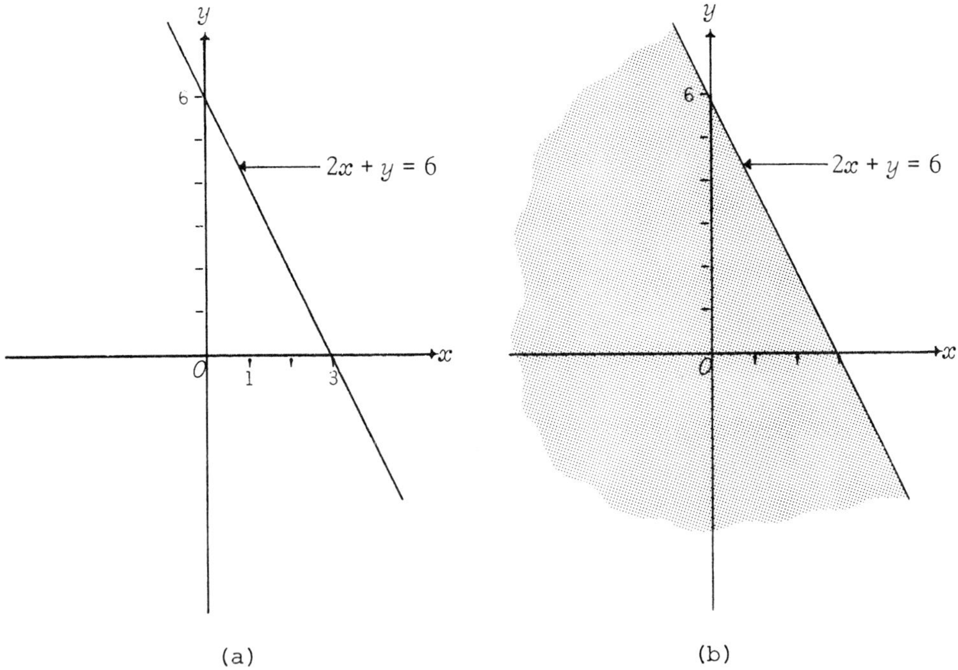

Figure 18.2

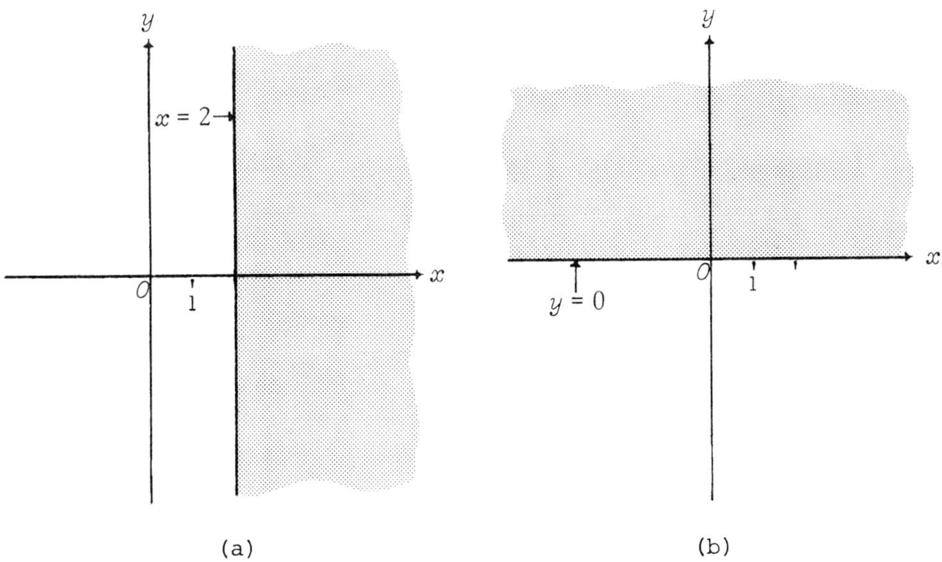

Figure 18.3

GRAPH OF A SYSTEM OF LINEAR INEQUALITIES IN TWO UNKNOWNS

The graph of a system of inequalities consists of those points that satisfy *all* the inequalities. The method of graphing is best presented by example.

Example 18-4

Graph the system

$$\begin{aligned} x + 3y &\leqslant 6 \\ 2x - y &\geqslant 4 \end{aligned}$$

On a single set of axes graph both lines. Then shade in turn each half-plane defined by the given inequalities. This is shown in figure 18.4(a). Each point in the doubly shaded region in the figure, or on its bounding half-lines, satisfies both of the given inequalities. The graph of the system is therefore this region, together with the half-lines, as shown in Figure 18.4(b). Note that this two-sided region is unbounded (that is, although it stops at the two half-lines, below them it extends without limit).

Example 18-5

Graph the system

$$\begin{aligned} 2x - 3y &\leqslant 0 \\ y &\geqslant 2 \end{aligned}$$

The desired graph is the two-sided shaded region shown in figure 18.5(b).

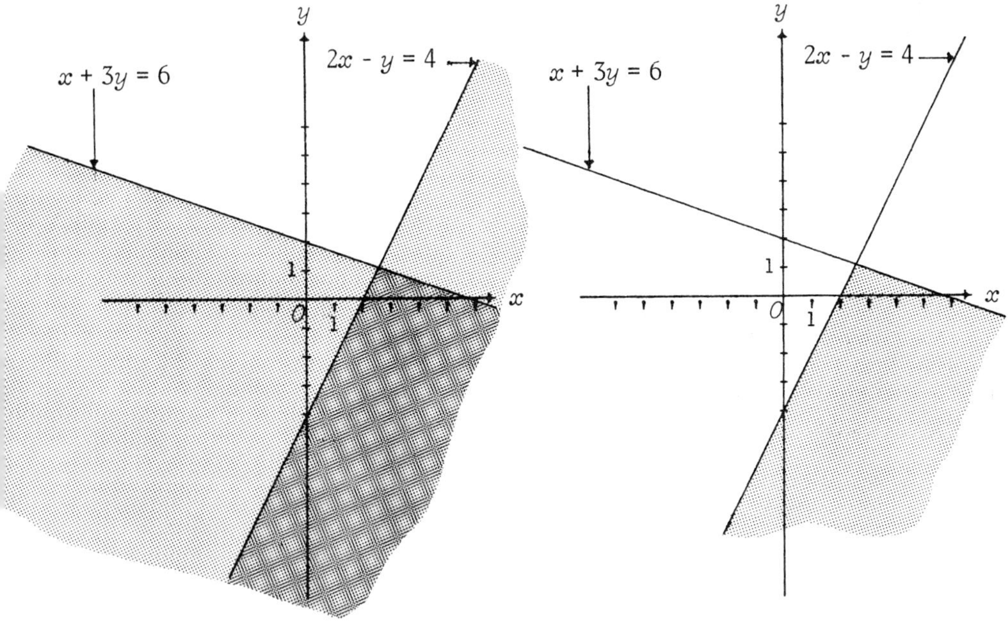

(a) (b)

Figure 18.4

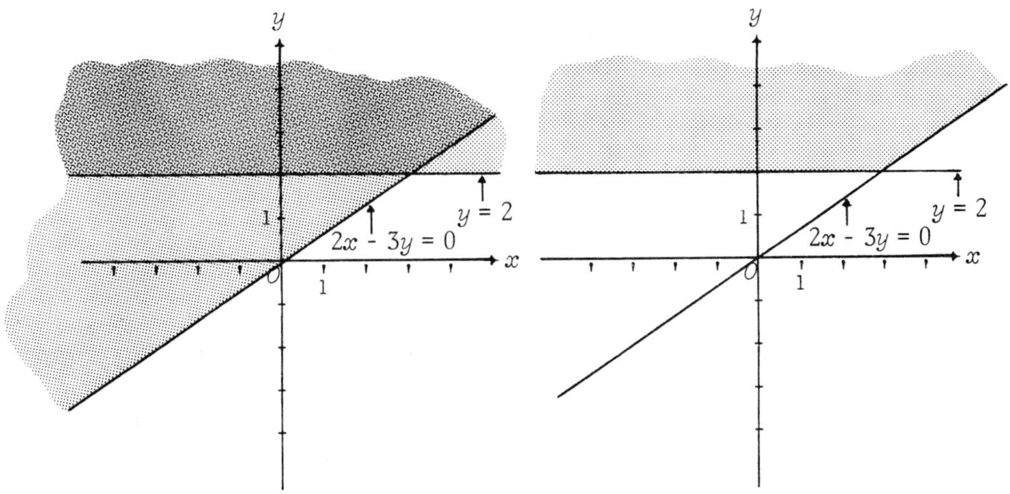

(a) (b)

Figure 18.5

Example 18-6

Graph the solution of the system

$$\begin{aligned} x \quad\quad &\geq 0 \\ y &\geq 0 \\ x + \ y &\geq 1 \\ x - 2y &\leq 4 \end{aligned}$$

Most applied problems in linear programming include, as this one does, the constraints $x \geq 0$ and $y \geq 0$, restricting the graphical solution to the first quadrant. This is so because prices, quantities of products or other items, number of bacteria or of people, and so on, cannot be negative.

In figure 18.6, the four-sided shaded region, entirely in the first quadrant, and its bounding half-lines and line segments, comprise the graph of the given system.

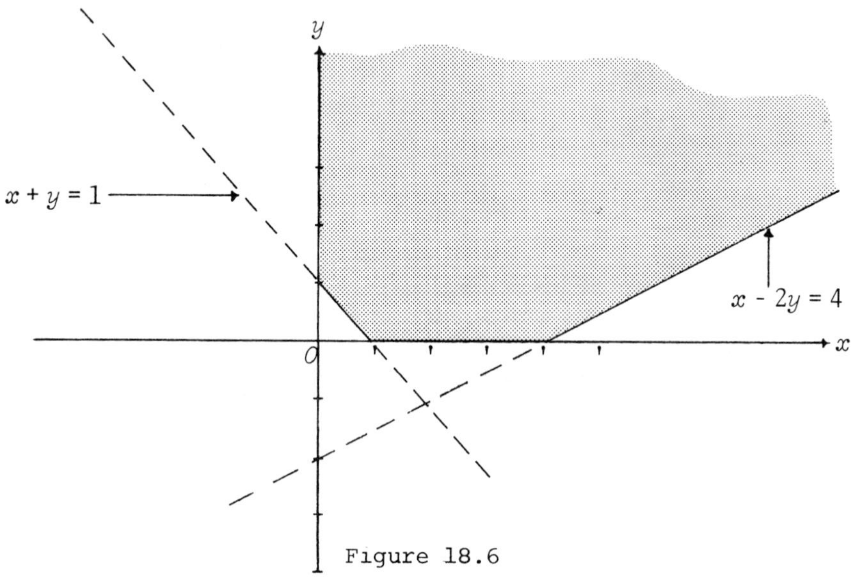

Figure 18.6

FINDING SYSTEMS OF LINEAR INEQUALITIES FROM WORD PROBLEMS

Each of the following problems can be modeled by a system of linear inequalities. For those which involve only two variables, we graph the intersection as we did above.

Example 18-7

The Rolleezy Company makes two models of skateboards. One requires two hours of hand labor and one hour of machine time; the other calls for $1\frac{1}{2}$ hours of hand labor and $1\frac{1}{2}$ hours of machine time. The company has 3000 man-hours and 1800 hours of machine time available each week. Set up and graph the associated system of linear inequalities for the weekly production levels Rolleezy can maintain.

We let x and y be respectively the numbers of the two models of skateboards produced weekly. Then $x \geqslant 0$ and $y \geqslant 0$, because negative numbers of skateboards make no sense. The problem tells us that it takes $2x$ hours of hand labor for x units of one model and $1.5y$ hours of hand labor for y units of the other, and that the total number of man-hours cannot exceed 3000. Thus

$$2x + 1.5y \leqslant 3000$$

The conditions on the need for and the availability of machine time per week lead to the inequality

$$x + 1.5y \leqslant 1800$$

The required system of linear inequalities is therefore

$$\begin{aligned} x \qquad &\geqslant 0 \\ y &\geqslant 0 \\ 2x + 1.5y &\leqslant 3000 \\ x + 1.5y &\leqslant 1800 \end{aligned}$$

The graph of this system is the shaded region and its four-sided boundary shown in Figure 18.7. Every point in the shaded region or

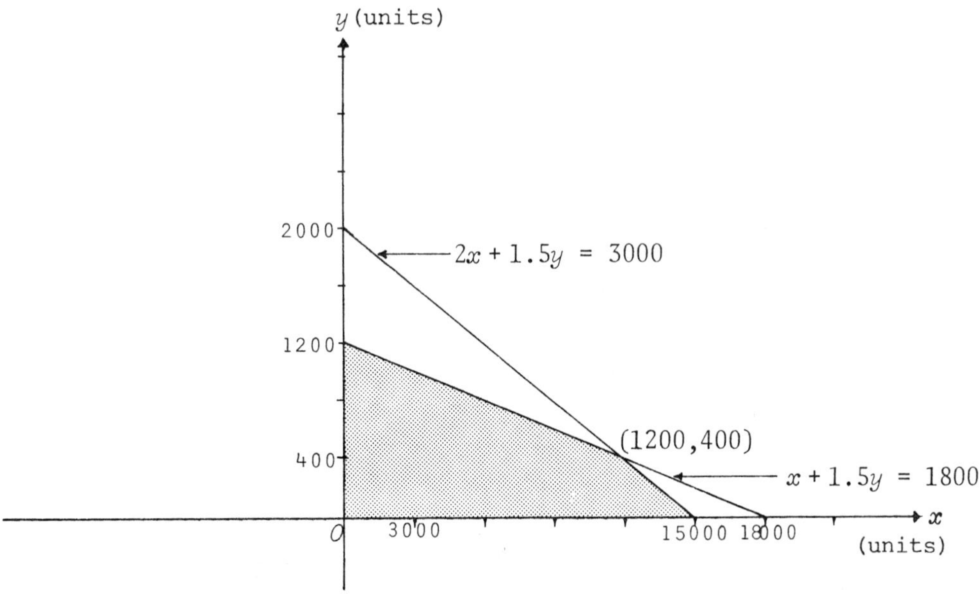

Figure 18.7

on its boundary is a possible pair of production levels for the two models of skateboards that Rolleezy can make subject to all the conditions specified. We note in passing that each of the four "corners" of the four-sided figure, (0,0), (0,1200), (1200,400), and (15 000, 0), satisfies these conditions.

(It is also worth noting that, practically speaking, x and y in this real-world problem must be integers.)

Example 18-8

A nutritionist who has purchased two food substitutes notes that one ounce of the first provides three units of protein and two units of carbohydrate, while one ounce of the second provides five units of protein but only one unit of carbohydrate. The nutritionist knows that an adult male needs at least 50 units of protein and 25 units of carbohydrate daily. Find and graph the resulting system of linear inequalities for the daily intake level of an adult male of these two food substitutes. (Assume no other sources of protein or carbohydrate.)

We let x and y be the number of ounces respectively of the two food substitutes to be consumed by an adult male. Then $x \geqslant 0$ and $y \geqslant 0$ because negative numbers of ounces have no meaning here. Since

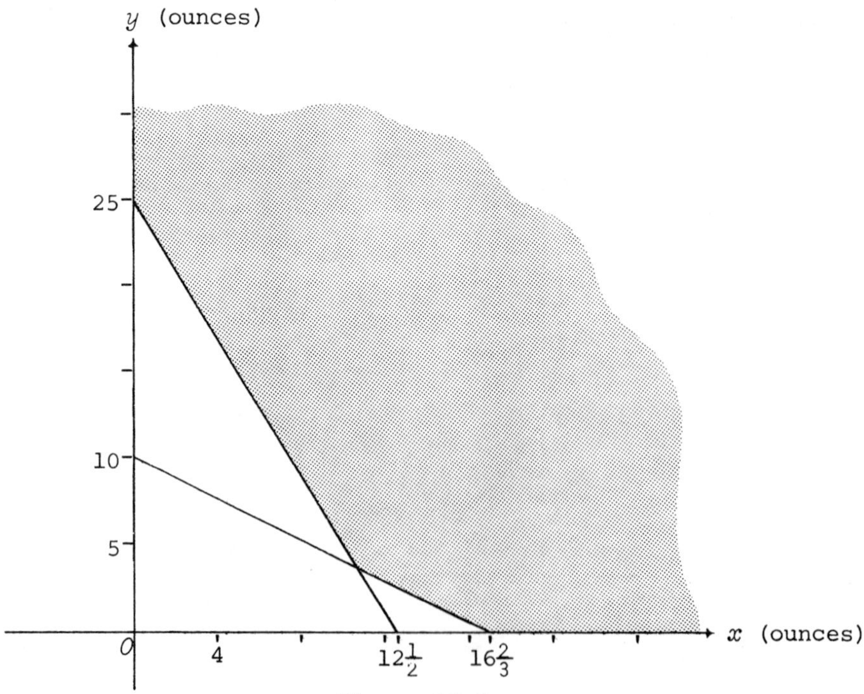

Figure 18.8

the first substitute provides $3x$ units of protein, the second $5y$ units, and at least 50 units are needed daily, we have the inequality $3x + 5y \geqslant 50$. Similarly, the carbohydrate requirement leads to the inequality $2x + y \geqslant 25$. The system is therefore

$$\begin{aligned} x \qquad &\geqslant 0 \\ y &\geqslant 0 \\ 3x + 5y &\geqslant 50 \\ 2x + \ y &\geqslant 25 \end{aligned}$$

The graph is the shaded region in Figure 18.8. Each point in this region or on its boundary represents a pair of daily intake levels (in ounces) of the two substitutes that is guaranteed to satisfy the minimum requirements of an adult male for protein and carbohydrate.

Example 18-9

An oil company leases two kinds of tankers. One can carry 10,000 barrels of crude oil, 10,000 barrels of gasoline, and 1000 barrels of airplane fuel, while the other can carry 7500 barrels of crude, 3000 of gasoline, and 3000 of plane fuel. A city wishes to import at least 60,000 barrels of crude, at least 30,000 barrels of gasoline, and at least 9000 barrels of plane fuel. Find and graph the resulting set of inequalities for the number of each kind of tanker the oil company leases to supply the city.

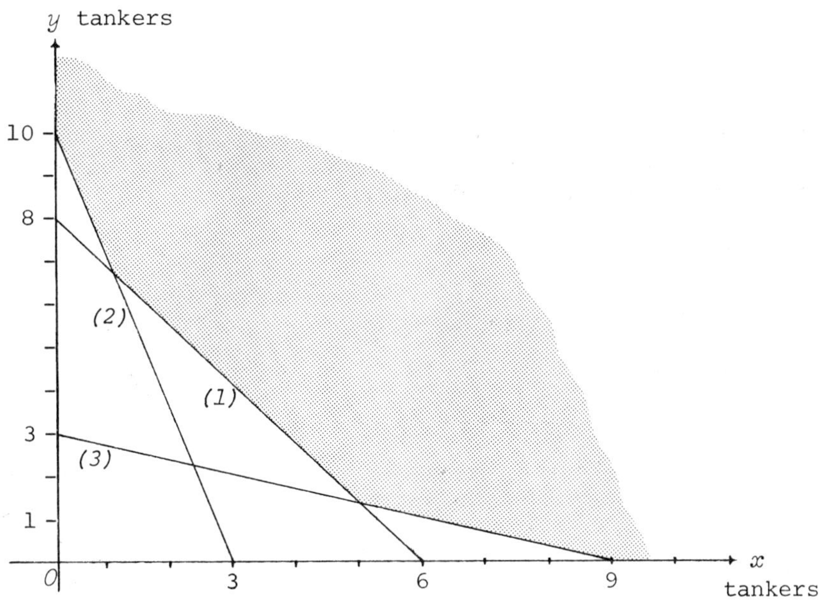

Figure 18.9

If x and y are the number of tankers of each type the oil company leases, then the descriptive set of inequalities is

$$\begin{aligned} x \qquad\qquad &\geqslant 0 \\ y &\geqslant 0 \\ 10{,}000x + 7500y &\geqslant 60{,}000 \qquad (1) \\ 10{,}000x + 3000y &\geqslant 30{,}000 \qquad (2) \\ 1000x + 3000y &\geqslant 9000 \qquad (3) \end{aligned}$$

Figure 18.9 (page 153) shows the graph of the given system of inequalities. The lines related to the inequalities (*1*), (*2*), and (*3*) are labeled correspondingly on the graph.

Example 18-10

Ad advertising agency promoting a new product plans to spend a total of at least \$50,000 but not more than \$100,000, distributing this among TV commercials, radio spots, and magazine ads. The amount to be spent on TV must be at least twice that for radio, and more money is to be spent on magazine ads than on radio spots. What set of inequalities governs the distribution of advertising funds for the new product?

If x, y, and z are the amounts allotted respectively to TV, radio, and magazines, then the constraining set of inequalities is

$$\begin{aligned} x \geqslant 0 \qquad y &\geqslant 0 \qquad z \geqslant 0 \\ x + y + z &\geqslant 50{,}000 \\ x + y + z &\leqslant 100{,}000 \\ x &\geqslant 2y \\ z &\geqslant y \end{aligned}$$

Since three variables are involved, the graph of this system is in 3-space. We omit a sketch, since in cases involving three or more variables sketches are not helpful in seeking the solution.

EXERCISES

1. Graph each of the following:

(a) $2x - y \geqslant 6$ (b) $y + 3x \leqslant 6$ (c) $2x + 3y \leqslant 10$
(d) $x \geqslant y + 2$ (e) $x - 2y \geqslant 0$ (f) $3x \leqslant 4y$
(g) $x \leqslant 0$ (h) $y \geqslant 2$

2. Graph the regions represented by each of the following systems:

(a) $x + y \geqslant 2$, $2y + 3x \leqslant 6$ (b) $2x - 3y \leqslant 12$, $x + 5y \leqslant 5$ (c) $2x + 3y \leqslant 0$, $x - y \geqslant 1$

(d) $x \leqslant 2y$, $y \geqslant 3x$ (e) $y \leqslant 2x + 1$, $y \geqslant 2x - 1$ (f) $y \leqslant 3$, $x \geqslant 1$

(g)	(h)	(i)
$x \geqslant 0,\ y \geqslant 0$	$x \geqslant 0,\ y \geqslant 0$	$x \geqslant 0,\ y \geqslant 0$
$x + y \geqslant 1$	$x + 3y \leqslant 6$	$2x + 5y \geqslant 10$
$x \leqslant 2$	$x + y \leqslant 3$	$3x + 4y \geqslant 12$

3. An automobile plant produces two models of cars. The first requires five hours of assembly-line time and three hours of finishing work; the second requires seven hours on the assembly line and three hours of finishing work. Each week a maximum of 2000 hours of assembly-line time and a maximum of 1000 hours of finishing-work time are available. Find the system of inequalities imposed by these conditions on the numbers of each model to be produced.
4. A can of mixed nuts contains two ounces of cashews, three ounces of walnuts, and seven ounces of peanuts. A bag of mixed nuts contains respectively three, five, and eight ounces of these kinds of nuts. The Nuts-To-You company has the option of buying a maximum of 1000 ounces of cashews, up to 2000 ounces of walnuts, and no more than 4000 ounces of peanuts. If the company buys its nuts only in these cans or bags, what set of inequalities describes the constraints on the quantities?
5. One fat-soluble pill contains 1000 units of vitamin A and 100 units of vitamin D; a second brand contains 1500 units of A and 75 units of D. An adult needs at least 5000 units of A and a minimum of 400 units of D daily. Find the system of inequalities for an adult's daily intake level of these two kinds of pills, assuming that these are the only sources used of vitamins A and D.
6. A supplier is willing to ship five pounds of pork, three pounds of chicken, and two pounds of steak per ten-pound box of meat. Another supplier furnishes four pounds of pork, three of chicken, and three of steak per ten-pound box. A restaurant owner plans to stock at least 100 pounds of pork, no less than 75 of chicken, and a minimum of 50 of steak. What system of linear inequalities is specified for the number of boxes of meat the restaurant owner must buy from each supplier?
7. A man wishes to invest at least \$5000 but not more than \$10,000 in stocks, bonds, and certificates of deposit. He plans to put at least as much money in stocks as in bonds, and at least twice as much in certificates as in bonds. Write a set of inequalities for the amounts to be invested in stocks, bonds, and certificates of deposit.

TEST A

1. Graph each of the following: (a) $x - 7y \leqslant 7$; (b) $3x \geqslant 5y$.

2. Graph each system:

(a) $x - y \geqslant 2$
$2x + y \leqslant 4$

(b) $x \geqslant 0,\ y \geqslant 0$
$x + y \leqslant 3$
$2x + y \geqslant 4$

3. A company produces black-and-white television sets, which require one hour of hand labor and two hours of machine time, and color sets, which require two hours of hand labor and three hours of machine time. The company has a maximum of 1000 man-hours and up to 1500 hours of machine time available each week. What set of inequalities must the quantities of the two kinds of TV sets satisfy?

TEST B

1. Graph each of the following: (a) $5x + y \geqslant 5$; (b) $2x + y \leqslant 0$.
2. Graph each system:

(a) $x - 6y \leqslant 6$
$-x + y \geqslant 1$

(b) $x \geqslant 0,\ y \geqslant 0$
$x + y \geqslant 4$
$2x + 5y \leqslant 10$

3. Food A yields 25 units of vitamins, 50 units of carbohydrates, and 40 units of proteins per pound; food B yields 30, 20, and 35 units per pound respectively of these three ingredients. A hospital needs at least 500 units of vitamins, 600 units of carbohydrates, and 700 units of protein from these two foods per week. What set of inequalities on the weekly amounts of these foods results from these conditions?

SUMMARY OF §§14-18

You should now be able to

(1) distinguish functions from nonfunctions;
(2) find the domain of a function;
(3) given $f(x)$, evaluate $f(a)$, $f(-x)$, and so forth;
(4) find the slope and intercepts of a line from its equation;
(5) graph a line from its equation;
(6) recognize when lines are parallel or perpendicular from their equations;
(7) find and graph linear functions described in words;
(8) derive a system (or set) of linear equations from a word problem;
(9) solve a system of linear equations by the method of substitution or by addition and subtraction;
(10) interpret solutions of systems of equations geometrically;
(11) graph a linear inequality in two unknowns;
(12) find and graph a system of linear inequalities defined by a word problem.

REVIEW EXERCISES

1. Which of these define functions with domain S and range T?
   (a) S is the set of all current United States senators and T is the set of their ages rounded to the nearest year.
   (b) S is the set of all nonnegative integers $\{0,1,2,...\}$ and T is the set of their square roots.
2. Find the domain fo each function:

   (a) $f(x) = 3x^2 - 3$ (b) $y = \sqrt{x - 4}$

   (c) $g(x) = \dfrac{x}{x^2 + 9}$ (d) $y = \dfrac{3}{x^2 - 2x}$
3. For each of the following functions, find $f(0)$, $f(2)$, and $f(-1)$.

   (a) $f(x) = 3x - 1$ (b) $f(x) = \dfrac{4}{x + 2}$

   (c) $f(x) = x^2 + 2$ (d) $f(x) = 1 - 2x^2$
4. Find the slope, if it exists, of the line through

   (a) $(0,3)$ and $(-2,0)$ (b) $(4,1)$ and $(2,-3)$

   (c) $(a, 1 - 2a)$ and $(a + h, 1 - 2(a + h))$

   (d) $(-5,1)$ and $(-5,4)$
5. Find the equation of, and graph, the line
   (a) whose slope is 3 and whose y-intercept is -2.
   (b) passing through the points $(4,3)$ and $(0,5)$.
   (c) parallel to $3x - 2y = 6$ and passing through $(-2,-2)$.
6. A farm tractor costing \$10,000 can be expected to be usable for eight years, after which its scrap value will be \$2000. Find its value after four years, assuming straight-line depreciation.
7. If 100 motel units can be rented at \$25 a night but only 70 units at \$30 each, find the demand function, assuming it is linear.
8. U-Drive rents its cars for \$12 a day. If it has fixed costs of \$180 per day and estimates depreciation and other costs at \$3 per rented car per day, what set of equations has the break-even point as its solution?
9. Determine whether the lines in each system below are identical, parallel, or intersect in a single point:

(a) $3x + y = 2$, $2y - 6x = 4$ (b) $y = 1 - x$, $x + y = 3$ (c) $y = 3x - 1$, $6x - 2y - 2 = 0$

10. Mr. Gambol invested $10,000 in a total of 60 shares of three different stocks, buying them at $50, $100, and $600 per share. What system of equations for the numbers of shares of each stock defines his investment?
11. Find the solution of each system of equations:

(a) $2x - y = 5$, $3x - 2y = 6$ (b) $4x + 2y - 3z = 3$, $2x + 3y + 6z = 7$, $6x - y - 6z = 7$ (c) $x - 2y = -6$, $y + z = 7$, $2x + z = 1$

12. Solve the system of equations of exercise 8 above.
13. Suppose you know that Mr. Gambol (in exercise 10 above) bought as many shares at $100 as he did of the other two stocks combined. Solve the new enlarged system of equations.
14. A chamber orchestra sold a total of 184 subscription tickets last season. Regular season tickets cost $25 each, but students and senior citizens paid only $15 each. If subscription sales totaled $3800, how many tickets were sold at each price?

15. Graph each inequality on a separate set of axes:

(a) $y - 2x \geqslant 3$ (b) $3x + 2y \leqslant 0$

16. Graph each of the following systems of inequalities:

(a) $x - 2y \leqslant 2$, $-x + 7y \leqslant 3$ (b) $x \geqslant 0$, $y \geqslant 0$, $2x + 3y \geqslant 6$, $5x - 2y \geqslant 10$

17. A widow who has collected $20,000 of insurance benefits decides to invest at least half in stocks, treasury notes, and money-market certificates. She decides to put at least as much in stocks as in the other two investments combined, but not more than half as much in notes as in certificates. Write a system of inequalities for the amounts of each type of investment that satisfy the widow's wishes.

CHAPTER TEST

1. Find the domain of $f(x) = \dfrac{x - 2}{x^2 - 3x}$.
2. If $f(x) = x^2 + 2x$, find and simplify $f(0)$ and $f(-2)$.
3. Find the equation of the line through $(1,-3)$ and $(2,-5)$.
4. Find the line parallel to $x + 2y = 5$ and with x-intercept equal to -3. Graph it.
5. A rock-group manager knows from experience that 3000 will attend a concert at \$6 apiece but only 2000 if the price is \$8. (a) Find the demand function, assuming it is linear. (b) How many tickets can be sold if the price is dropped to \$5?
6. Only one of the following systems of equations has a unique solution. Find the solution of that system.

   (a) $\begin{aligned} y &= 2 + x \\ 2x - 2y &= 4 \end{aligned}$ (b) $\begin{aligned} 3x - y &= 7 \\ 2y &= 6x + 7 \end{aligned}$ (c) $\begin{aligned} 3x + y &= -1 \\ 4x + 3y &= 7 \end{aligned}$

7. A manufacturer is willing to produce $800p$ calculators at a selling price of p dollars per calculator. At this price consumers are willing to buy $9500 - 150p$ calculators. What is the market price?
8. The publisher of a popular mathematics text retails the paperback edition at \$12, the hard-cover edition at \$20. One year he sells 2000 copies altogether, for a total on retail sales of \$30,000. How many copies of each edition were sold that year?
9. Find the solution of the following system:

$$\begin{aligned} x + y - 2z &= 0 \\ 2x - 3y - z &= 8 \\ 3x + 5y + 5z &= 9 \end{aligned}$$

10. Graph the inequality $3x - 4y \leqslant 12$.
11. Graph the system of inequalities

$$\begin{aligned} x \geqslant 0,\ y &\geqslant 0 \\ 2x + y &\leqslant 4 \\ -x + 3y &\leqslant 6 \end{aligned}$$

12. Products A and B are made by a company with 24 hours per week available on machine I and 36 on machine II. Each product requires four hours on I; A requires 10 on II, B only two hours on II. What system of inequalities describes the contraints on the weekly quantities of each product the company can make?

Chapter 4

Matrices

In this chapter we introduce mathematical objects known as matrices. We will see how important they are in organizing and displaying data. After developing the properties of matrices and finding out how to operate on them algebraically, we will use them to solve systems of linear equations. One common technique for solving such systems involves using the inverse of a matrix, which we will also consider. This technique is readily adapted to computer use.

Matrices have pervaded almost every field of inquiry and have been of great value in the solutions of numerous important real-life problems. We will consider some of these applications here. In particular, we will see how matrices may be applied to input-output analysis in the field of economics and to sequences of experiments in probability problems called Markov processes.

SECTION 19. BASIC CONCEPTS OF MATRICES

In this section we give the basic definitions of elementary matrix theory and show how we organize given data in matrix form and how we interpret data given in a matrix. (Note that the plural of "matrix" is "matrices.")

DEFINITIONS

A *matrix* is a rectangular array of numbers; thus

Example 19-1

$$R = \begin{pmatrix} 0.18 & 0.35 & 2.05 & 2.22 \\ 0.15 & 0.33 & 1.19 & 1.39 \\ 0.18 & 0.35 & 1.36 & 1.81 \end{pmatrix}$$

As shown in this example, a matrix is usually denoted by a capital letter.

Matrix R was obtained from the information tabulated below about postal rates early in 1981 for different classes of mail:

	WEIGHT			
CLASS OF MAIL	1 oz	2 oz	12 oz	1 lb
First	0.18	0.35	2.05	2.22*
Second	0.15	0.33	1.19	1.39
Third	0.18	0.35	1.36	1.81

**To most zones in the U.S.*

We note that R has three *rows* (horizontal) and four *columns* (vertical); we call R a "three-by-four" matrix (usually written "3×4"). The numbers in the matrix are called *elements* or *entries*. The elements in the second row of R are 0.15, 0.33, 1.19, and 1.39 (representing the cost, in dollars, via second-class, of mail weighing respectively 1 oz, 2 oz, 12 oz, and 1 lb, early in 1981). The entries in the fourth column of matrix R are 2.22, 1.39, and 1.81 (telling the cost for a one-pound package via first-, second-, and third-class mail respectively, early in 1981).

The *size* of a matrix depends on how many rows and columns it has and is usually called its *dimension*. A matrix with m rows and n columns is an $m \times n$ matrix and has dimension $m \times n$. If m and n are equal, the matrix is called a *square matrix*.

Example 19-2

What is the dimension of each of these matrices?

$$A = \begin{pmatrix} 1 & 0 & 3 & 2 & -5 \\ 4 & -1 & 0 & 0 & 2 \end{pmatrix} \qquad B = \begin{pmatrix} 0 & -2 \\ -3 & 5 \\ 1 & 4 \end{pmatrix}$$

$$C = \begin{pmatrix} 0 & -3 & 1 \\ -2 & 5 & 4 \end{pmatrix} \qquad D = \begin{pmatrix} \frac{1}{2} & \frac{1}{4} & 0 & \frac{1}{4} \end{pmatrix} \qquad E = \begin{pmatrix} 1 \\ 0 \\ 1 \end{pmatrix}$$

$$F = \begin{pmatrix} 1 & 3 \\ 2 & 4 \end{pmatrix} \qquad G = \begin{pmatrix} \frac{1}{2} & 0 & \frac{1}{2} \\ \frac{1}{3} & \frac{2}{3} & 0 \\ 1 & 0 & 0 \end{pmatrix}$$

A is 2×5; B is 3×2; C is 2×3; D is 1×4; E is 3×1; F is 2×2; G is 3×3. Note that matrices B and C do *not* have the same dimension. Only F and G are square matrices; F is a 2×2 square matrix while G has dimension 3×3.

A matrix of dimension $1 \times n$ (consisting of only one row) is called a *row vector*; one of dimension $m \times 1$ (with only one column) is called a *column vector*. In Example 19-2, D is a row vector and E is a column vector.

If the elements of a vector (either row or column) are all nonnegative and sum to 1, the vector is called a *probability vector*. Thus

$$\begin{pmatrix} \frac{1}{2} & \frac{1}{3} & \frac{1}{6} \end{pmatrix}, \qquad \begin{pmatrix} \frac{1}{4} \\ \frac{3}{4} \end{pmatrix}, \qquad \begin{pmatrix} \frac{1}{8} \\ 0 \\ \frac{7}{8} \end{pmatrix}, \qquad \begin{pmatrix} 0 & 1 & 0 & 0 \end{pmatrix}$$

are all probability vectors. But neither of the following is:

$$\begin{pmatrix} \frac{1}{2} & \frac{1}{3} & \frac{1}{3} \end{pmatrix}, \qquad \begin{pmatrix} \frac{1}{2} \\ \frac{3}{2} \\ -1 \end{pmatrix}$$

Are any of the matrices in example 19-2 probability vectors?

MORE ON SQUARE MATRICES

A *transition matrix* is a square matrix in which each row is a probability vector. Two examples are

$$\begin{pmatrix} \frac{1}{2} & \frac{1}{2} \\ \frac{1}{3} & \frac{2}{3} \end{pmatrix} \qquad \text{and} \qquad \begin{pmatrix} 1 & 0 & 0 \\ 0 & \frac{1}{4} & \frac{3}{4} \\ 0 & 1 & 0 \end{pmatrix}$$

Matrix G of example 19-2 is also a transition matrix. As we shall see later in this chapter, transition matrices (also called *Markov matrices*) are important in predicting long-run trends in certain kinds of experiments.

A *permutation matrix* is a transition matrix with exactly one 1 in each row and one 1 in each column, zeros everywhere else; for example

$$\begin{pmatrix} 0 & 1 \\ 1 & 0 \end{pmatrix} \quad \text{and} \quad \begin{pmatrix} 0 & 1 & 0 \\ 1 & 0 & 0 \\ 0 & 0 & 1 \end{pmatrix}$$

The *main* (or *major* or *principal*) *diagonal* of a square matrix is the set of elements whose row and column are the same (the elements along the line from the upper left corner to the lower right corner). A *diagonal matrix* is a square matrix in which the only nonzero elements, if any, appear on the main diagonal. Two examples are

$$\begin{pmatrix} 3 & 0 \\ 0 & 1 \end{pmatrix} \quad \text{and} \quad \begin{pmatrix} 2 & 0 & 0 \\ 0 & -1 & 0 \\ 0 & 0 & 0 \end{pmatrix}$$

In a later section, in which matrices are used to solve systems of linear equations, we will be interested in transforming a given square matrix into a diagonal permutation matrix.

A *symmetric matrix* is a square matrix in which the element in row i column j is the same as the entry in row j column i, for every i and j. The following two are examples:

$$\begin{pmatrix} 1 & 5 \\ 5 & 1 \end{pmatrix} \quad \text{and} \quad \begin{pmatrix} 1 & 4 & 6 \\ 4 & 0 & 9 \\ 6 & 9 & 2 \end{pmatrix}$$

MATRIX NOTATION

The most general notation for a matrix A of dimension $m \times n$ is

$$A = \begin{pmatrix} a_{11} & a_{12} & a_{13} \cdots a_{1n} \\ a_{21} & a_{22} & a_{23} \cdots a_{2n} \\ \vdots & \vdots & \vdots \quad\quad \vdots \\ a_{m1} & a_{m2} & a_{m3} \cdots a_{mn} \end{pmatrix}$$

Here the element in row i and column j is denoted by the symbol "a_{ij}". Thus, for example, the element a_{52} of a matrix is the one in its fifth row and second column; the element in the third row and fourth column of a matrix is designated by "a_{34}". An arbitrary element in the third row of matrix A is denoted as "a_{3j}", where j gives the entry's column; an arbitrary element in the second column is represented by "a_{i2}", where i gives the row.

The general display notation for matrix A given above is also often condensed to

$$A = (a_{ij}) \qquad i = 1,2,\ldots,m;\ j = 1,2,\ldots,n$$

Example 19-3

Suppose

$$A = \begin{pmatrix} 2 & -7 & 0 \\ 0 & 3 & -1 \\ -4 & -2 & 5 \\ 1 & 0 & 6 \end{pmatrix}$$

Then A is a 4×3 matrix. Here a_{23} (the entry in the second row and third column) is -1; $a_{41} = 1$; $a_{13} = 0$; $a_{31} = -4$; and so forth.

NOTE. The definition of symmetric matrix given earlier can now be stated more simply: a symmetric matrix is one in which $a_{ij} = a_{ji}$ for all i and j.

EQUALITY OF MATRICES

Two matrices are *equal* if they have the same dimension and if corresponding elements are equal.

Thus, since both are square matrices of dimension 2×2,

$$\begin{pmatrix} a & b \\ c & d \end{pmatrix} = \begin{pmatrix} 2 & -1 \\ 0 & 3 \end{pmatrix}$$

if and only if $a = 2$, $b = -1$, $c = 0$, and $d = 3$. Similarly, since the following matrices both have dimension 2×1, if it is given that

$$\begin{pmatrix} 2x + 3y \\ x - 4y \end{pmatrix} = \begin{pmatrix} 3 \\ -5 \end{pmatrix}$$

it follows that

$$2x + 3y = 3$$

and that

$$x - 4y = -5$$

conversely, if it is given that these elements are equal, it follows that the matrices are.

APPLICATIONS: USING MATRICES TO DISPLAY DATA

We illustrate with the following examples, keeping an eye out for any of the special kinds of matrices defined above.

Example 19-4

A certain vitamin pill contains 10,000 units of vitamin A, 400 units of D, and 100 units of E. The row vector that describes the specific amounts of these three vitamins in this pill is

$$(10{,}000 \quad 400 \quad 100)$$

Example 19-5

The probabilities of throwing two, three, ..., twelve on a pair of honest dice are displayed in the row vector

$$\left(\tfrac{1}{36} \quad \tfrac{1}{18} \quad \tfrac{1}{12} \quad \tfrac{1}{9} \quad \tfrac{5}{36} \quad \tfrac{1}{6} \quad \tfrac{5}{36} \quad \tfrac{1}{9} \quad \tfrac{1}{12} \quad \tfrac{1}{18} \quad \tfrac{1}{36}\right)$$

Example 19-6

To determine the charges for long-distance phone calls, the United States is divided into three zones. In 1980 the charges for calls from the Virgin Islands were as displayed in the following matrix:

	WEEKEND		EVENING		DAYTIME	
TO ZONE	1st min.	add. min.	1st min.	add. min.	1st min.	add. min.
A	0.53	0.45	0.87	0.73	1.34	1.13
B	0.57	0.49	0.93	0.79	1.44	1.23
C	0.61	0.53	1.00	0.86	1.54	1.33

Example 19-7

A rat is in a maze with six compartments, as shown at the right. Suppose that the rat is equally likely to leave the compartment it is in by any of the exits from that compartment. The following matrix then gives the probability of entering any compartment from any other:

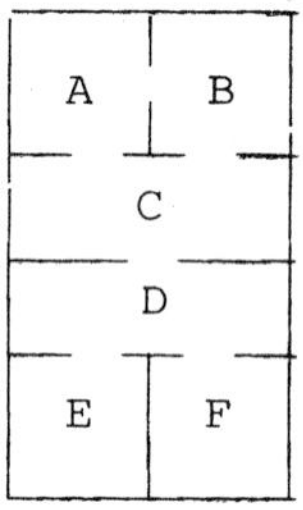

	TO A	B	C	D	E	F
FROM A	0	$\frac{1}{2}$	$\frac{1}{2}$	0	0	0
B	$\frac{1}{2}$	0	$\frac{1}{2}$	0	0	0
C	$\frac{1}{3}$	$\frac{1}{3}$	0	$\frac{1}{3}$	0	0
D	0	0	$\frac{1}{3}$	0	$\frac{1}{3}$	$\frac{1}{3}$
E	0	0	0	1	0	0
F	0	0	0	1	0	0

Since every row of this matrix is a probability vector, it is a transition matrix.

Example 19-8

The following matrix gives the distances in miles between pairs of the three principal islands of the U. S. Virgin Islands:

	ST. THOMAS	ST. CROIX	ST. JOHN
ST. THOMAS	0	40	3
ST. CROIX	40	0	39
ST. JOHN	3	39	0

For every i and j, the entry in the ith row and jth column is the same as that in the jth row and ith column. (Why?) The matrix is therefore a symmetric matrix.

Example 19-9

Suppose four guests at a hotel are all unhappy with their rooms and agree to shift as follows: the guests in rooms 1 and 3 will trade rooms, and so will those in rooms 2 and 4. This exchange can be represented by the following matrix:

	TO ROOM			
FROM ROOM	1	2	3	4
1	0	0	1	0
2	0	0	0	1
3	1	0	0	0
4	0	1	0	0

Since every row has exactly one 1 and every column has exactly one 1 and all other entries are 0, this is a permutation matrix.

Example 19-10

On a certain campus there are separate dormitories for freshmen, sophomores, juniors, and seniors. The numbers of students of each class living in each dormitory last year are given by this matrix:

		DORMS:			
		F	So	J	Se
STUDENTS:	F	600	0	0	0
	So	0	500	0	0
	J	0	0	450	0
	Se	0	0	0	400

Since the only nonzero entries are along the major diagonal, this is a diagonal matrix.

Example 19-11

Burning equivalent amounts of coal, oil, and gas produces four pollutants in the relative amounts given in this matrix:

		POLLUTANT:			
		A	B	C	D
FUEL:	COAL	2	3	1	5
	OIL	0	2	10	3
	GAS	4	1	5	2

This is an example of an *input-output* matrix, to be discussed in more detail later in this chapter.

EXERCISES

1. What is the dimension of each of the following matrices?

$$A = \begin{pmatrix} 2 & 0 & 3 \\ 1 & -1 & 1 \end{pmatrix} \qquad B = \begin{pmatrix} 1 & 0 & 2 \\ 2 & 1 & 0 \\ 3 & -1 & 1 \\ 0 & 2 & 5 \end{pmatrix}$$

$$C = (1 \quad 4 \quad 0 \quad 5) \qquad D = \begin{pmatrix} 3 \\ 2 \\ 0 \\ 1 \\ -1 \end{pmatrix}$$

2. Which of the following are probability vectors?

$$A = \begin{pmatrix} 1 \\ 0 \end{pmatrix} \qquad B = (0 \quad \tfrac{1}{3} \quad \tfrac{2}{3})$$

$$C = (\tfrac{2}{5} \quad -\tfrac{3}{5} \quad \tfrac{4}{5} \quad \tfrac{1}{5} \quad \tfrac{1}{5}) \qquad D = \begin{pmatrix} \frac{1}{3} \\ \frac{1}{6} \\ \frac{2}{5} \\ \frac{1}{15} \end{pmatrix}$$

3. Which of the following are transition matrices? permutation matrices? diagonal matrices? symmetric matrices?

(a) $\begin{pmatrix} 1 & 0 & 0 \\ 0 & 0 & 1 \\ 0 & 1 & 0 \end{pmatrix}$ (b) $\begin{pmatrix} 3 & 0 \\ 0 & 5 \end{pmatrix}$ (c) $\begin{pmatrix} \frac{1}{2} & 0 & \frac{1}{2} \\ 0 & 1 & 0 \\ \frac{1}{2} & \frac{1}{3} & \frac{1}{6} \end{pmatrix}$

(d) $\begin{pmatrix} 0 & 1 & 2 \\ 2 & 1 & 0 \\ 0 & 2 & 1 \end{pmatrix}$ (e) $\begin{pmatrix} 1 & -1 & 3 \\ -1 & 2 & 0 \\ 3 & 0 & 3 \end{pmatrix}$ (f) $\begin{pmatrix} 1 & 0 \\ 0 & 1 \end{pmatrix}$

(g) $\begin{pmatrix} 1 & 0 \\ 0 & 1 \\ 1 & 0 \end{pmatrix}$ (h) $\begin{pmatrix} 0 & 8 \\ 7 & 2 \end{pmatrix}$

In 4 through 9, write a matrix that presents the data given:

4. A sailboat charter service has hourly, daily, and weekly rental

rates of \$20, \$100, and \$500 respectively.

5. The climate on a certain Pacific island is such that on any given day the probabilities that it will be completely sunny, partly cloudy, or totally overcast are respectively $\frac{3}{4}$, $\frac{1}{5}$, and $\frac{1}{20}$.
6. Each of three students took five examinations in a nursing-qualification course, with the following results: the first student scored 425, 300, 350, 325, and 375; the second scored 500, 475, 525, 450, and 475; the third scored 375, 250, 325, 275, and 300.
7. During a certain week one employee worked 8 hours, $7\frac{1}{2}$ hrs, 8 hrs, 8 hrs, and 10 hrs respectively on Monday through Friday; a second employee worked 6 hrs, 4 hrs, 5 hrs, $4\frac{1}{2}$ hrs, and 8 hrs; a third worked 11, 8, 8, 0, and 0 hrs.
8. The manufacture of an automobile of a certain model costs \$1500 in raw materials, takes 5 hours of work in an assembly plant, and requires 3 hours in a finishing shop; one automobile of a second model uses 2100 dollars' worth of raw materials, 6 hours of assembly-plant time, and 5 hours of finishing time.
9. Of the consumers now using Brand A, three fourths will still be using it six months from now but one sixth will have switched to Brand B and the rest to Brand C. Half of those now using Brand B will still be using it in six months, while one fifth will have switched to Brand C and the rest to A. Of those now using Brand C, all will switch in six months, in equal numbers to Brands A and B.

   Is the matrix for these data a transition matrix?

10. There are eight blood types: A^+, A^-, B^+, B^-, AB^+, AB^-, 0^+, 0^-. The plusses and minuses in these labels refer to the so-called "Rh factor." A person who is Rh-negative cannot receive blood from an Rh-positive donor. A person with type A blood cannot receive blood from an individual of type B or AB; one of type B cannot receive from a type A or AB donor; and one of type 0 cannot receive from A, B, or AB.

    If these are the only restrictions, write a matrix that shows all the safe donors for each type of recipient. Use a "1" to indicate a safe donor-recipient pair, a "0" for all other pairs.
11. Suppose

$$\begin{pmatrix} 2x + 3y \\ 3x - 2y \end{pmatrix} = \begin{pmatrix} 4 \\ -7 \end{pmatrix}$$

Evaluate x and y. (Hint: solve a system of linear equations.)

TEST A

1. Give an example of (a) a probability vector; (b) a transition matrix.

2. What is the dimension of each of the following matrices?

$$A = \begin{pmatrix} 2 & 1 & 5 & 7 \end{pmatrix} \qquad B = \begin{pmatrix} 3 & 1 \\ 1 & 2 \\ 0 & 1 \end{pmatrix} \qquad C = \begin{pmatrix} 5 + 2x & 6 \\ 0 & -1 + 3x \end{pmatrix}$$

3. Five publishers (I, II, III, IV, V) offer books that are suitable for this course. Each has both a hard-cover and a soft-cover edition available. The retail prices for the hard-cover editions are respectively \$13, \$12, \$19, \$14, and \$15; for the soft-cover editions the prices are \$11, \$8, \$12, \$8, and \$9. Display this information in matrix form.

TEST B

1. Give an example of (a) a permutation matrix; (b) a diagonal matrix.
2. What is the dimension of each of the following matrices?

$$A = \begin{pmatrix} 2 \\ 3 \end{pmatrix} \qquad B = \begin{pmatrix} 4 & 1 \\ 0 & -1 \end{pmatrix} \qquad C = \begin{pmatrix} 1 & 5 & 6 & 3 \\ 2 & 0 & -1 & \frac{1}{2} \end{pmatrix}$$

3. A fast-food restaurant takes orders one evening from three groups of students. Dorm I orders 10 hamburgers, 12 bottles of beer, and 2 pizzas; Dorm II orders 6 hamburgers, 20 bottles of beer, and 12 hot dogs; the Garden Apartments order 8 bottles of beer and 4 pizzas. Present these data in a matrix.

SECTION 20. MATRIX OPERATIONS

One very important use of matrices is in solving systems of linear equations. This is the subject of §22 below. To solve such systems, and to use matrices in other applications, we need to acquire skill in performing the basic operations on matrices. In this section and the next we define and practice these operations.

MATRIX ADDITION AND SUBTRACTION

We can *add* two matrices, or *subtract* one from another, only if they have the same dimension. To do so we just add, or subtract, corresponding elements. The *sum*, or the *difference*, is a matrix of the same dimension as those added or subtracted.

Example 20-1

(a) $\begin{pmatrix} 1 & 2 \\ 3 & 4 \end{pmatrix} + \begin{pmatrix} 0 & -3 \\ 2 & 6 \end{pmatrix} = \begin{pmatrix} 1 & -1 \\ 5 & 10 \end{pmatrix}$

(b) $\begin{pmatrix} 2 & 3 & 8 \\ -5 & 0 & 6 \end{pmatrix} + \begin{pmatrix} 1 & 2 & 5 \\ 3 & 4 & -2 \end{pmatrix} = \begin{pmatrix} 3 & 5 & 13 \\ -2 & 4 & 4 \end{pmatrix}$

(c) $\begin{pmatrix} 5 \\ 8 \\ -1 \end{pmatrix} + \begin{pmatrix} 13 \\ 4 \\ -4 \end{pmatrix} = \begin{pmatrix} 18 \\ 12 \\ -5 \end{pmatrix}$

(d) $\begin{pmatrix} 6 & 2 \\ -5 & 1 \\ 0 & 4 \\ 1 & 3 \end{pmatrix} - \begin{pmatrix} 1 & 4 \\ 0 & -3 \\ -2 & 4 \\ 1 & 3 \end{pmatrix} = \begin{pmatrix} 5 & -2 \\ -5 & 4 \\ 2 & 0 \\ 0 & 0 \end{pmatrix}$

Generalizing, we see that

> If $A = (a_{ij})$ is an $m\times n$ matrix
> and $B = (b_{ij})$ is an $m\times n$ matrix
> then $A + B = (a_{ij} + b_{ij})$
> and $A - B = (a_{ij} - b_{ij})$

Note again that we can add (or subtract) two matrices only if they have the same dimension or shape.

Example 20-2

The matrices below show how the prices of four items changed at three different stores last year; the amounts are in cents:

	CHANGE IN PRICE FROM JUNE 1 TO JULY 1			CHANGE IN PRICE FROM JULY 1 TO AUGUST 1		
	STORE			STORE		
	A	B	C	A	B	C
ITEM I	5	3	7	-3	0	-1
II	-1	2	6	1	-1	4
III	12	9	10	6	4	6
IV	-4	-1	-3	-1	0	3

The sum of these two matrices is

	STORE		
	A	B	C
ITEM I	2	3	6
II	0	1	10
III	18	13	16
IV	-5	-1	0

The sum gives the net change in the prices of the four items at the three stores from June 1 to August 1 last year. For example, item III changed in price most at store A (+18¢), least at store B (+13¢); item IV cost 5¢ less at store A on August 1 than it did there on June 1, but this item was the same price on both dates at store C.

SCALAR MULTIPLICATION

In the context of matrices, ordinary real numbers are called *scalars*.

Example 20-3

Last year an automobile dealer employing three salesmen recorded the following average monthly sales:

	SUB-COMPACTS	COMPACTS
SALESMAN I	28	10.6
II	20	8
III	24.2	13

For this year the dealer expects a 10% increase in all sales. If we let A be the matrix shown just above, then the matrix for the expected increase in sales this year is computed as follows:

$$0.10A = 0.10 \times \begin{pmatrix} 28 & 10.6 \\ 20 & 8 \\ 24.2 & 13 \end{pmatrix} = \begin{pmatrix} 0.10\times 28 & 0.10\times 10.6 \\ 0.10\times 20 & 0.10\times 8 \\ 0.10\times 24.2 & 0.10\times 13 \end{pmatrix} = \begin{pmatrix} 2.8 & 1.06 \\ 2 & 0.8 \\ 2.42 & 1.3 \end{pmatrix}$$

This example illustrates *scalar multiplication* (or *multiplication by a scalar*): to multiply a matrix by a real number, we multiply each element of the matrix by that real number. Thus

> If $A = (a_{ij})$ and c is a scalar, then $cA = (ca_{ij})$ is the product of matrix A by the scalar c.

Example 20-4

(a) $3\times\begin{pmatrix} 4 \\ 1 \\ 5 \end{pmatrix} = \begin{pmatrix} 12 \\ 3 \\ 15 \end{pmatrix}$ (b) $8\times\begin{pmatrix} 1 & 0 \\ -2 & 3 \end{pmatrix} = \begin{pmatrix} 8 & 0 \\ -16 & 24 \end{pmatrix}$

Example 20-5

Display in matrix form the hoped-for average monthly sales for this year of the dealer in example 20-3.

The answer is

$$1.10A = \begin{pmatrix} 30.8 & 11.66 \\ 22.0 & 8.8 \\ 26.62 & 14.3 \end{pmatrix}$$

Note that this answer is the sum $A + 0.10A$, which can also be written $1A + 0.10A$.

ZERO MATRICES

A *zero matrix*, designated by O, is one all of whose elements are zero. A zero matrix of the proper shape serves as an *additive identity*: thus

$$\begin{pmatrix} 2 & 3 & 1 \\ -4 & -1 & 5 \end{pmatrix} + \begin{pmatrix} 0 & 0 & 0 \\ 0 & 0 & 0 \end{pmatrix} = \begin{pmatrix} 2 & 3 & 1 \\ -4 & -1 & 5 \end{pmatrix}$$

and, in general, if A is any matrix and O is the zero matrix of the same dimension, then $A + O = A$.

THE ALGEBRA OF MATRICES

Convince yourself of the truth of the following rules about matrices:

If A, B, and C have the same dimension, if r is a real number, and if O is the zero matrix of the proper size, then

(1)	$A + B = B + A$	(addition is commutative)
(2)	$(A + B) + C = A + (B + C)$	(addition is associative)
(3)	$r(A + B) = rA + rB$	(scalar multiplication distributes over addition)
(4)	$A + O = O + A = A$	(O is the additive identity)
(5)	$A + (-A) = (-A) + A = O$	($-A = -1A$ is the additive inverse of A)

(Show that subtraction of matrices is neither commutative nor associative.)

Note how the rules are verified in the following example.

Example 20-6

(a) $(1\ \ 2\ \ 5) + (0\ \ 0\ \ 0) = (1\ \ 2\ \ 5)$.

(b) $\begin{pmatrix} 8 & -1 \\ 2 & 0 \end{pmatrix} + \begin{pmatrix} 0 & 0 \\ 0 & 0 \end{pmatrix} = \begin{pmatrix} 8 & -1 \\ 2 & 0 \end{pmatrix}$

(c) If $A = \begin{pmatrix} 1 & -2 & 3 \\ -3 & 0 & -4 \end{pmatrix}$, then $-A = (-1)\times\begin{pmatrix} 1 & -2 & 3 \\ -3 & 0 & -4 \end{pmatrix} =$

$\begin{pmatrix} -1 & 2 & -3 \\ 3 & 0 & 4 \end{pmatrix}$ and $A + (-A) = \begin{pmatrix} 0 & 0 & 0 \\ 0 & 0 & 0 \end{pmatrix} = O$.

Example 20-7

Person-to-person weekday rates in October 1980 from Myrtle Beach, South Carolina, to Fairbanks, Alaska and to Honolulu are given (in dollars) by the matrix

	FIRST THREE MINUTES	EACH ADD'L MINUTE
TO FAIRBANKS	6.60	2.20
HONOLULU	7.50	2.50

Suppose the phone company decided to increase the initial-three-minute rate to Fairbanks and Honolulu by \$1.10 and \$1.25 respectively, and to reduce the additional-minute charges respectively by \$0.40 and \$0.50. (a) What matrix represents the rate changes? (b) What matrix represents the new rates?

(a) The rate-change matrix is $\begin{pmatrix} 1.10 & -0.40 \\ 1.25 & -0.50 \end{pmatrix}$

(b) The new rates are given by the sum

$$\begin{pmatrix} 6.60 & 2.20 \\ 7.50 & 2.50 \end{pmatrix} + \begin{pmatrix} 1.10 & -0.40 \\ 1.25 & -0.50 \end{pmatrix} = \begin{pmatrix} 7.70 & 1.80 \\ 8.75 & 2.00 \end{pmatrix}$$

EXERCISES

Perform the indicated operations whenever they are defined.

1. $(1 \quad 2) + (-1 \quad 3)$
2. $\begin{pmatrix} 3 \\ 4 \end{pmatrix} + (1 \quad 6)$
3. $5\times\begin{pmatrix} 0 \\ 1 \end{pmatrix}$
4. $6\times(1 \quad 2 \quad 0)$
5. $7\times\begin{pmatrix} 1 & 2 \\ 0 & -1 \end{pmatrix} + \begin{pmatrix} 2 & 3 \\ 4 & 1 \end{pmatrix}$
6. $4\times\begin{pmatrix} 5 & 1 \\ 0 & 2 \end{pmatrix} + \begin{pmatrix} 1 \\ 3 \end{pmatrix} + (2 \quad 4)$
7. $8\times\begin{pmatrix} 1 & -1 \\ 2 & 6 \end{pmatrix} + \begin{pmatrix} 3 & 2 \\ 4 & 1 \end{pmatrix}$
8. $5\times\begin{pmatrix} 3 & 0 & 1 \\ 2 & 6 & 4 \end{pmatrix} - 2\times\begin{pmatrix} 1 & 8 & 7 \\ 0 & 1 & -1 \end{pmatrix}$
9. $\begin{pmatrix} 4 & 2 \\ 1 & 1 \\ 3 & 0 \end{pmatrix} + 6\times\begin{pmatrix} 8 & 2 \\ 7 & 0 \\ 1 & 1 \end{pmatrix}$
10. $(1 \quad 4 \quad 5) + (2 \quad 7)$

TEST A

Perform the indicated operations whenever they are defined.

1. $\begin{pmatrix} 2 & 1 \\ 6 & 7 \end{pmatrix} + \begin{pmatrix} 1 & -1 \\ 0 & 2 \end{pmatrix}$
2. $\begin{pmatrix} 3 \\ 4 \end{pmatrix} + 8\times\begin{pmatrix} 1 \\ 6 \end{pmatrix}$
3. $\begin{pmatrix} 5 & 1 & 2 \\ 0 & 3 & 4 \end{pmatrix} + 2\times\begin{pmatrix} 1 & 0 \\ -1 & 0 \end{pmatrix}$

TEST B

Perform the indicated operations whenever they are defined.

1. $\begin{pmatrix} 3 & 1 & 0 \\ 2 & 6 & 5 \end{pmatrix} + 9\times\begin{pmatrix} 2 & 1 & 0 \\ -1 & 2 & 3 \end{pmatrix}$
2. $\begin{pmatrix} 1 \\ 2 \end{pmatrix} + (3 \quad 4)$
3. $18\times\begin{pmatrix} 0 & 0 \\ 0 & 0 \end{pmatrix} + \begin{pmatrix} 0 \\ 0 \end{pmatrix}$

SECTION 21. MULTIPLICATION OF MATRICES

In §20 we defined scalar multiplication: the product of a real number and a matrix. In this section we define another type of multiplication for matrices, the product of two matrices. We start with an application so as to motivate the definition of this new and --at first--rather strange operation.

Example 21-1

Suppose four different plants manufacture chairs, tables, and desks daily as shown below:

	NUMBER OF CHAIRS	TABLES	DESKS
PLANT I	50	10	20
II	60	15	5
III	30	10	10
IV	10	5	10

Suppose, further, that the amounts of wood and of metal (the raw materials) needed in the manufacture of each chair, table, and desk are given by the following:

	NEEDED UNITS OF WOOD	METAL
CHAIR	3	1
TABLE	10	2
DESK	9	3

We now ask a series of questions.

QUESTION (1). How much wood is needed daily by plant I?

Since plant I makes 50 chairs daily and each chair requires 3 units of wood, it needs $50\cdot 3 = 150$ units of wood for its chairs; since plant I makes 10 tables and each table needs 10 units of wood, it requires $10\cdot 10 = 100$ units of wood for its tables; finally, for its desks, plant I needs $20\cdot 9 = 180$ units of wood. So altogether the wood needed daily by plant I is

$$50\cdot 3 + 10\cdot 10 + 20\cdot 9 = 430 \text{ units}$$

If we display the given information in matrices, with the articles-per-plant matrix on the left and the raw-materials-per-article matrix at the right, we have

$$\begin{pmatrix} 50 & 10 & 20 \\ 60 & 15 & 5 \\ 30 & 10 & 10 \\ 10 & 5 & 10 \end{pmatrix} \begin{pmatrix} 3 & 1 \\ 10 & 2 \\ 9 & 3 \end{pmatrix}$$

--where we have shaded the entries in the first row of the left-hand matrix and in the first column of the right-hand matrix. The answer to question (1) was obtained by multiplying the entries in the first row of the left-hand matrix by the corresponding entries in the first column of the right-hand matrix and then adding the products.

QUESTION (2). How much metal does plant I need altogether daily? For the chairs plant I needs 50·1 units of metal; for tables, 10·2 units of metal; and for desks, 20·3 units. So altogether the metal needed daily by plant I is

$$50\cdot1 + 10\cdot2 + 20\cdot3 = 130 \text{ units}$$

This answer was gotten by multiplying the entries in the first row of the left-hand matrix by the corresponding entries in the *second* column of the right-hand matrix, then adding those products.

QUESTION (3). How much wood does plant II need daily?
In the first matrix we find the number of chairs, tables, and desks produced daily by plant II in the second row. The amount of wood each article needs is given in the first column of the second matrix. The answer is thus

$$60\cdot3 + 15\cdot10 + 5\cdot9 = 375 \text{ units}$$

QUESTION (4). How about plant II's metal needs? These are obtained similarly, by multiplying the entries in row number two of the matrix on the left by the corresponding entries in column two of the matrix at the right, then summing:

$$60\cdot1 + 15\cdot2 + 5\cdot3 = 105 \text{ units}$$

QUESTIONS (5) through (8). These are now obvious, and we can merely give the answers. Plant III requires wood as follows:

$$30\cdot3 + 10\cdot10 + 10\cdot9 = 280 \text{ units}$$

and metal as follows:

$$30\cdot1 + 10\cdot2 + 10\cdot3 = 80 \text{ units}$$

Finally, the needs of plant IV for wood and metal respectively are

$$10\cdot3 + 5\cdot10 + 10\cdot9 = 170 \text{ units}$$

and

$$10\cdot1 + 5\cdot2 + 10\cdot3 = 50 \text{ units}$$

These results can be displayed in matrix form as follows:

	NEEDED UNITS OF	
	WOOD	METAL
PLANT I	430	130
II	375	105
III	280	80
IV	170	50

If we let A be the articles-per-plant matrix, B the raw-materials-per-article matrix, and C the raw-materials-per-plant matrix given just above, then C is called the MATRIX PRODUCT AB. So we have

$$AB = C$$

Writing out the matrices yields

$$\underset{A}{\begin{pmatrix} 50 & 10 & 20 \\ 60 & 15 & 5 \\ 30 & 10 & 10 \\ 10 & 5 & 10 \end{pmatrix}} \cdot \underset{B}{\begin{pmatrix} 3 & 1 \\ 10 & 2 \\ 9 & 3 \end{pmatrix}} = \underset{C}{\begin{pmatrix} 430 & 130 \\ 375 & 105 \\ 280 & 80 \\ 170 & 50 \end{pmatrix}}$$

Notice the dimensions of these matrices: A is 4×3, B is 3×2, and C is 4×2.

Let us review how we obtained the various entries of C. The entry of C in row 3 and column 2, for example, was obtained by multiplying the entries in row 3 of A by the corresponding entries in column 2 of B and summing these products. Using subscripts, we can write

$$c_{32} = a_{31}b_{12} + a_{32}b_{22} + a_{33}b_{32}$$

We show this below, shading the pertinent data:

$$\begin{pmatrix} a_{11} & a_{12} & a_{13} \\ a_{21} & a_{22} & a_{23} \\ a_{31} & a_{32} & a_{33} \\ a_{41} & a_{42} & a_{43} \end{pmatrix} \cdot \begin{pmatrix} b_{11} & b_{12} \\ b_{21} & b_{22} \\ b_{31} & b_{32} \end{pmatrix} = \begin{pmatrix} c_{11} & c_{12} \\ c_{21} & c_{22} \\ c_{31} & c_{32} \\ c_{41} & c_{42} \end{pmatrix}$$

DEFINITION OF MATRIX MULTIPLICATION

We see from the example and description above that we can multiply two matrices A and B, with A on the left and B on the right, *only* if A has the same number of columns as B has rows. Thus

If A is m by n, B must be n by something;
if A is $m \times n$ and B is $n \times p$, then AB is defined. If the product $AB = C$, then C is an $m \times p$ matrix.

> Let A by $m \times n$ and B be $n \times p$. If $A = (a_{ij})$, $B = (b_{ij})$, and $AB = C$ where $C = (c_{ij})$, then
>
> $$c_{ij} = a_{i1}b_{1j} + a_{i2}b_{2j} + \ldots + a_{in}b_{nj}$$

It is helpful to think of matrix multiplication as *row-by-column multiplication*.

Example 21-2

$$\begin{pmatrix} 3 & 1 \\ 0 & 2 \\ -4 & 1 \\ 5 & -2 \end{pmatrix} \cdot \begin{pmatrix} 2 \\ -5 \end{pmatrix} = \begin{pmatrix} 3\cdot 2 + 1\cdot -5 \\ 0\cdot 2 + 2\cdot -5 \\ -4\cdot 2 + 1\cdot -5 \\ 5\cdot 2 + -2\cdot -5 \end{pmatrix} = \begin{pmatrix} 1 \\ -10 \\ -13 \\ 20 \end{pmatrix}$$

dimensions: $(4\times 2)\cdot(2\times 1) = (4\times 1) = (4\times 1)$

Example 21-3

$$\begin{pmatrix} -1 & 6 & 7 \end{pmatrix} \cdot \begin{pmatrix} -2 & 0 \\ 5 & 3 \\ 4 & -1 \end{pmatrix} = \begin{pmatrix} -1\cdot -1 + 6\cdot 5 + 7\cdot 4 & -1\cdot 0 + 6\cdot 13 + 7\cdot -1 \end{pmatrix} = \begin{pmatrix} 60 & -25 \end{pmatrix}$$

$(1\times 3) \cdot (3\times 2) = (1\times 2) = (1\times 2)$

Example 21-4

$$\begin{pmatrix} -4 \\ 3 \end{pmatrix} \cdot (5 \quad 7) = \begin{pmatrix} -4\cdot 5 & -4\cdot 7 \\ 3\cdot 5 & 3\cdot 7 \end{pmatrix} = \begin{pmatrix} -20 & -28 \\ 15 & 21 \end{pmatrix}$$

$(2\times 1)\cdot(1\times 2) = (2\times 2) = (2\times 2)$

NOTES ON MATRIX MULTIPLICATION

We can think of the dimensions of two matrices as follows:

$$A \; (m\times n) \cdot (n\times p) \; B$$

(OUTER: m and p; INNER: n and n)

(1) The product of A and B is defined only if their *inner* dimensions are equal.

(2) The dimension of the product matrix AB is obtained from their *outer* dimensions: AB has dimension $m\times p$.

(3) If a row vector A and a column vector B have the same number of elements then the product AB is defined to be the scalar (real number) obtained by multiplying corresponding elements and adding the products (as shown in the next example).

Example 21-5

(a) $$(3 \quad -1 \quad 2)\cdot \begin{pmatrix} -2 \\ 4 \\ 1 \end{pmatrix} = 3\cdot -2 + -1\cdot 4 + 2\cdot 1 = -8$$

(b) $$(4 \quad 1 \quad -5 \quad 2)\cdot\begin{pmatrix}-1\\0\\-2\\3\end{pmatrix} = 4\cdot -1 + 1\cdot 0 + -5\cdot -2 + 2\cdot 3 = 12$$

(4) If A is a square matrix, then $A^2 = AA$, $A^3 = A^2A = AAA$, and so on.

(5) Note that matrix multiplication is not commutative. If we reverse the order of the matrices in example 21-3, for instance, the display looks like this:

$$\begin{pmatrix}-2 & 0\\5 & -3\\4 & -1\end{pmatrix}(-1 \quad 6 \quad 7)$$

Suppose we label these two matrices Q and S respectively. But the dimension of Q is 3×2, the dimension of S is 1×3. So the "inner" dimensions, respectively 2 and 1, are different, and the product QS is not defined. But in the original order SQ the "inner" dimensions, respectively 3 and 3, *are* the same, and the product SQ was obtained in Example 21-3.

Even when both AB and BA are defined they are not in general equal, as we now show.

Example 21-6

Let $A = \begin{pmatrix}1 & -2\\3 & -1\end{pmatrix}$ and $B = \begin{pmatrix}4 & 5\\2 & 0\end{pmatrix}$. Then

$$AB = \begin{pmatrix}1 & -2\\3 & -1\end{pmatrix}\cdot\begin{pmatrix}4 & 5\\2 & 0\end{pmatrix} = \begin{pmatrix}0 & 5\\10 & 15\end{pmatrix}$$

$$BA = \begin{pmatrix}4 & 5\\2 & 0\end{pmatrix}\cdot\begin{pmatrix}1 & -2\\3 & -1\end{pmatrix} = \begin{pmatrix}19 & -13\\2 & -4\end{pmatrix}$$

So $AB \neq BA$.

PROPERTIES OF MATRIX MULTIPLICATION

The following rules supplement those given in §20 (p. 174) about matrix addition and multiplication by a scalar. A, B, and C are matrices; r is a scalar:

(6) $(AB)C = A(BC)$ (matrix multiplication is associative)
(7) $A(B + C) = AB + AC$ (matrix multiplication distributes over addition)
(8) $r(AB) = (rA)B = A(rB)$

IDENTITY MATRICES

A square matrix with 1's along the principal diagonal (from upper left to lower right) and with 0's everywhere else is called an *identity matrix* and is denoted by I. The dimension of I can be indicated by a subscript when not clear from context; for example

$$I_2 = \begin{pmatrix} 1 & 0 \\ 0 & 1 \end{pmatrix} \qquad I_4 = \begin{pmatrix} 1 & 0 & 0 & 0 \\ 0 & 1 & 0 & 0 \\ 0 & 0 & 1 & 0 \\ 0 & 0 & 0 & 1 \end{pmatrix} \qquad I_n = \begin{pmatrix} 1 & 0 & \dots & 0 \\ 0 & 1 & \dots & 0 \\ \vdots & \vdots & & \vdots \\ 0 & 0 & \dots & 1 \end{pmatrix}$$

--where the last has dimension $n \times n$.

The student can easily verify that if A is an $m \times n$ matrix then

$$A \cdot I_n = A$$
$$I_m \cdot A = A$$

and if A is square, of dimension $m \times m$, and I is the identity matrix of the same dimension, then

$$AI = IA = A$$

Thus an identity matrix of the proper dimension serves as the multiplicative identity in matrix multiplication (just as the number 1 does in ordinary multiplication of real numbers).

The next two examples illustrate some of the properties and definitions given above.

Example 21-7

If $M = \begin{pmatrix} 1 & 3 \\ 0 & -2 \end{pmatrix}$, then

$$M^2 = \begin{pmatrix} 1 & 3 \\ 0 & -2 \end{pmatrix} \cdot \begin{pmatrix} 1 & 3 \\ 0 & -2 \end{pmatrix} = \begin{pmatrix} 1 & -3 \\ 0 & 4 \end{pmatrix}$$

$$M^3 = \begin{pmatrix} 1 & 3 \\ 0 & -2 \end{pmatrix} \cdot \begin{pmatrix} 1 & 3 \\ 0 & -2 \end{pmatrix} \cdot \begin{pmatrix} 1 & 3 \\ 0 & -2 \end{pmatrix} = \begin{pmatrix} 1 & -3 \\ 0 & 4 \end{pmatrix} \cdot \begin{pmatrix} 1 & 3 \\ 0 & -2 \end{pmatrix} = \begin{pmatrix} 1 & 9 \\ 0 & -8 \end{pmatrix}$$

Property (6) above implies that $A^2A = AA^2$. (Why?) Verify this for matrix M.

Example 21-8

(a) $\begin{pmatrix} 1 & 0 \\ 0 & 1 \end{pmatrix} \begin{pmatrix} 2 & 3 \\ 5 & -6 \end{pmatrix} = \begin{pmatrix} 2 & 3 \\ 5 & -6 \end{pmatrix}$; $\begin{pmatrix} 2 & 3 \\ 5 & -6 \end{pmatrix} \begin{pmatrix} 1 & 0 \\ 0 & 1 \end{pmatrix} = \begin{pmatrix} 2 & 3 \\ 5 & -6 \end{pmatrix}$

(b) $\begin{pmatrix} 1 & 2 & 6 \\ 4 & 2 & 0 \\ -1 & 1 & 5 \end{pmatrix} \cdot \begin{pmatrix} 1 & 0 & 0 \\ 0 & 1 & 0 \\ 0 & 0 & 1 \end{pmatrix} = \begin{pmatrix} 1 & 2 & 6 \\ 4 & 2 & 0 \\ -1 & 1 & 5 \end{pmatrix}$

These cases show the effect of multiplying by an identity matrix.

Example 21-9

The following shows the amount of time required at each of two production stages in the manufacture of three models of cars:

$$\begin{array}{ll} & \text{HOURS NEEDED IN} \\ & \text{ASSEMBLY} \quad \text{FINISHING} \\ \text{MODEL I} & \\ \text{II} & \\ \text{III} & \end{array} \begin{pmatrix} 5 & 3 \\ 4 & 4 \\ 6 & 7 \end{pmatrix} \qquad \text{(matrix } A\text{)}$$

If each hour in the assembly shop costs \$400 and each hour in the finishing shop costs \$250, find the total cost of these production stages for each model.

The hourly costs, in dollars, are displayed below:

$$\begin{array}{l} \text{COST FOR} \\ \text{ONE HR} \end{array}$$

$$\begin{array}{l} \text{ASSEMBLY} \\ \text{FINISHING} \end{array} \begin{pmatrix} 400 \\ 250 \end{pmatrix} \qquad \text{(matrix } B\text{)}$$

Matrix A is $3{\times}2$ and gives the hours-in-production-stage per model; matrix B is $2{\times}1$ and gives the cost-per-hour per production stage. The product matrix AB therefore will be $3{\times}1$ and will give the cost per model:

$$\begin{pmatrix} 5 & 3 \\ 4 & 4 \\ 6 & 7 \end{pmatrix} \cdot \begin{pmatrix} 400 \\ 250 \end{pmatrix} = \begin{pmatrix} 5\cdot400 + 3\cdot250 \\ 4\cdot400 + 4\cdot250 \\ 6\cdot400 + 7\cdot250 \end{pmatrix} = \begin{pmatrix} 2750 \\ 2600 \\ 4150 \end{pmatrix}$$

So we have (in dollars):

$$\begin{array}{l} \text{COST PER} \\ \text{MODEL} \end{array}$$

$$\begin{array}{l} \text{MODEL I} \\ \text{II} \\ \text{III} \end{array} \begin{pmatrix} 2750 \\ 2600 \\ 4150 \end{pmatrix}$$

ROW-BY-COLUMN MULTIPLICATION

When we multiply matrix A by matrix B and get the product $AB = C$ we know that

If A is $m{\times}n$ and B is $n{\times}p$ then C is $m{\times}p$.

Just as the common inner dimension "disappears" in the dimension of the product matrix,

$$(m{\times}n) \cdot (n{\times}p) = (m{\times}p)$$

so also does the common category of the two matrices disappear in their product. Thus, in Example 21-1 we multiplied the given matrices as follows:

$$\begin{pmatrix}\text{articles}\\ \text{per}\\ \text{plant}\end{pmatrix}\cdot\begin{pmatrix}\text{raw-material units}\\ \text{per}\\ \text{article}\end{pmatrix} = \begin{pmatrix}\text{raw-material units}\\ \text{per}\\ \text{plant}\end{pmatrix}$$

The category common to the two matrices multiplied is the articles category; it does not appear in their matrix product.

In Example 21-9 we multiplied as follows:

$$\begin{pmatrix}\text{production-stage}\\ \text{hours per}\\ \text{model}\end{pmatrix}\cdot\begin{pmatrix}\text{cost per}\\ \text{production-stage}\\ \text{hour}\end{pmatrix} = \begin{pmatrix}\text{cost}\\ \text{per}\\ \text{model}\end{pmatrix}$$

The hours category disappears in the product.

EXERCISES

1. Calculate any of the following products that are defined.

(a) $\begin{pmatrix}1 & 3\end{pmatrix}\begin{pmatrix}4\\-1\end{pmatrix}$ (b) $\begin{pmatrix}6 & 2 & 1\end{pmatrix}\begin{pmatrix}1\\0\\2\end{pmatrix}$ (c) $\begin{pmatrix}4\\5\end{pmatrix}\begin{pmatrix}1 & 3\end{pmatrix}$

(d) $\begin{pmatrix}1\\0\end{pmatrix}\begin{pmatrix}1 & 0 & 0\end{pmatrix}$ (e) $\begin{pmatrix}3 & 4\end{pmatrix}\begin{pmatrix}1 & 2 & -1\\0 & 3 & 5\end{pmatrix}$ (f) $\begin{pmatrix}2 & 1 & 0\\3 & 0 & 4\end{pmatrix}\begin{pmatrix}2\\6\end{pmatrix}$

(g) $\begin{pmatrix}1\\0\end{pmatrix}\begin{pmatrix}2 & 3\\6 & 1\end{pmatrix}$ (h) $\begin{pmatrix}3 & 1\\0 & 2\\-1 & 4\end{pmatrix}\begin{pmatrix}2\\5\end{pmatrix}$ (i) $\begin{pmatrix}1 & 0 & -1\end{pmatrix}\begin{pmatrix}3 & 4 & 0\\1 & 1 & 0\\2 & 6 & 1\end{pmatrix}$

(j) $\begin{pmatrix}1 & 0\\2 & 3\end{pmatrix}\begin{pmatrix}4 & 5\\-1 & 2\end{pmatrix}$ (k) $\begin{pmatrix}1 & 0 & 2\\3 & 4 & 5\end{pmatrix}\begin{pmatrix}0 & -1\\1 & 2\\2 & 3\end{pmatrix}$ (l) $\begin{pmatrix}1 & 0\\2 & 3\\-1 & 4\end{pmatrix}\begin{pmatrix}1 & 3\\4 & 1\\2 & -1\end{pmatrix}$

(m) $\begin{pmatrix}2 & 0 & -1\\1 & -2 & 0\\0 & 3 & 1\end{pmatrix}\begin{pmatrix}4 & 0 & 1\\2 & -1 & 0\\1 & 0 & -2\end{pmatrix}$

2. Multiply as indicated, if possible.

(a) $\begin{pmatrix}3 & 1 & 5\end{pmatrix}\begin{pmatrix}a\\b\\c\end{pmatrix}$ (b) $\begin{pmatrix}1\\4\end{pmatrix}\begin{pmatrix}x\\y\end{pmatrix}$ (c) $\begin{pmatrix}2 & 3\end{pmatrix}\begin{pmatrix}u\\v\end{pmatrix}$ (d) $\begin{pmatrix}1 & 4\\0 & 2\end{pmatrix}\begin{pmatrix}x\\y\end{pmatrix}$

(e) $\begin{pmatrix}1 & 3 & 5\\2 & 0 & 1\end{pmatrix}\begin{pmatrix}p\\q\\r\end{pmatrix}$ (f) $\begin{pmatrix}3 & 4 & 1\\5 & 2 & 6\\0 & 1 & -1\end{pmatrix}\begin{pmatrix}x\\y\\z\end{pmatrix}$

3. In each of the following, show that $A(BC) = (AB)C$.

(a) $A = \begin{pmatrix}3 & 2\\1 & -1\end{pmatrix}$ $B = \begin{pmatrix}0 & 2\\4 & 1\end{pmatrix}$ $C = \begin{pmatrix}-1 & 0\\1 & 2\end{pmatrix}$

(b) $A = \begin{pmatrix}2 & -1\\0 & 1\end{pmatrix}$ $B = \begin{pmatrix}3 & 0 & 1\\2 & 4 & -1\end{pmatrix}$ $C = \begin{pmatrix}1 & 2 & 5\\0 & 3 & 1\\1 & 4 & -1\end{pmatrix}$

4. Let

$$A = \begin{pmatrix} 1 & 0 \\ 2 & -1 \\ 4 & 3 \end{pmatrix}, \qquad B = \begin{pmatrix} 5 & 0 \\ 1 & -1 \end{pmatrix}, \qquad C = \begin{pmatrix} 2 & 1 \\ -1 & 0 \end{pmatrix}$$

Show that $A(B + C) = AB + AC$.

5. Calculate A^2 and A^3 if A equals

(a) $\begin{pmatrix} 1 & 0 \\ 0 & 2 \end{pmatrix}$ (b) $\begin{pmatrix} 0 & 1 \\ 1 & 0 \end{pmatrix}$ (c) $\begin{pmatrix} 2 & 0 \\ 0 & 3 \end{pmatrix}$ (d) $\begin{pmatrix} 1 & 0 \\ 3 & 4 \end{pmatrix}$

(e) $\begin{pmatrix} 3 & 2 \\ 1 & 4 \end{pmatrix}$ (f) $\begin{pmatrix} 1 & 0 & 0 \\ 0 & 0 & 1 \\ 0 & 1 & 0 \end{pmatrix}$ (g) $\begin{pmatrix} 3 & 0 & 0 \\ 0 & 2 & 0 \\ 0 & 0 & 4 \end{pmatrix}$ (h) $\begin{pmatrix} 1 & 2 & 0 \\ 0 & -1 & 3 \\ 4 & 0 & 5 \end{pmatrix}$

6. Refer to the cost-of-mail-by-class matrix in Example 19-1 (page 162). Suppose a small business needs to mail 10 one-oz letters, five two-ounce letters, two eight-ounce packages, and one one-pound package. (a) Which class of mail costs the least? (b) How much can be saved using this class of mail rather than first-class mail?

7. Suppose two stores owned by the same proprietor recorded sales during one week as follows:

	TV SETS	RADIOS	PHONOGRAPHS
STORE I	10	7	3
STORE II	4	15	8

If the selling price was \$350 per TV set, \$50 per radio, and \$100 per phonograph, how much did each store take in on these items during the week?

8. On the basis of one week's observations in a laboratory, data were obtained and summarized in matrix form as follows: Matrix R tells how many servings of each of three different foods were eaten during the week by John Jones and Sally Silver. Matrix S gives the number of units of vitamins A, B, and D supplied by a serving of each of the three foods. (a) Assign categories to rows and columns in R and S so that matrix RS tells the vitamin intake of each person. (b) How must matrices R and S be organized so that matrix SR, instead of matrix RS, displays this information?

TEST A

1. If defined, calculate the following:

(a) $(2 \quad 3 \quad -1)\begin{pmatrix} 0 \\ 1 \\ 1 \end{pmatrix}$ (b) $\begin{pmatrix} 2 & 3 \\ 1 & 4 \\ 0 & 1 \end{pmatrix}\begin{pmatrix} x \\ y \\ z \end{pmatrix}$ (c) $\begin{pmatrix} 2 \\ 3 \end{pmatrix}\begin{pmatrix} 1 & 4 \\ 0 & 1 \end{pmatrix}$

(d) $\begin{pmatrix} 1 & 0 & 1 \\ 2 & 0 & 3 \end{pmatrix}\begin{pmatrix} a \\ b \\ c \end{pmatrix}$ (e) A^2, where $A = \begin{pmatrix} 2 & 0 \\ 1 & -1 \end{pmatrix}$.

2. The following tables give information about the sales one day of four books at two stores and about the selling prices of these books:

	NO. COPIES SOLD OF BOOK NO.			
	I	II	III	IV
STORE X	20	5	40	5
Y	10	3	30	8

	PRICE IN $$
BOOK I	10
II	12
III	9
IV	20

(a) Write a matrix that displays the revenue at each store from each book. (b) How much did store Y take in altogether?

TEST B

1. If defined, calculate the following:

(a) $(1\ 4)\begin{pmatrix} 2 \\ 6 \end{pmatrix}$ (b) $\begin{pmatrix} 2 & 1 & 3 \\ 0 & -1 & 4 \end{pmatrix}\begin{pmatrix} a \\ b \\ c \end{pmatrix}$ (c) $\begin{pmatrix} 4 & 6 \\ 7 & 1 \end{pmatrix}(x\ y)$

(d) $\begin{pmatrix} 3 \\ 2 \end{pmatrix}(1\ 0)$ (e) $\begin{pmatrix} 1 & 0 \\ 0 & 1 \end{pmatrix}\begin{pmatrix} 2 & 19 \\ 32 & 4 \end{pmatrix}$ (c) $\begin{pmatrix} 1 & -1 \\ 2 & 3 \\ 0 & 4 \end{pmatrix}\begin{pmatrix} 2 & 1 & 0 \\ 3 & 4 & -1 \end{pmatrix}$

2. Four chain stores sell TV sets during one month as recorded below in the left-hand table, and at prices shown on the right:

	NO. SOLD OF MODEL		
	DELUXE	MEDIUM	ECONOMY
STORE A	2	1	5
B	0	3	2
C	4	2	1
D	3	0	4

	PRICE IN $$
DELUXE	500
MEDIUM	400
ECONOMY	300

(a) Write a matrix for the revenues for TV sets during the month at these stores. (b) Which store took in the most? the least?

SECTION 22

USING MATRICES TO SOLVE SYSTEMS OF LINEAR EQUATIONS

In this section we will see how closely matrices are related to systems of linear equations, then show how we can use matrices to solve such systems more efficiently than we did earlier. Although there are many ways of using matrices for this purpose, the technique introduced here, called the *Gauss-Jordan* method, is especially useful because it enables us to determine easily whether a system has a solution and, if so, to find any solutions. This method is also readily adapted to computers.

MATRIX EQUATIONS

We start with a system of two equations in two unknowns.

Example 22-1

The system

$$\begin{aligned} 5x + 4y &= 3 \\ 2x + y &= -1 \end{aligned}$$

can we written as follows, using the definition of equality of matrices:

$$\begin{pmatrix} 5x + 4y \\ 2x + y \end{pmatrix} = \begin{pmatrix} 3 \\ -1 \end{pmatrix}$$

or it can be written as follows, using also multiplication of matrices:

$$\begin{pmatrix} 5 & 4 \\ 2 & 1 \end{pmatrix} \begin{pmatrix} x \\ y \end{pmatrix} = \begin{pmatrix} 3 \\ -1 \end{pmatrix}$$

If we represent the matrices in this equation by

$$A = \begin{pmatrix} 5 & 4 \\ 2 & 1 \end{pmatrix} \qquad X = \begin{pmatrix} x \\ y \end{pmatrix} \qquad B = \begin{pmatrix} 3 \\ -1 \end{pmatrix}$$

then we get an even neater, more compact notation for the *matrix equation*

$$AX = B$$

Here A is called the *coefficient matrix* of the system, the column vector X is called the *solution vector*, and the column vector B is called the *vector of constants*.

Example 22-2

Write the following system as a matrix equation:

$$\begin{aligned} x - y + z &= 2 \\ 3x + 4z &= 1 \\ x + 2y &= 0 \end{aligned}$$

Using matrices, we get

$$\begin{pmatrix} 1 & -1 & 1 \\ 3 & 0 & 4 \\ 1 & 2 & 0 \end{pmatrix} \begin{pmatrix} x \\ y \\ z \end{pmatrix} = \begin{pmatrix} 2 \\ 1 \\ 0 \end{pmatrix}$$

Since this system involves three unknowns, the solution vector X has dimension 3×1.

Example 22-3

Write the following matrix equation as a system of linear equations:

$$\begin{pmatrix} 2 & 0 & -1 & 3 \\ 1 & 1 & 0 & 0 \\ -3 & 2 & 1 & 4 \\ 0 & -3 & 0 & 5 \end{pmatrix} \begin{pmatrix} x_1 \\ x_2 \\ x_3 \\ x_4 \end{pmatrix} = \begin{pmatrix} 5 \\ -4 \\ 0 \\ 3 \end{pmatrix}$$

The system of equations in four unknowns defined by this matrix equation is

$$\begin{aligned} 2x_1 - x_3 + 3x_4 &= 5 \\ x_1 + x_2 &= -4 \\ -3x_1 + 2x_2 + x_3 + 4x_4 &= 0 \\ -3x_2 + 5x_4 &= 3 \end{aligned}$$

AUGMENTED MATRICES

When solving a system of equations by matrices it is convenient to omit the solution vector and to incorporate the vector of constants into the coefficient matrix, with a vertical bar to set it off; the result is called the *augmented matrix* for the system.

The augmented matrices for the systems in examples 22-1, -2, and -3 are shown here:

EXAMPLE 22-1

$$\left(\begin{array}{cc|c} 5 & 4 & 3 \\ 2 & 1 & -1 \end{array}\right)$$

EXAMPLE 22-2

$$\left(\begin{array}{ccc|c} 1 & -1 & 1 & 2 \\ 3 & 0 & 4 & 1 \\ 1 & 2 & 0 & 0 \end{array}\right)$$

EXAMPLE 22-3

$$\left(\begin{array}{cccc|c} 2 & 0 & -1 & 3 & 5 \\ 1 & 1 & 0 & 0 & -4 \\ -3 & 2 & 1 & 4 & 0 \\ 0 & -3 & 0 & 5 & 3 \end{array}\right)$$

The change to an augmented matrix is obviously just a change in notation. The augmented matrix displays only the essential features of the system of equations; that is, only the features that are involved in the solution process. We will return to this in just a moment.

CONVENTIONAL SOLUTION OF A SYSTEM OF EQUATIONS

We will now carefully analyze how a system of equations is solved by the conventional method presented in §17 (page 137). Then we will use this analysis to develop the method of solving the same system using matrices.

Example 22-4

Solve the system

$$\begin{aligned} x + 2y - z &= 0 \\ 2x - y + 3z &= 15 \\ -3x + y + 2z &= 3 \end{aligned}$$

Our aim is to derive a sequence of equivalent systems (systems with the same solutions as the original system) until we obtain one of the form

$$\begin{aligned} x \qquad\quad &= a \\ y \quad\; &= b \\ z &= c \end{aligned}$$

We will then have solved the original system.

There are three operations we can perform on a system of equations that yield an equivalent system. We denote these operations as "Op1," "Op2," and "Op3" for easy reference:

Op1: interchange any two equations;
Op2: multiply any equation by any constant other than zero;
Op3: replace any equation by the sum of that equation and a (nonzero) multiple of any other equation.

We now show how we apply these operations to the given system of equations, systematically working first on the terms in x, then on those in y, and finally on those in z. Keep in mind at all times the shape of the terminal system at which we are aiming.

Since the coefficient of x in the first equation is 1, we are off to a quick start. To eliminate the term in x from the second equation, we use Op3: we multiply the first equation by -2 and add the result to the second equation:

$$(-2)(x + 2y - z = 0) + (2x - y + 3z = 15) \rightarrow -5y + 5z = 15$$

Replacing the second equation of the original system by this, our new (and equivalent) system is

$$\begin{aligned} x + 2y - z &= 0 \\ -5y + 5z &= 15 \\ -3x + y + 2z &= 3 \end{aligned}$$

Now we want to eliminate the term in x from the third equation. We apply Op3 again, this time multiplying the first equation by 3 and adding the result to the third equation:

$$3(x + 2y - z = 0) + (-3x + y + 2z = 3) \rightarrow 7y - z = 3$$

So the new, but still equivalent, system is

$$\begin{aligned} x + 2y - z &= 0 \\ -5y + 5z &= 15 \\ 7y - z &= 3 \end{aligned}$$

We turn next to the terms in y, being careful, as we operate on them, not to disturb the terms in x, which are now in the desired position. Since our goal is for the coefficient of y in the second equation to be 1, we use Op2 to multiply the second equation by $-\frac{1}{5}$, obtaining the new equivalent system

$$\begin{aligned} x + 2y - z &= 0 \\ y - z &= -3 \\ 7y - z &= 3 \end{aligned}$$

To eliminate the term in y from the first equation, we multiply the second equation by -2 and add the result to the first (Op3):

$$\begin{aligned} x + z &= 6 \\ y - z &= -3 \\ 7y - z &= 3 \end{aligned}$$

To eliminate the y-term from the third equation, we again use Op3, this time adding -7 times the second equation to the third:

$$\begin{aligned} x + z &= 6 \\ y - z &= -3 \\ 6z &= 24 \end{aligned}$$

Finally we turn to the terms in z (taking care not to disturb the terms in x and y which are now as we want them). Since we want the coefficient of z in the third equation to be 1, we invoke Op2 to multiply the third equation by $\frac{1}{6}$:

$$\begin{aligned} x + z &= 6 \\ y - z &= -3 \\ z &= 4 \end{aligned}$$

To eliminate the term in z from the first equation, we add -1 times the third equation to the first (Op3):

$$\begin{aligned} x &= 2 \\ y - z &= -3 \\ z &= 4 \end{aligned}$$

To eliminate the z-term from the second equation we add 1 times the third equation to the second equation (again Op3):

$$\begin{aligned} x &= 2 \\ y &= 1 \\ z &= 4 \end{aligned}$$

--and finally we have the solution to the original system: it is $x = 2$, $y = 1$, and $z = 4$.

Since it is easy to make careless errors, you should always check your solution in the original system. In this example we have

$$\begin{aligned} (2) + 2(1) - (4) &\stackrel{\checkmark}{=} 0 \\ 2(2) - (1) + 3(4) &\stackrel{\checkmark}{=} 15 \\ -3(2) + (1) + 2(4) &\stackrel{\checkmark}{=} 3 \end{aligned}$$

A Note on Operation Op3

Since this operation often confuses people, we suggest that you devote some extra attention to it. Op3 is used to *eliminate* the term in a particular variable in a particular equation. When we use Op3 we *replace* that equation by a new one in which the coefficient of the variable in question is zero. For example, in the system

$$\begin{aligned} x + 2y - z &= 0 \\ y - z &= -3 \\ 7y - z &= 3 \end{aligned}$$

we eliminate the y-term in the first equation by *replacing* the first equation by a new one, namely

(-2 times equation two) + (equation one)

Similarly, we *replace* equation three by a new equation, namely

(-7 times equation two) + (equation three)

We recommend that you reread example 22-4 before going on, to be sure you understand how we obtained each new equivalent system from the preceding system.

SOLUTION BY MATRICES: GAUSS-JORDAN ROW OPERATIONS

The sequence of equivalent systems of equations just demonstrated can be expressed neatly in terms of matrix equations, and even more compactly using augmented matrices. Thus, at the beginning of example 22-4 we have:

SYSTEM OF EQUATIONS	MATRIX EQUATION	AUGMENTED MATRIX
$\begin{aligned} x + 2y - z &= 0 \\ 2x - y + 3z &= 15 \\ -3x + y + 2z &= 3 \end{aligned}$	$\begin{pmatrix} 1 & 2 & -1 \\ 2 & -1 & 3 \\ -3 & 1 & 2 \end{pmatrix} \begin{pmatrix} x \\ y \\ z \end{pmatrix} = \begin{pmatrix} 0 \\ 15 \\ 3 \end{pmatrix}$	$\left(\begin{array}{ccc\|c} 1 & 2 & -1 & 0 \\ 2 & -1 & 3 & 15 \\ -3 & 1 & 2 & 3 \end{array}\right)$

There is a set of three operations we can perform on the rows of an augmented matrix, called the *Gauss-Jordan row operations*, that precisely parallel the three operations we have been using on systems of equations. We denote these matrix operations by "Ma1," "Ma2," and "Ma3," and note that each is obtained just by replacing the word "equation," wherever it appears below on the left, by "row":

SYSTEM OF EQUATIONS		AUGMENTED MATRIX	
Op1.	Interchange any two equations.	Ma1.	Interchange any two rows.
Op2.	Multiply any equation by a nonzero constant.	Ma2.	Multiply any row by a nonzero constant.
Op3.	Replace any equation by the sum of that equation and a nonzero multiple of any other equation.	Ma3.	Replace any row by the sum of that row and a nonzero multiple of any other row.

In the display that follows we show at the left the sequence of equivalent systems of equations obtained in example 22-4; at the right we show the corresponding sequence of augmented matrices obtained by applying the parallel Gauss-Jordan row operations:

$$\begin{aligned} x + 2y - z &= 0 \\ 2x - y + 3z &= 15 \\ -3x + y + 2z &= 3 \end{aligned} \qquad \left(\begin{array}{ccc|c} 1 & 2 & -1 & 0 \\ 2 & -1 & 3 & 15 \\ -3 & 1 & 2 & 3 \end{array}\right)$$

$$\downarrow \text{(by Op3)} \qquad \downarrow \text{(by Ma3)}$$

$$\begin{aligned} x + 2y - z &= 0 \\ -5y + 5z &= 15 \\ -3x + y + 2z &= 3 \end{aligned} \qquad \left(\begin{array}{ccc|c} 1 & 2 & -1 & 0 \\ 0 & -5 & 5 & 15 \\ -3 & 1 & 2 & 3 \end{array}\right)$$

$$\downarrow \text{(by Op3)} \qquad \downarrow \text{(by Ma3)}$$

$$\begin{aligned} x + 2y - z &= 0 \\ -5y + 5z &= 15 \\ 7y - z &= 3 \end{aligned} \qquad \left(\begin{array}{ccc|c} 1 & 2 & -1 & 0 \\ 0 & -5 & 5 & 15 \\ 0 & 7 & -1 & 3 \end{array}\right)$$

$$\downarrow \text{(by Op2)} \qquad \downarrow \text{(by Ma2)}$$

$$\begin{aligned} x + 2y - z &= 0 \\ y - z &= -3 \\ 7y - z &= 3 \end{aligned} \qquad \left(\begin{array}{ccc|c} 1 & 2 & -1 & 0 \\ 0 & 1 & -1 & -3 \\ 0 & 7 & -1 & 3 \end{array}\right)$$

$$\downarrow \text{(by Op3)} \qquad \downarrow \text{(by Ma3)}$$

$$\begin{aligned} x \quad + z &= 6 \\ y - z &= -3 \\ 7y - z &= 3 \end{aligned} \qquad \left(\begin{array}{ccc|c} 1 & 0 & 1 & 6 \\ 0 & 1 & -1 & -3 \\ 0 & 7 & -1 & 3 \end{array}\right)$$

$$\downarrow \text{(by Op3)} \qquad \downarrow \text{(by Ma3)}$$

$$\begin{aligned} x \quad + z &= 6 \\ y - z &= -3 \\ 6z &= 24 \end{aligned} \qquad \left(\begin{array}{ccc|c} 1 & 0 & 1 & 6 \\ 0 & 1 & -1 & -3 \\ 0 & 0 & 6 & 24 \end{array}\right)$$

(turn page)

(continued)

$$\Big\downarrow \text{(by Op2)} \qquad\qquad\qquad\qquad \Big\downarrow \text{(by Ma2)}$$

$$\begin{aligned} x \quad\; + z &= 6 \\ y - z &= -3 \\ z &= 4 \end{aligned} \qquad\qquad \left(\begin{array}{ccc|c} 1 & 0 & 1 & 6 \\ 0 & 1 & -1 & -3 \\ 0 & 0 & 1 & 4 \end{array}\right)$$

$$\Big\downarrow \text{(by Op3)} \qquad\qquad\qquad\qquad \Big\downarrow \text{(by Ma3)}$$

$$\begin{aligned} x \quad\quad\; &= 2 \\ y - z &= -3 \\ z &= 4 \end{aligned} \qquad\qquad \left(\begin{array}{ccc|c} 1 & 0 & 0 & 2 \\ 0 & 1 & -1 & -3 \\ 0 & 0 & 1 & 4 \end{array}\right)$$

$$\Big\downarrow \text{(by Op3)} \qquad\qquad\qquad\qquad \Big\downarrow \text{(by Op3)}$$

$$\begin{aligned} x \quad\quad\; &= 2 \\ y \quad &= 1 \\ z &= 4 \end{aligned} \qquad\qquad \left(\begin{array}{ccc|c} 1 & 0 & 0 & 2 \\ 0 & 1 & 0 & 1 \\ 0 & 0 & 1 & 4 \end{array}\right)$$

For a system of three equations in three unknowns (call them x, y, and z) the desired terminations of the respective solution sequences are of the form

$$\begin{aligned} x \quad\quad\; &= a \\ y \quad &= b \\ z &= c \end{aligned} \qquad \text{and} \qquad \left(\begin{array}{ccc|c} 1 & 0 & 0 & a \\ 0 & 1 & 0 & b \\ 0 & 0 & 1 & c \end{array}\right)$$

Note that the coefficient matrix on the right--the part of the augmented matrix to the left of the vertical bar--is the identity matrix. Later in this section we will consider examples in which the coefficient matrix cannot be reduced to the identity matrix.

Example 22-5

Solve the system of equations

$$\begin{aligned} 2y - z &= -2 \\ x + y + z &= 3 \\ 2x - y - z &= 0 \end{aligned}$$

by applying Gauss-Jordan row operations to augmented matrices.

The original system yields the augmented matrix

$$\left(\begin{array}{ccc|c} 0 & 2 & -1 & -2 \\ 1 & 1 & 1 & 3 \\ 2 & -1 & -1 & 0 \end{array}\right)$$

Our goal, remember, is to reduce the coefficient matrix to the multiplicative identity:

$$\left(\begin{array}{ccc|c} 1 & 0 & 0 & a \\ 0 & 1 & 0 & b \\ 0 & 0 & 1 & c \end{array}\right)$$

Since the first row in our original augmented matrix starts out with 0 and we want a 1, we begin by interchanging the first and second rows (Ma1). We then continue to operate on the rows as indicated below, to obtain the desired terminal matrix:

$$\left(\begin{array}{ccc|c} 0 & 2 & -1 & -2 \\ 1 & 1 & 1 & 3 \\ 2 & -1 & -1 & 0 \end{array}\right) \begin{array}{c}\text{interchange}\\ \text{rows 1 and 2}\\ \text{(Ma1)} \longrightarrow\end{array} \left(\begin{array}{ccc|c} 1 & 1 & 1 & 3 \\ 0 & 2 & -1 & -2 \\ 2 & -1 & -1 & 0 \end{array}\right) \begin{array}{c}(-2)\cdot\text{(row 1)}\\ +\ \text{(row 3)}\\ \text{(Ma3)} \longrightarrow\end{array} \left(\begin{array}{ccc|c} 1 & 1 & 1 & 3 \\ 0 & 2 & -1 & -2 \\ 0 & -3 & -3 & -6 \end{array}\right)$$

$$\begin{array}{c}(\tfrac{1}{2})\cdot\text{(row 2)}\\ \text{(Ma2)} \longrightarrow\end{array} \left(\begin{array}{ccc|c} 1 & 1 & 1 & 3 \\ 0 & 1 & -\frac{1}{2} & -1 \\ 0 & -3 & -3 & -6 \end{array}\right) \begin{array}{c}-\ \text{(row 2)}\\ +\ \text{(row 1)}\\ \text{(Ma3)} \longrightarrow\end{array} \left(\begin{array}{ccc|c} 1 & 0 & \frac{3}{2} & 4 \\ 0 & 1 & -\frac{1}{2} & -1 \\ 0 & -3 & -3 & -6 \end{array}\right) \begin{array}{c}(3)\cdot\text{(row 2)}\\ +\ \text{(row 3)}\\ \text{(Ma3)} \longrightarrow\end{array}$$

$$\left(\begin{array}{ccc|c} 1 & 0 & \frac{3}{2} & 4 \\ 0 & 1 & -\frac{1}{2} & -1 \\ 0 & 0 & -\frac{9}{2} & -9 \end{array}\right) \begin{array}{c}(-\tfrac{2}{9})\cdot\text{(row 3)}\\ \text{(Ma3)} \longrightarrow\end{array} \left(\begin{array}{ccc|c} 1 & 0 & \frac{3}{2} & 4 \\ 0 & 1 & -\frac{1}{2} & -1 \\ 0 & 0 & 1 & 2 \end{array}\right) \begin{array}{c}(-\tfrac{3}{2})\cdot\text{(row 3)}\\ +\ \text{(row 1)}\\ \text{(Ma3)} \longrightarrow\end{array} \left(\begin{array}{ccc|c} 1 & 0 & 0 & 1 \\ 0 & 1 & -\frac{1}{2} & -1 \\ 0 & 0 & 1 & 2 \end{array}\right)$$

$$\begin{array}{c}(\tfrac{1}{2})\cdot\text{(row 3)}\\ +\ \text{(row 2)}\\ \text{(Ma3)} \longrightarrow\end{array} \left(\begin{array}{ccc|c} 1 & 0 & 0 & 1 \\ 0 & 1 & 0 & 0 \\ 0 & 0 & 1 & 2 \end{array}\right)$$

We can now read off the solution of the given system of equations: $x = 1$, $y = 0$, $z = 2$. Of course we check (in the original system!):

$$\begin{aligned} 2(0) - (2) &\overset{\checkmark}{=} -2 \\ (1) + (0) + (2) &\overset{\checkmark}{=} 3 \\ 2(1) - (0) - (2) &\overset{\checkmark}{=} 0 \end{aligned}$$

STEPS TO FOLLOW IN USING THE GAUSS-JORDAN TECHNIQUE

If the coefficient matrix is 3×3,

STEP 1. Get 1 in the upper left-hand corner. Then use operation Ma3, as necessary, to get 0 everywhere else in column one.

STEP 2. Get 1 in row two, column two. Then use Ma3 to get 0 everywhere else in column 2.

STEP 3. Get 1 in row three, column three. Then use Ma3 to get 0 everywhere else in column 3.

(For a coefficient matrix of other dimensions the modification of the foregoing is obvious.) To summarize,

Our ultimate goal is to reduce the coefficient matrix to the multiplicative identity, if possible.

SYSTEMS OF n EQUATIONS IN n UNKNOWNS WHERE $n \neq 3$

The Gauss-Jordan technique can be applied to any system of n equations in n unknowns, as we now demonstrate.

Example 22-6

Using augmented matrices and Gauss-Jordan row operations, solve the system

$$\begin{aligned} 2x - 3y &= 3 \\ 3x + 4y &= 13 \end{aligned}$$

The steps leading to the solution are given below. You should justify each arrow in the reduction of the coefficient matrix to the multiplicative identity:

$$\left(\begin{array}{cc|c} 2 & -3 & 2 \\ 3 & 4 & 13 \end{array}\right) \longrightarrow \left(\begin{array}{cc|c} 1 & -\frac{3}{2} & \frac{3}{2} \\ 3 & 4 & 13 \end{array}\right) \longrightarrow \left(\begin{array}{cc|c} 1 & -\frac{3}{2} & \frac{3}{2} \\ 0 & \frac{17}{2} & \frac{17}{2} \end{array}\right) \longrightarrow \left(\begin{array}{cc|c} 1 & -\frac{3}{2} & \frac{3}{2} \\ 0 & 1 & 1 \end{array}\right) \longrightarrow \left(\begin{array}{cc|c} 1 & 0 & 3 \\ 0 & 1 & 1 \end{array}\right)$$

The given system therefore has the solution $x = 3$, $y = 1$. We check:

$$2(3) - 3(1) \overset{\checkmark}{=} 3 \qquad\qquad 3(3) + 4(1) \overset{\checkmark}{=} 13$$

Example 22-7

Using the methods of this section, solve the system

$$\begin{aligned} w + x + y + z &= 2 \\ x \qquad - z &= 1 \\ 2w \qquad + 3y \quad z &= 6 \\ 2x + y \qquad &= 1 \end{aligned}$$

Once again, you should justify each step in the derivation:

$$\left(\begin{array}{cccc|c} 1 & 1 & 1 & 1 & 2 \\ 0 & 1 & 0 & -1 & 1 \\ 2 & 0 & 3 & 1 & 6 \\ 0 & 2 & 1 & 0 & 1 \end{array}\right) \to \left(\begin{array}{cccc|c} 1 & 1 & 1 & 1 & 2 \\ 0 & 1 & 0 & -1 & 1 \\ 0 & -2 & 1 & -1 & 2 \\ 0 & 2 & 1 & 0 & 1 \end{array}\right) \to \left(\begin{array}{cccc|c} 1 & 0 & 1 & 2 & 1 \\ 0 & 1 & 0 & -1 & 1 \\ 0 & -2 & 1 & -1 & 2 \\ 0 & 2 & 1 & 0 & 1 \end{array}\right) \to \left(\begin{array}{cccc|c} 1 & 0 & 1 & 2 & 1 \\ 0 & 1 & 0 & -1 & 1 \\ 0 & 0 & 1 & -3 & 4 \\ 0 & 2 & 1 & 0 & 1 \end{array}\right) \to$$

$$\left(\begin{array}{cccc|c} 1 & 0 & 1 & 2 & 1 \\ 0 & 1 & 0 & -1 & 1 \\ 0 & 0 & 1 & -3 & 4 \\ 0 & 0 & 1 & 2 & -1 \end{array}\right) \to \left(\begin{array}{cccc|c} 1 & 0 & 0 & 5 & -3 \\ 0 & 1 & 0 & -1 & 1 \\ 0 & 0 & 1 & -3 & 4 \\ 0 & 0 & 1 & 2 & -1 \end{array}\right) \to \left(\begin{array}{cccc|c} 1 & 0 & 0 & 5 & -3 \\ 0 & 1 & 0 & -1 & 1 \\ 0 & 0 & 1 & -3 & 4 \\ 0 & 0 & 0 & 5 & -5 \end{array}\right) \to \left(\begin{array}{cccc|c} 1 & 0 & 0 & 5 & -3 \\ 0 & 1 & 0 & -1 & 1 \\ 0 & 0 & 1 & -3 & 4 \\ 0 & 0 & 0 & 1 & -1 \end{array}\right) \to$$

$$\left(\begin{array}{cccc|c} 1 & 0 & 0 & 0 & 2 \\ 0 & 1 & 0 & -1 & 1 \\ 0 & 0 & 1 & -3 & 4 \\ 0 & 0 & 0 & 1 & -1 \end{array}\right) \to \left(\begin{array}{cccc|c} 1 & 0 & 0 & 0 & 2 \\ 0 & 1 & 0 & 0 & 0 \\ 0 & 0 & 1 & -3 & 4 \\ 0 & 0 & 0 & 1 & -1 \end{array}\right) \to \left(\begin{array}{cccc|c} 1 & 0 & 0 & 0 & 2 \\ 0 & 1 & 0 & 0 & 0 \\ 0 & 0 & 1 & 0 & 1 \\ 0 & 0 & 0 & 1 & -1 \end{array}\right)$$

We thus have $w = 2$, $x = 0$, $y = 1$, $z = -1$.

Checking yields

$$(2) + (0) + (1) + (-1) \overset{\checkmark}{=} 2$$
$$(0) + (-1) \overset{\checkmark}{=} 1$$
$$2(2) + 3(1) + (-1) \overset{\checkmark}{=} 6$$
$$2(0) + (1) \overset{\checkmark}{=} 1$$

A NOTE ON PROCEDURE

As you acquire skill in performing row operations, you will see that several operations not affecting each other can be performed at once. For instance, in Example 22-7 we can simultaneously perform the final three operations and move immediately

$$\text{from} \quad \left(\begin{array}{cccc|c} 1 & 0 & 0 & 5 & -3 \\ 0 & 1 & 0 & -1 & 1 \\ 0 & 0 & 1 & -3 & 4 \\ 0 & 0 & 0 & 1 & -1 \end{array}\right) \quad \text{to} \quad \left(\begin{array}{cccc|c} 1 & 0 & 0 & 0 & 2 \\ 0 & 1 & 0 & 0 & 0 \\ 0 & 0 & 1 & 0 & 1 \\ 0 & 0 & 0 & 1 & -1 \end{array}\right)$$

(Of course, in applying Gauss-Jordan row operations on a blackboard or on paper one does not have to rewrite the whole augmented matrix with each operation, since one can erase the row to be changed and insert the new entries.)

SYSTEMS OF LINEAR EQUATIONS WITHOUT A UNIQUE SOLUTION

Each system of equations considered so far in this section has had a unique solution. As noted in §16, however, a system of n equations in n unknowns may have no solution at all or it may have more than one. Fortunately, when we use the Gauss-Jordan row operations on an augmented matrix it is easy to tell which of the three cases applies:

I.	The system has a unique solution.	The terminal coefficient matrix is a multiplicative identity.
II.	The system has no solution.	Some row in the terminal augmented matrix has zeros everywhere to the left of the vertical bar and a nonzero constant to the right.
III.	The system has more than one solution.	There is no row of the type defined for Case II, but there is a row all of whose elements are zero.

Example 22-8

Use the methods of this section to solve the system

$$\begin{aligned} x - y + z &= 4 \\ 2x + y + 2z &= 11 \\ x \quad + z &= 6 \end{aligned}$$

We obtain the following sequence of augmented matrices; here we do only one step at a time to avoid obscuring the point being made:

$$\left(\begin{array}{ccc|c} 1 & -1 & 1 & 4 \\ 2 & 1 & 2 & 11 \\ 1 & 0 & 1 & 6 \end{array}\right) \to \left(\begin{array}{ccc|c} 1 & -1 & 1 & 4 \\ 0 & 3 & 0 & 3 \\ 1 & 0 & 1 & 6 \end{array}\right) \to \left(\begin{array}{ccc|c} 1 & -1 & 1 & 4 \\ 0 & 3 & 0 & 3 \\ 0 & 1 & 0 & 2 \end{array}\right) \to \left(\begin{array}{ccc|c} 1 & -1 & 1 & 4 \\ 0 & 1 & 0 & 1 \\ 0 & 1 & 0 & 2 \end{array}\right) \to$$

$$\left(\begin{array}{ccc|c} 1 & 0 & 1 & 5 \\ 0 & 1 & 0 & 1 \\ 0 & 1 & 0 & 2 \end{array}\right) \to \left(\begin{array}{ccc|c} 1 & 0 & 1 & 5 \\ 0 & 1 & 0 & 1 \\ 0 & 0 & 0 & 1 \end{array}\right)$$

We stop with the rightmost matrix above because a row (the third) has all zeros to the left of the vertical bar and 1 to the right of it. It should be clear that there is *no* set of values for x, y, and z that will satisfy the condition implicit in row three of the last matrix, namely

$$0 \cdot x + 0 \cdot y + 0 \cdot z = 1$$

This system illustrates Case II: it has no solution.

[Closer inspection of the original system reveals its *inconsistency*; that is, its having no solution. The sum of the first and second equations, taken with the third equation, form the following equivalent system:

$$\begin{aligned} 3x + 3z &= 15 \\ x + z &= 6 \end{aligned}$$

If we now divide the first of these two equations by 3, then we have the inconsistent requirement that the sum of x and z must be both 5 and 6!]

Example 22-9

Solve the system

$$\begin{aligned} x + y + z &= 4 \\ 3x + y + 3z &= 6 \\ 2x - y + 2z &= -1 \end{aligned}$$

We get the following sequence of augmented matrices (performing some obvious steps at the same time):

$$\left(\begin{array}{ccc|c} 1 & 1 & 1 & 4 \\ 3 & 1 & 3 & 6 \\ 2 & -1 & 2 & -1 \end{array}\right) \to \left(\begin{array}{ccc|c} 1 & 1 & 1 & 4 \\ 0 & -2 & 0 & -6 \\ 0 & -3 & 0 & -9 \end{array}\right) \to \left(\begin{array}{ccc|c} 1 & 1 & 1 & 4 \\ 0 & 1 & 0 & 3 \\ 0 & 1 & 0 & 3 \end{array}\right) \to \left(\begin{array}{ccc|c} 1 & 0 & 1 & 1 \\ 0 & 1 & 0 & 3 \\ 0 & 0 & 0 & 0 \end{array}\right)$$

Since the bottom row of the augmented matrix now has all zeros (but no other row of this matrix has all zeros to the left of the vertical bar with a nonzero number to the right), we stop. We note that *any* set of values for x, y, z satisfies the condition

$$0 \cdot x + 0 \cdot y + 0 \cdot z = 0$$

Indeed, the last matrix tells us more:

$$\begin{aligned} y &= 3 \\ x + z &= 1 \end{aligned}$$

For example, (0, 3, 1), (8, 3, -7), and (-96, 3, 97) are all solutions. If t is a real number, every set $(t, 3, 1 - t)$ is a solution. The given system, which is Case III, has an infinite number of solutions.

[Again, let us inspect the original system, this time to expose its *redundancy* (that is, that it has more than one solution). Note that if we add the first and third equations, and also the second and third equations, we obtain

$$\begin{aligned} 3x + 3z &= 3 \\ 5x + 5z &= 5 \end{aligned}$$

But both of these equations tell us that x and z add up to 1! Using this fact in any of the given equations immediately yields that $y = 3$.]

EXERCISES

1. Solve each of the following systems using augmented matrices and Gauss-Jordan row operations. (Check your answers!)

(a) $\begin{aligned} x - y + 2z &= 7 \\ y + 3z &= 7 \\ 2x - y + z &= 9 \end{aligned}$ (b) $\begin{aligned} x + 2y - z &= 4 \\ -x + y + 3z &= 7 \\ 2y + z &= 7 \end{aligned}$ (c) $\begin{aligned} 2x - y + z &= 7 \\ 2y + 3z &= 5 \\ 3x + y - 4z &= -11 \end{aligned}$

(d) $\begin{aligned} -3x + 2y - z &= 4 \\ 2x - y + 2z &= -3 \\ -x + z &= -1 \end{aligned}$ (e) $\begin{aligned} y + 2z &= 2 \\ x + 2y - z &= 1 \\ 2x - y + 3z &= 7 \end{aligned}$ (f) $\begin{aligned} y + z &= 1 \\ 2x - y &= 3 \\ 3x + z &= 8 \end{aligned}$

(g) $\begin{aligned} 2x - 3y &= 1 \\ 5x + 2y &= 12 \end{aligned}$ (h) $\begin{aligned} -x + y &= 5 \\ 3x - 2y &= -12 \end{aligned}$

(i) $\begin{aligned} w + x - y &= -2 \\ 2x + y + z &= 3 \\ -w - y - z &= -2 \\ w + 3z &= 5 \end{aligned}$ (j) $\begin{aligned} w + y &= 6 \\ 2x - y &= 4 \\ y + 4z &= 6 \\ 2w - z &= 7 \end{aligned}$

2. Using Gauss-Jordan row operations determine whether each of the following has a unique solution, no solution, or many solutions.

(a) $\begin{aligned} 2x + y - z &= 3 \\ x - y - 2z &= 6 \\ x - z &= 3 \end{aligned}$ (b) $\begin{aligned} x + z &= 5 \\ x - y &= 2 \\ y + 2z &= 4 \end{aligned}$ (c) $\begin{aligned} x - y - z &= 1 \\ x + 2y + z &= 4 \\ 4x + 2y &= 5 \end{aligned}$

(d) $\begin{aligned} y - z &= -1 \\ x + y + z &= 0 \\ 2x - y &= -2 \end{aligned}$ (e) $\begin{aligned} 2x - y + z &= 3 \\ x + y + z &= 4 \\ 4x + y + 3z &= 10 \end{aligned}$ (f) $\begin{aligned} x + y &= 5 \\ y + z &= 3 \\ x - z &= 2 \end{aligned}$

(g) $\begin{aligned} x - y + z &= 2 \\ 2x - 2y + 2z &= 4 \\ 3x - 3y + 3z &= 6 \end{aligned}$ (h) $\begin{aligned} x + 2y - z &= 1 \\ 2x + 4y - 2z &= 2 \\ 3x + 6y - 3z &= 4 \end{aligned}$ (i) $\begin{aligned} y + z &= 2 \\ y - z &= 2 \\ x + y + z &= 2 \end{aligned}$

TEST A

Using Gauss-Jordan row operations find the unique solution if it exists; if not, tell whether the system has no solution or many solutions.

1. $\begin{aligned} x - y + 3z &= 7 \\ 2x + z &= 11 \\ y - 2z &= -1 \end{aligned}$
2. $\begin{aligned} 2y + z &= 3 \\ x + y + z &= 2 \\ x - y &= 1 \end{aligned}$
3. $\begin{aligned} y - z &= 4 \\ 2x + y &= 3 \\ x + z &= -1 \end{aligned}$
4. $\begin{aligned} 2x + y - z &= 1 \\ x - y + z &= 2 \\ 3x + y - z &= 2 \end{aligned}$

TEST B

Using Gauss-Jordan row operations find the unique solution if it exists; if not, tell whether the system has no solution or many solutions.

1. $\begin{aligned} 3y - z &= 4 \\ x + y + z &= 1 \\ x + 7y - z &= 9 \end{aligned}$
2. $\begin{aligned} x + 2y - z &= 2 \\ -y + z &= 0 \\ 2x + y - z &= 2 \end{aligned}$
3. $\begin{aligned} y + z &= 1 \\ 2x - y + 3z &= -1 \\ x + 2y - z &= 2 \end{aligned}$
4. $\begin{aligned} x - z &= 0 \\ y + z &= 1 \\ x + y &= 2 \end{aligned}$

SECTION 23. INVERSE OF A MATRIX

Every real number, other than 0, has a multiplicative inverse. If $b \neq 0$ then the number b^{-1} such that $bb^{-1} = b^{-1}b = 1$ is the multiplicative inverse of b. In this section we define and compute the (multiplicative) inverse of a square matrix (if it has one). We also show how to use matrix inversion in solving systems of linear equations.

In the next section we will apply the methods introduced here.

DEFINITIONS

If A is a square matrix, then A^{-1} is its *inverse* if $AA^{-1} = I$, where I is the identity matrix.

Example 23-1

If $A = \begin{pmatrix} 2 & 3 \\ 3 & 5 \end{pmatrix}$, then $A^{-1} = \begin{pmatrix} 5 & -3 \\ -3 & 2 \end{pmatrix}$ is its inverse, since

$$AA^{-1} = \begin{pmatrix} 2 & 3 \\ 3 & 5 \end{pmatrix}\begin{pmatrix} 5 & -3 \\ -3 & 2 \end{pmatrix} = \begin{pmatrix} 1 & 0 \\ 0 & 1 \end{pmatrix}$$

Example 23-2

If $A = \begin{pmatrix} 1 & 0 & 0 \\ 3 & 0 & 1 \\ 0 & 1 & -1 \end{pmatrix}$, then $A^{-1} = \begin{pmatrix} 1 & 0 & 0 \\ -3 & 1 & 1 \\ -3 & 1 & 0 \end{pmatrix}$, since $AA^{-1} = \begin{pmatrix} 1 & 0 & 0 \\ 0 & 1 & 0 \\ 0 & 0 & 1 \end{pmatrix}$.

(Verify this!)

FINDING THE INVERSE OF A MATRIX

Suppose $A = \begin{pmatrix} 3 & 5 \\ 1 & 2 \end{pmatrix}$. To find A^{-1}, let $A^{-1} = \begin{pmatrix} x & u \\ y & v \end{pmatrix}$. Then it follows that

$$\begin{pmatrix} 3 & 5 \\ 1 & 2 \end{pmatrix}\begin{pmatrix} x & u \\ y & v \end{pmatrix} = \begin{pmatrix} 1 & 0 \\ 0 & 1 \end{pmatrix}$$

Multiplying the matrices on the left yields

$$\begin{pmatrix} 3x + 5y & 3u + 5v \\ x + 2y & u + 2v \end{pmatrix} = \begin{pmatrix} 1 & 0 \\ 0 & 1 \end{pmatrix}$$

From this matrix equation we obtain the two pairs of equations

$$\begin{aligned} 3x + 5y &= 1 \\ x + 2y &= 0 \end{aligned} \quad \text{and} \quad \begin{aligned} 3u + 5v &= 0 \\ u + 2v &= 1 \end{aligned}$$

Rewritten as augmented matrices, these are

$$\left(\begin{array}{cc|c} 3 & 5 & 1 \\ 1 & 2 & 0 \end{array}\right) \quad \text{and} \quad \left(\begin{array}{cc|c} 3 & 5 & 0 \\ 1 & 2 & 1 \end{array}\right)$$

Since matrix A appears at the left of the bar in both of these augmented matrices, any Gauss-Jordan operation performed on one matrix will have the same effect on the coefficient matrix A in the other. We can therefore compress the notation even more, and work on the solution of both pairs of equations at the same time, by combining the two augmented matrices into the following single form:

$$\left(\begin{array}{cc|cc} 3 & 5 & 1 & 0 \\ 1 & 2 & 0 & 1 \end{array}\right)$$

We now use the techniques of §22 to reduce matrix A (which appears in this new augmented matrix as the part to the left of the bar) to the identity matrix. When this has been done, the matrix that results on the right of the bar is the desired inverse of A.

$$\left(\begin{array}{cc|cc} 3 & 5 & 1 & 0 \\ 1 & 2 & 0 & 1 \end{array}\right)$$

interchange the rows: $$\left(\begin{array}{cc|cc} 1 & 2 & 0 & 1 \\ 3 & 5 & 1 & 0 \end{array}\right)$$

replace the second row by the second minus 3 times the first: $$\left(\begin{array}{cc|cc} 1 & 2 & 0 & 1 \\ 0 & -1 & 1 & -3 \end{array}\right)$$

multiply the second row by -1: $$\left(\begin{array}{cc|cc} 1 & 2 & 0 & 1 \\ 0 & 1 & -1 & 3 \end{array}\right)$$

replace the first row by the first minus 2 times the second: $$\left(\begin{array}{cc|cc} 1 & 0 & 2 & -5 \\ 0 & 1 & -1 & 3 \end{array}\right)$$

Thus the inverse of $\begin{pmatrix} 3 & 5 \\ 1 & 2 \end{pmatrix}$ is $\begin{pmatrix} 2 & -5 \\ -1 & 3 \end{pmatrix}$. Check to verify that

$$\begin{pmatrix} 3 & 5 \\ 1 & 2 \end{pmatrix}\begin{pmatrix} 2 & -5 \\ -1 & 3 \end{pmatrix} = \begin{pmatrix} 1 & 0 \\ 0 & 1 \end{pmatrix}$$

Example 23-3

Find the inverse of $B = \begin{pmatrix} 2 & 0 & 5 \\ -5 & 4 & 0 \\ 0 & 1 & 3 \end{pmatrix}$.

Using letters with subscripts for the elements of the inverse matrix B^{-1}, we have that

$$BB^{-1} = \begin{pmatrix} 2 & 0 & 5 \\ -5 & 4 & 0 \\ 0 & 1 & 3 \end{pmatrix} \begin{pmatrix} x_1 & y_1 & z_1 \\ x_2 & y_2 & z_2 \\ x_3 & y_3 & z_3 \end{pmatrix} = \begin{pmatrix} 1 & 0 & 0 \\ 0 & 1 & 0 \\ 0 & 0 & 1 \end{pmatrix}$$

In parallel with the explanation given above for a 2×2 matrix, we see that the elements of B^{-1} must satisfy the following three matrix equations:

$$\begin{pmatrix} 2 & 0 & 5 \\ -5 & 4 & 0 \\ 0 & 1 & 3 \end{pmatrix} \begin{pmatrix} x_1 \\ x_2 \\ x_3 \end{pmatrix} = \begin{pmatrix} 1 \\ 0 \\ 0 \end{pmatrix} \qquad \begin{pmatrix} 2 & 0 & 5 \\ -5 & 4 & 0 \\ 0 & 1 & 3 \end{pmatrix} \begin{pmatrix} y_1 \\ y_2 \\ y_3 \end{pmatrix} = \begin{pmatrix} 0 \\ 1 \\ 0 \end{pmatrix} \qquad \begin{pmatrix} 2 & 0 & 5 \\ -5 & 4 & 0 \\ 0 & 1 & 3 \end{pmatrix} \begin{pmatrix} z_1 \\ z_2 \\ z_3 \end{pmatrix} = \begin{pmatrix} 0 \\ 0 \\ 1 \end{pmatrix}$$

Since the sequence of row operations needed to transform the coefficient matrix B to an identity matrix is the same for all three systems of equations, we can--again just as in our handling of the 2×2 matrix--construct a single augmented matrix that enables us to solve the three systems at the same time:

$$\left(\begin{array}{ccc|ccc} 2 & 0 & 5 & 1 & 0 & 0 \\ -5 & 4 & 0 & 0 & 1 & 0 \\ 0 & 1 & 3 & 0 & 0 & 1 \end{array}\right)$$

multiply top row by $\frac{1}{2}$:

$$\left(\begin{array}{ccc|ccc} 1 & 0 & \frac{5}{2} & \frac{1}{2} & 0 & 0 \\ -5 & 4 & 0 & 0 & 1 & 0 \\ 0 & 1 & 3 & 0 & 0 & 1 \end{array}\right)$$

replace the second row by the second plus 5 times the first:

$$\left(\begin{array}{ccc|ccc} 1 & 0 & \frac{5}{2} & \frac{1}{2} & 0 & 0 \\ 0 & 4 & \frac{25}{2} & \frac{5}{2} & 1 & 0 \\ 0 & 1 & 3 & 0 & 0 & 1 \end{array}\right)$$

interchange second and bottom rows:

$$\left(\begin{array}{ccc|ccc} 1 & 0 & \frac{5}{2} & \frac{1}{2} & 0 & 0 \\ 0 & 1 & 3 & 0 & 0 & 1 \\ 0 & 4 & \frac{25}{2} & \frac{5}{2} & 1 & 0 \end{array}\right)$$

replace the bottom row by the bottom minus 4 times the second:

$$\left(\begin{array}{ccc|ccc} 1 & 0 & \frac{5}{2} & \frac{1}{2} & 0 & 0 \\ 0 & 1 & 3 & 0 & 0 & 1 \\ 0 & 0 & \frac{1}{2} & \frac{5}{2} & 1 & -4 \end{array}\right)$$

multiply the bottom row by 2:
replace the top row by the top row minus 5/2 times the bottom:
replace the middle row by the middle row minus 3 times the bottom:

$$\left(\begin{array}{ccc|ccc} 1 & 0 & 0 & -12 & -5 & 20 \\ 0 & 1 & 0 & -15 & -6 & 25 \\ 0 & 0 & 1 & 5 & 2 & -8 \end{array}\right)$$

Thus we have

$$B^{-1} = \begin{pmatrix} -12 & -5 & 20 \\ -15 & -6 & 25 \\ 5 & 2 & -8 \end{pmatrix}$$

Check, by showing that

$$\begin{pmatrix} 2 & 0 & 5 \\ -5 & 4 & 0 \\ 0 & 1 & 3 \end{pmatrix}\begin{pmatrix} -12 & -5 & 20 \\ -15 & -6 & 25 \\ 5 & 2 & -8 \end{pmatrix} = \begin{pmatrix} 1 & 0 & 0 \\ 0 & 1 & 0 \\ 0 & 0 & 1 \end{pmatrix}$$

SUMMARY OF PROCEDURE FOR FINDING THE INVERSE

We shall see shortly that some square matrices do not have inverses; that is, they are not *invertible*. But if a square matrix A does have an inverse A^{-1} it can be computed as follows:

(1) Construct an augmented matrix with A to the left of the vertical bar and with the identity matrix I of the same dimension on the right.

(2) Use appropriate Gauss-Jordan row operations to transform A into an identity matrix, applying each operation at the same time to the matrix to the right of the vertical bar.

When the part to the left has thus been transformed into an identity matrix, the part to the right of the vertical bar is A^{-1}, the inverse of A.

Example 23-4

Find the inverse of $\begin{pmatrix} 1 & 0 & 1 & 0 \\ 0 & 2 & 3 & 0 \\ 1 & 0 & -1 & 1 \\ 5 & 1 & 0 & 3 \end{pmatrix}$.

Each step in the following derivation should be justified, and the answer (to the right of the vertical bar in the last augmented matrix) should be checked:

$$\left(\begin{array}{cccc|cccc} 1 & 0 & 1 & 0 & 1 & 0 & 0 & 0 \\ 0 & 2 & 3 & 0 & 0 & 1 & 0 & 0 \\ 1 & 0 & -1 & 1 & 0 & 0 & 1 & 0 \\ 5 & 1 & 0 & 3 & 0 & 0 & 0 & 1 \end{array}\right) \longrightarrow \left(\begin{array}{cccc|cccc} 1 & 0 & 1 & 0 & 1 & 0 & 0 & 0 \\ 0 & 2 & 3 & 0 & 0 & 1 & 0 & 0 \\ 0 & 0 & -2 & 1 & -1 & 0 & 1 & 0 \\ 0 & 1 & -5 & 3 & -5 & 0 & 0 & 1 \end{array}\right)$$

$$\longrightarrow \left(\begin{array}{cccc|cccc} 1 & 0 & 1 & 0 & 1 & 0 & 0 & 0 \\ 0 & 1 & -5 & 3 & -5 & 0 & 0 & 1 \\ 0 & 0 & -2 & 1 & -1 & 0 & 1 & 0 \\ 0 & 2 & 3 & 0 & 0 & 1 & 0 & 0 \end{array}\right) \longrightarrow \left(\begin{array}{cccc|cccc} 1 & 0 & 1 & 0 & 1 & 0 & 0 & 0 \\ 0 & 1 & -5 & 3 & -5 & 0 & 0 & 1 \\ 0 & 0 & -2 & 1 & -1 & 0 & 1 & 0 \\ 0 & 0 & 13 & -6 & 10 & 1 & 0 & -2 \end{array}\right)$$

$$\longrightarrow \left(\begin{array}{cccc|cccc} 1 & 0 & 1 & 0 & 1 & 0 & 0 & 0 \\ 0 & 1 & -5 & 3 & -5 & 0 & 0 & 1 \\ 0 & 0 & 1 & -\frac{1}{2} & \frac{1}{2} & 0 & -\frac{1}{2} & 0 \\ 0 & 0 & 13 & -6 & 10 & 1 & 0 & -2 \end{array}\right) \longrightarrow \left(\begin{array}{cccc|cccc} 1 & 0 & 1 & \frac{1}{2} & \frac{1}{2} & 0 & \frac{1}{2} & 0 \\ 0 & 1 & 0 & \frac{1}{2} & -\frac{5}{2} & 0 & -\frac{5}{2} & 1 \\ 0 & 0 & 1 & -\frac{1}{2} & \frac{1}{2} & 0 & -\frac{1}{2} & 0 \\ 0 & 0 & 0 & \frac{1}{2} & \frac{7}{2} & 1 & \frac{13}{2} & -2 \end{array}\right)$$

$$\longrightarrow \left(\begin{array}{cccc|cccc} 1 & 0 & 0 & \frac{1}{2} & \frac{1}{2} & 0 & \frac{1}{2} & 0 \\ 0 & 1 & 0 & \frac{1}{2} & -\frac{5}{2} & 0 & -\frac{5}{2} & 1 \\ 0 & 0 & 1 & -\frac{1}{2} & \frac{1}{2} & 0 & -\frac{1}{2} & 0 \\ 0 & 0 & 0 & 1 & 7 & 2 & 13 & -4 \end{array}\right) \longrightarrow \left(\begin{array}{cccc|cccc} 1 & 0 & 0 & 0 & -3 & -1 & -6 & 2 \\ 0 & 1 & 0 & 0 & -6 & -1 & -9 & 3 \\ 0 & 0 & 1 & 0 & 4 & 1 & 6 & -2 \\ 0 & 0 & 0 & 1 & 7 & 2 & 13 & -4 \end{array}\right)$$

NONINVERTIBLE MATRICES

Not every square matrix has an inverse. If matrix A has no inverse then the procedure outlined above will come to a halt before A can be transformed into an identity matrix.

Example 23-5

Find the inverse, if it exists, of $\begin{pmatrix} 1 & 2 & 3 \\ 0 & 1 & -2 \\ 2 & 5 & 4 \end{pmatrix}$.

Applying appropriate row operations to the relevant augmented matrix yields this sequence:

$$\left(\begin{array}{ccc|ccc} 1 & 2 & 3 & 1 & 0 & 0 \\ 0 & 1 & -2 & 0 & 1 & 0 \\ 2 & 5 & 4 & 0 & 0 & 1 \end{array}\right) \rightarrow \left(\begin{array}{ccc|ccc} 1 & 2 & 3 & 1 & 0 & 0 \\ 0 & 1 & -2 & 0 & 1 & 0 \\ 0 & 1 & -2 & -2 & 0 & 1 \end{array}\right) \rightarrow \left(\begin{array}{ccc|ccc} 1 & 0 & 7 & 1 & -2 & 0 \\ 0 & 1 & -2 & 0 & 1 & 0 \\ 0 & 0 & 0 & -2 & -1 & 0 \end{array}\right)$$

But we can go no farther. So we conclude that the given matrix has no inverse.

PROPERTIES OF INVERSES

The following rules supplement those given in §20 (p. 174) and §21 (p. 180):

(9) If $AA^{-1} = I$, then $A^{-1}A = I$. (a matrix commutes with its inverse)

(10) If $AB = AC = I$, then $B = C$. (if a matrix has an inverse, that inverse is unique)

FORMULA FOR THE INVERSE OF A 2×2 MATRIX

When we apply the procedure outlined earlier for finding the inverse of the general 2×2 matrix $A = \begin{pmatrix} a & b \\ c & d \end{pmatrix}$ we obtain

$$A^{-1} = \begin{pmatrix} \dfrac{d}{ad-bc} & \dfrac{-b}{ad-bc} \\[2ex] \dfrac{-c}{ad-bc} & \dfrac{a}{ad-bc} \end{pmatrix} = \frac{1}{ad-bc}\begin{pmatrix} d & -b \\ -c & a \end{pmatrix}$$

The number $ad - bc$ is called the *determinant* of the matrix A, usually denoted by $|A|$. If $|A| \neq 0$, then A has an inverse; if $|A| = 0$, then A is not invertible.

Example 23-6

The inverse of $\begin{pmatrix} 2 & 5 \\ 4 & 3 \end{pmatrix}$ is $\dfrac{1}{2\cdot3 - 5\cdot4}\begin{pmatrix} 3 & -5 \\ -4 & 2 \end{pmatrix} = \begin{pmatrix} -\frac{3}{14} & \frac{5}{14} \\ \frac{4}{14} & -\frac{2}{14} \end{pmatrix} = \begin{pmatrix} -\frac{3}{14} & \frac{5}{14} \\ \frac{2}{7} & -\frac{1}{7} \end{pmatrix}$.

Verify this result by showing that the product of the given matrix and the proposed inverse is in fact the identity matrix.

USING INVERSES TO SOLVE LINEAR SYSTEMS

Inverse of matrices can be used to solve systems of n linear equations in n unknowns as follows.

If A is the $n\times n$ coefficient matrix, X the column vector of variables, and B the column vector of real numbers (scalars), then we can express the system of equations in the matrix form

$$AX = B$$

Then, if A has an inverse A^{-1} and we multiply both sides of this equation by A^{-1}, we have

$$A^{-1}(AX) = A^{-1}B$$

But since $A^{-1}(AX) = (A^{-1}A)X = IX = X$, it follows that

$$X = A^{-1}B$$

Note how we apply this in the following.

Example 23-7

Using matrix inversion, solve the system $\begin{aligned} 2x - 3y &= 7 \\ x - y &= 2 \end{aligned}$

Note that

$$\begin{pmatrix} 2 & -3 \\ 1 & -1 \end{pmatrix}^{-1} = \frac{1}{(2)(-1) - (1)(-3)}\begin{pmatrix} -1 & 3 \\ -1 & 2 \end{pmatrix} = \begin{pmatrix} -1 & 3 \\ -1 & 2 \end{pmatrix}$$

Therefore

$$\begin{pmatrix} x \\ y \end{pmatrix} = \begin{pmatrix} -1 & 3 \\ -1 & 2 \end{pmatrix}\begin{pmatrix} 7 \\ 2 \end{pmatrix} = \begin{pmatrix} -1 \\ -3 \end{pmatrix}$$

Check the solution $x = -1$, $y = -3$ in the original system of equations.

In the next example we demonstrate the special usefulness of inverses for solving systems of equations. Note that each of the following systems has precisely the same coefficient matrix; only the vector of constants is different from one system to another.

Example 23-8

Solve the systems

(a)
$$\begin{aligned} x \quad\;\; - 2z &= -1 \\ y + 3z &= 11 \\ x + y \qquad &= 7 \end{aligned}$$

(b)
$$\begin{aligned} x \quad\;\; - 2z &= -5 \\ y + 3z &= 6 \\ x + y \qquad &= -1 \end{aligned}$$

(c)
$$\begin{aligned} x \quad\;\; - 2z &= 16 \\ y + 3z &= 3 \\ x + y \qquad &= 23 \end{aligned}$$

We begin by finding the inverse of the coefficient matrix:

$$\left(\begin{array}{ccc|ccc} 1 & 0 & -2 & 1 & 0 & 0 \\ 0 & 1 & 3 & 0 & 1 & 0 \\ 1 & 1 & 0 & 0 & 0 & 1 \end{array}\right) \to \left(\begin{array}{ccc|ccc} 1 & 0 & -2 & 1 & 0 & 0 \\ 0 & 1 & 3 & 0 & 1 & 0 \\ 0 & 1 & 2 & -1 & 0 & 1 \end{array}\right) \to \left(\begin{array}{ccc|ccc} 1 & 0 & -2 & 1 & 0 & 0 \\ 0 & 1 & 3 & 0 & 1 & 0 \\ 0 & 0 & -1 & -1 & -1 & 1 \end{array}\right)$$

$$\to \left(\begin{array}{ccc|ccc} 1 & 0 & -2 & 1 & 0 & 0 \\ 0 & 1 & 3 & 0 & 1 & 0 \\ 0 & 0 & 1 & 1 & 1 & -1 \end{array}\right) \to \left(\begin{array}{ccc|ccc} 1 & 0 & 0 & 3 & 2 & -2 \\ 0 & 1 & 0 & -3 & -2 & 3 \\ 0 & 0 & 1 & 1 & 1 & -1 \end{array}\right)$$

Now we take the vector of constants in each system and multiply it on the left by the inverse we have found:

(a) $$\begin{pmatrix} x \\ y \\ z \end{pmatrix} = \begin{pmatrix} 3 & 2 & -2 \\ -3 & -2 & 3 \\ 1 & 1 & -1 \end{pmatrix} \begin{pmatrix} -1 \\ 11 \\ 7 \end{pmatrix} = \begin{pmatrix} 5 \\ 2 \\ 3 \end{pmatrix} \to x = 5,\ y = 2,\ z = 3$$

(b) $$\begin{pmatrix} x \\ y \\ z \end{pmatrix} = \begin{pmatrix} 3 & 2 & -2 \\ -3 & -2 & 3 \\ 1 & 1 & -1 \end{pmatrix} \begin{pmatrix} -5 \\ 6 \\ -1 \end{pmatrix} = \begin{pmatrix} -1 \\ 0 \\ 2 \end{pmatrix} \to x = -1,\ y = 0,\ z = 2$$

(c) $$\begin{pmatrix} x \\ y \\ z \end{pmatrix} = \begin{pmatrix} 3 & 2 & -2 \\ -3 & -2 & 3 \\ 1 & 1 & -1 \end{pmatrix} \begin{pmatrix} 16 \\ 3 \\ 23 \end{pmatrix} = \begin{pmatrix} 8 \\ 15 \\ -4 \end{pmatrix} \qquad x = 8,\ y = 15,\ z = -4$$

EXERCISES

1. Using Gauss-Jordan row operations find the inverse (if it exists) of each of the following square matrices. Check your answers!

(a) $\begin{pmatrix} 2 & 3 \\ 3 & 5 \end{pmatrix}$ (b) $\begin{pmatrix} 3 & 4 \\ 2 & 1 \end{pmatrix}$ (c) $\begin{pmatrix} 2 & 4 \\ 3 & 6 \end{pmatrix}$ (d) $\begin{pmatrix} 0 & 1 \\ 4 & 3 \end{pmatrix}$

(e) $\begin{pmatrix} -1 & 0 \\ 1 & 2 \end{pmatrix}$ (f) $\begin{pmatrix} 1 & 0 & 2 \\ 0 & 1 & -1 \\ 2 & 3 & 2 \end{pmatrix}$ (g) $\begin{pmatrix} 2 & 4 & 10 \\ 1 & -1 & 0 \\ 0 & 1 & 2 \end{pmatrix}$

(h) $\begin{pmatrix} 3 & -1 & 1 \\ 1 & 2 & 5 \\ 5 & 3 & 11 \end{pmatrix}$ (i) $\begin{pmatrix} 1 & 0 & 0 & 2 \\ 0 & 1 & 1 & -1 \\ 2 & 0 & -2 & 1 \\ 3 & 1 & -1 & -2 \end{pmatrix}$

2. Using the formula

$$\begin{pmatrix} a & b \\ c & d \end{pmatrix}^{-1} = \frac{1}{ad - bc}\begin{pmatrix} d & -b \\ -c & a \end{pmatrix}$$

find the inverse (if it exists) of each of the following:

(a) $\begin{pmatrix} 3 & 5 \\ 1 & 2 \end{pmatrix}$ (b) $\begin{pmatrix} 2 & 8 \\ 1 & 4 \end{pmatrix}$ (c) $\begin{pmatrix} 0 & -1 \\ 2 & 1 \end{pmatrix}$ (d) $\begin{pmatrix} -2 & 0 \\ 1 & 3 \end{pmatrix}$

3. Using inverses, solve each system of linear equations in each set of three systems.

(a) $\begin{aligned} 3x + y &= 7 \\ 2x + y &= 6 \end{aligned}$ $\quad\begin{aligned} 3x + y &= -3 \\ 2x + y &= -2 \end{aligned}$ $\quad\begin{aligned} 3x + y &= 7 \\ 2x + y &= 2 \end{aligned}$

(b) $\begin{aligned} 2x - y &= 0 \\ 5x + 3y &= 11 \end{aligned}$ $\quad\begin{aligned} 2x - y &= 21 \\ 5x + 3y &= 47 \end{aligned}$ $\quad\begin{aligned} 2x - y &= -1 \\ 5x + 3y &= 3 \end{aligned}$

(c) $\begin{aligned} -x + 2y &= 0 \\ 3x - 5y &= 3 \end{aligned}$ $\quad\begin{aligned} -x + 2y &= 1 \\ 3x - 5y &= -2 \end{aligned}$ $\quad\begin{aligned} -x + 2y &= -5 \\ 3x - 5y &= 20 \end{aligned}$

(d) $\begin{aligned} x + y \phantom{{}- z} &= 3 \\ 2y - z &= 1 \\ 3x \phantom{{}+y} + 2z &= 9 \end{aligned}$ $\quad\begin{aligned} x + y \phantom{{}- z} &= 1 \\ 2y - z &= 4 \\ 3x \phantom{{}+y} + 2z &= -4 \end{aligned}$ $\quad\begin{aligned} x + y \phantom{{}- z} &= 19 \\ 2y - z &= 9 \\ 3x \phantom{{}+y} + 2z &= 26 \end{aligned}$

(e) $\begin{aligned} -x \phantom{{}+3y} + z &= 2 \\ 2x + y \phantom{{}+2z} &= 1 \\ 3x + 3y + 2z &= 7 \end{aligned}$ $\quad\begin{aligned} -x \phantom{{}+3y} + z &= 0 \\ 2x + y \phantom{{}+2z} &= -3 \\ 3x + 3y + 2z &= -8 \end{aligned}$ $\quad\begin{aligned} -x \phantom{{}+3y} + z &= 0 \\ 2x + y \phantom{{}+2z} &= 4 \\ 3x + 3y + 2z &= 15 \end{aligned}$

TEST A

1. Using Gauss-Jordan row operations find the inverse of each of the following square matrices.

(a) $\begin{pmatrix} 1 & 3 \\ 2 & -1 \end{pmatrix}$ (b) $\begin{pmatrix} 0 & -1 \\ 2 & 0 \end{pmatrix}$ (c) $\begin{pmatrix} 2 & 0 & -1 \\ 1 & 0 & -1 \\ 0 & 2 & 3 \end{pmatrix}$

2. Using inverses, solve the systems of linear equations in each set.

(a) $\begin{aligned} 3x + 7y &= -1 \\ x + 2y &= 0 \end{aligned}$ $\quad\begin{aligned} 3x + 7y &= 33 \\ x + 2y &= 10 \end{aligned}$ $\quad\begin{aligned} 3x + 7y &= -10 \\ x + 2y &= -3 \end{aligned}$

(b) $\begin{aligned} x - y \phantom{{}+2z} &= 0 \\ y + 2z &= 4 \\ 3x \phantom{{}-y} - z &= 5 \end{aligned}$ $\quad\begin{aligned} x - y \phantom{{}+2z} &= -5 \\ y + 2z &= 25 \\ 3x \phantom{{}-y} - z &= -10 \end{aligned}$

TEST B

1. Using Gauss-Jordan row operations find the inverse of each of the following square matrices.

(a) $\begin{pmatrix} -1 & 0 \\ 1 & -1 \end{pmatrix}$ (b) $\begin{pmatrix} 3 & 0 \\ 0 & 2 \end{pmatrix}$ (c) $\begin{pmatrix} 0 & 0 & 1 \\ 2 & 0 & 3 \\ 1 & 1 & 0 \end{pmatrix}$

2. Using inverses, solve the systems of linear equations in each set.

(a) $\begin{aligned} 5x - 3y &= 19 \\ 2x - y &= 9 \end{aligned}$ $\quad$ $\begin{aligned} 5x - 3y &= 24 \\ 2x - y &= 9 \end{aligned}$ $\quad$ $\begin{aligned} 5x - 3y &= 4 \\ 2x - y &= 3 \end{aligned}$

(b) $\begin{aligned} y + z &= 1 \\ x - 2y \quad &= 0 \\ x \quad - z &= -1 \end{aligned}$ $\quad$ $\begin{aligned} y + z &= 1 \\ x - 2y \quad &= -15 \\ x \quad - z &= -1 \end{aligned}$

SECTION 24. MATRICES IN INPUT-OUTPUT ANALYSIS

In this section and the next we present applications of matrices different from any given so far.

Here we show one way in which matrices are used in the field of economics and how the inverse of a matrix is used to solve certain sorts of economic problems.

INPUT-OUTPUT MATRICES

In 1973, Wassily Leontief was awarded the Nobel prize in economics for his application of mathematics to economic problems. Leontief used matrices to analyze the interaction of a set of interdependent industries. In an article in *Scientific American* for October 1951 he described his *input-output* analysis of the interaction in 1947 of a very large number of sectors of the American economy.

The underlying principle is this: The output of each industry is used, in part, as input for other industries, perhaps also by the industry itself. In most cases, there is also some use of an industry's output by nonindustrial consumers or for export; this is sometimes referred to as the demand of the "open sector," or as "consumer" demand, or as "final" or *external* demand.

Input-output analysis is used to ascertain how much (output) each industry must produce in order to meet both the demands of all industries (for input) and the external demand. The procedure has also been found valuable in estimating how changes in the prices of commodities and goods produced may affect the whole economy.

Example 24-1

Suppose that in a simple economic system there are only three industries: agriculture (food), transportation, and clothing. Each uses the output of all three: for example, the farmer needs clothes for his farm hands, transportation for his produce, and some of his own corn to feed the hogs he raises. Suppose, further, in our simple economy, that in order to produce \$1 worth of food the agricultural industry requires an input of \$0.40 worth of its own output, an input of \$0.30 worth from transportation, and an input of \$0.20 from clothing. For the transportation sector to produce \$1 worth of goods, the value of its inputs must be respectively \$0.30 from agriculture, \$0.30 from transportation, and \$0.10 from clothing. Finally, for each dollar of output from the clothing industry, that industry must receive inputs worth \$0.20 from agriculture, \$0.10 from transportation, and \$0.20 from itself.

These facts about our simple economic system are set forth succinctly in the following *input-output matrix*:

		TO		
		AGRICULTURE	TRANSPORTATION	CLOTHING
FROM	AGRICULTURE	.40	.30	.20
	TRANSPORTATION	.30	.30	.10
	CLOTHING	.20	.10	.20

Note that each entry is *output* from the industry listed for its row, and *input* for the industry listed for its column. Thus the third column, for example, tells the amounts required from agriculture, transportation, and clothing in order for the clothing industry to produce a dollar's worth of clothing.

Example 24-2

To the facts given for the preceding example, suppose we add now the specification that there are external (nonindustrial or consumer) demands as follows: for \$2,000,000 worth of agricultural products; for \$1,000,000 worth of transportation; and for \$1,500,000 worth of clothing. How much should each industry produce in order to meet the total demand?

Let us represent the external demand as a vertical vector, with entries in millions of dollars. For our fictitious economy we have

$$D = \begin{pmatrix} 2 \\ 1 \\ 1.5 \end{pmatrix}$$

Now let x, y, and z be respectively the total amounts of output from agriculture, transportation, and clothing respectively required to satisfy both internal (industrial) and external demands. We then have the following matrix equation (ignore for the moment the fact that some of the entries are shaded):

$$\begin{pmatrix} x \\ y \\ z \end{pmatrix} = \begin{pmatrix} .40 & .30 & .20 \\ .30 & .30 & .10 \\ .20 & .10 & .20 \end{pmatrix} \cdot \begin{pmatrix} x \\ y \\ z \end{pmatrix} + \begin{pmatrix} 2 \\ 1 \\ 1.5 \end{pmatrix}$$

or

$$X = R \cdot X + D$$

where R is the original input-output matrix. If we carry out the arithmetic implied by this matrix equation we obtain a system of three linear equations in x, y, and z. One of these equations, the one for which the terms are shaded in the matrix equation above, is for the *total* amount of clothing that must be produced in this economy to meet all demands. It is

$$z = 0.20x + 0.10y + 0.20z + 1.5$$

z ↑	$0.20x + 0.10y + 0.20z$ ↑	1.5 ↑
total amount of clothing (output) in millions of dollars	amts of clothing (inputs) needed for each of the three industries (in millions of dollars)	amt of clothing needed for consumer demand (in millions of dollars)

We explain the right side of the above equation as follows:

TO PRODUCE	REQUIRES
\$1 worth of food	\$0.20 worth of clothing
\$1 worth of transportation	\$0.10 worth of clothing
\$1 worth of clothing	\$0.20 worth of clothing

Therefore

TO PRODUCE	REQUIRES
\$$x$ million in food	\$$0.20x$ million of clothing
\$$y$ million in transportation	\$$0.10y$ million of clothing
\$$z$ million in clothing	\$$0.20z$ million of clothing

The *total* required output of clothing consists of the three amounts to the right directly above, required as inputs to the industries, plus \$1.5 million to satisfy external (consumer) demand.

We now solve the matrix equation for X:

$$\begin{aligned} X &= RX + D \\ X - RX &= D \\ IX - RX &= D \\ (I - R)X &= D \\ X &= (I - R)^{-1}D \end{aligned}$$

We have

$$I - R = \begin{pmatrix} 1 & 0 & 0 \\ 0 & 1 & 0 \\ 0 & 0 & 1 \end{pmatrix} - \begin{pmatrix} .4 & .3 & .2 \\ .3 & .3 & .1 \\ .2 & .1 & .2 \end{pmatrix} = \begin{pmatrix} .6 & -.3 & -.2 \\ -.3 & .7 & -.1 \\ -.2 & -.1 & .8 \end{pmatrix}$$

To find the inverse of $I - R$ we apply the methods of §23; the answer is

$$(I - R)^{-1} = \begin{pmatrix} 2.52 & 1.19 & 0.78 \\ 1.19 & 2.02 & 0.55 \\ 0.78 & 0.55 & 1.51 \end{pmatrix}$$

A hand calculator was used here, with computations made to three decimal places, rounded off to two places only in the final matrix.

Now, since $X = (I - R)^{-1}D$, we have

$$\begin{pmatrix} x \\ y \\ z \end{pmatrix} = \begin{pmatrix} 2.52 & 1.19 & 0.78 \\ 1.19 & 2.02 & 0.55 \\ 0.78 & 0.55 & 1.51 \end{pmatrix} \begin{pmatrix} 2 \\ 1 \\ 1.5 \end{pmatrix} = \begin{pmatrix} 7.40 \\ 5.22 \\ 4.38 \end{pmatrix}$$

So to meet the total demand our simple economy requires outputs valued at \$7.40 million from agriculture, at \$5.22 million from transportation, and at \$4.38 million from clothing.

Note that the outputs of the three industries required *for internal use only* are, respectively, in millions of dollars,

$$\begin{aligned} 7.40 - 2 &= 5.40 \text{ from the agricultural sector;} \\ 5.22 - 1 &= 4.22 \text{ from the transportation sector;} \\ 4.38 - 1.5 &= 2.88 \text{ from the clothing sector.} \end{aligned}$$

CONVENIENCE OF THE INVERSE IN INPUT-OUTPUT ANALYSIS

It is usually the case that the coefficient matrix, or input-output matrix (denoted by R in the above discussion), remains fixed in a particular economic system, whereas the external demand may change with time. Note thus that even if one has a series of problems in which the external demands are different, the inverse of $I - R$ need be computed only once.

Example 24-3

Suppose that the external demands on the three industries of example 24-3 increase as follows: to \$2,500,000 for food; to \$1,500,000 for transportation; and to \$2,000,000 for clothing. What production schedule is required to meet this increased demand?

With x, y, and z denoting the new total outputs required, we have

$$\begin{pmatrix} x \\ y \\ z \end{pmatrix} = \begin{pmatrix} 2.52 & 1.19 & 0.78 \\ 1.19 & 2.02 & 0.55 \\ 0.78 & 0.55 & 1.51 \end{pmatrix} \begin{pmatrix} 2.5 \\ 1.5 \\ 2 \end{pmatrix} = \begin{pmatrix} 9.65 \\ 7.10 \\ 5.80 \end{pmatrix}$$

To satisfy the increased demand, agriculture should produce goods worth \$9,650,000, transportation's output should be \$7,100,000, and the clothing industry's level should be \$5,800,000.

Example 24-4

Suppose a very simple economy has only two industries, A and B. Find the output levels required for each, given the following input-output matrix

$$R = \begin{pmatrix} 0.20 & 0.50 \\ 0.30 & 0.10 \end{pmatrix}$$

and external demands of (a) 20 units from industry A and 12 from B; (b) 24 units from A and 10 from B; (c) 10 from A and 15 from B. Assume the units are in millions of dollars.

We first compute the inverse of $I - R$:*

---

*The formula on page 203 for the inverse of a 2×2 matrix may also be used to obtain $(I - R)^{-1}$.

$$(I-R)^{-1} = \begin{pmatrix} .8 & -.5 \\ -.3 & .9 \end{pmatrix}^{-1} = \begin{pmatrix} 1.58 & .88 \\ .53 & 1.40 \end{pmatrix}$$

With x and y denoting respectively the required production levels for industries A and B, we have:

$$\text{(a)} \quad \begin{pmatrix} x \\ y \end{pmatrix} = \begin{pmatrix} 1.58 & .88 \\ .53 & 1.40 \end{pmatrix}\begin{pmatrix} 20 \\ 12 \end{pmatrix} = \begin{pmatrix} 42.16 \\ 27.40 \end{pmatrix}$$

$$\text{(b)} \quad \begin{pmatrix} x \\ y \end{pmatrix} = \begin{pmatrix} 1.58 & .88 \\ .53 & 1.40 \end{pmatrix}\begin{pmatrix} 24 \\ 10 \end{pmatrix} = \begin{pmatrix} 46.72 \\ 26.72 \end{pmatrix}$$

$$\text{(c)} \quad \begin{pmatrix} x \\ y \end{pmatrix} = \begin{pmatrix} 1.58 & .88 \\ .53 & 1.40 \end{pmatrix}\begin{pmatrix} 10 \\ 15 \end{pmatrix} = \begin{pmatrix} 29.00 \\ 26.30 \end{pmatrix}$$

Thus the production levels required to meet the total demand in the three cases are as follows:

	(a)	(b)	(c)
INDUSTRY A	\$42,160,000	\$46,720,000	\$29,000,000
INDUSTRY B	\$27,400,000	\$26,720,000	\$26,300,000

OTHER APPLICATIONS OF INPUT-OUTPUT ANALYSIS

The following example illustrates another use of the Leontief model.

Example 24-5

Suppose that for each dollar of operating costs of a county government, the major city of the county must pay the county \$0.25, while for each dollar of the city's operating costs the county must pay the city \$0.10. In addition to what it pays the city, the county has yearly operating expenses of \$1,500,000; in addition to what it must pay the county, the city has yearly operating expenses of \$600,000. Find the total yearly operating costs of the county and of the city.

The relevant input-output matrix R appears in the table

	COUNTY	CITY
COUNTY	0	0.10
CITY	0.25	0

where we have entered in each governmental unit's row the amount it must pay the other unit. The zeros show that neither governmental unit "pays" itself anything. With x and y representing the total operating costs of county and city respectively, in millions of dollars, we get the equation

$$X = \begin{pmatrix} x \\ y \end{pmatrix} = \begin{pmatrix} 0 & .10 \\ .25 & 0 \end{pmatrix}\begin{pmatrix} x \\ y \end{pmatrix} + \begin{pmatrix} 1.5 \\ .6 \end{pmatrix}$$

where the numbers in the final matrix D are also in millions of dollars. We see that

$$X = (I - R)^{-1}D$$
$$= \begin{pmatrix} 1 & -0.10 \\ -.25 & 1 \end{pmatrix}^{-1} D$$

To find the inverse of $I - R$ we will use the formula on page 203 for the inverse of a 2×2 matrix:

$$(I - R)^{-1} = \frac{1}{(1)(1) - (0.25)(0.10)} \begin{pmatrix} 1 & .10 \\ .25 & 1 \end{pmatrix}$$
$$= \frac{1}{0.975} \begin{pmatrix} 1 & .10 \\ .25 & 1 \end{pmatrix} = \begin{pmatrix} 1.026 & .102 \\ .256 & 1.026 \end{pmatrix}$$

where we have computed to three decimal places. Then

$$\begin{pmatrix} x \\ y \end{pmatrix} = \begin{pmatrix} 1.026 & .102 \\ .256 & 1.026 \end{pmatrix} \begin{pmatrix} 1.5 \\ .6 \end{pmatrix} = \begin{pmatrix} 1.600 \\ 1.000 \end{pmatrix}$$

Thus the county and city have total operating costs respectively of \$1,600,000 and \$1,000,000.

EXERCISES

1. In example 24-2 verify that $(I - R)^{-1}$ on page 210 is correct by showing that the product of $(I - R)$ and its inverse is I.
2. In example 24-2 calculate the production levels required if the external demand vector is $\begin{pmatrix} 3 \\ 2 \\ 2 \end{pmatrix}$.
3. Verify the correctness of $(I - R)^{-1}$ in example 24-4 on page 212 by showing that $(I - R)(I - R)^{-1} = I$.
4. A simple economy consists of three industries, I, II, and III. Suppose that the input-output matrix for this economy is

	I	II	III
I	0.5	0.2	0.6
II	0.2	0.4	0.1
III	0.1	0.3	0.2

   (a) How many units from industry II are used to produce one unit in industry III? (b) How many units from II are needed to produce 1 unit by I? (c) Which industry uses the most units from I to produce one of its own units?
5. Find the output levels required of each sector in the two-sector economy whose input-output and demand matrices are

$$R = \begin{pmatrix} .10 & .30 \\ .40 & .20 \end{pmatrix} \qquad D = \begin{pmatrix} 8 \\ 10 \end{pmatrix}$$

6. In example 24-5 calculate the new total operating costs if the county's yearly operating expenses drop to $1,427,000 and the city's to $550,000.
7. A manufacturing company has three factories, A, B, and C; each factory makes a single product. Internal demands for each dollar of output are as follows:

   A uses $0.20 worth of its own product,
   $0.30 worth of B's product,
   none of C's product;
   B uses $0.20 worth of A's product,
   none of its own product,
   $0.20 worth of C's product;
   C uses $0.20 worth of A's product,
   $0.50 worth of B's product,
   $0.20 worth of its own product.

   Outside demands for the products are as follows: $5000 for A's, $6000 for B's, and $4500 for C's.

   (a) Write the input-output matrix R and the external demand matrix D. [Hint: Express D in thosands of dollars.]

   (b) If X is the vector of production levels needed to satisfy the total demand, write the matrix equation involving X, R, and D.

   (c) Solve the equation in (b) for X; that is, find the total production levels required of the three factories.

   (d) Find the new production levels required if the external demands on A, B, and C change to $4000, $2000, and $5000 respectively.
8. Two power plants, one nuclear and one coal-burning, are not very far apart. For each unit of power produced by the nuclear plant it is necessary to use 0.1 units of its own power and 0.05 units of power from the coal-burning plant. To produce a unit of power by the coal-burning plant it takes 0.1 units of its own power plus 0.4 units of nuclear power. What should the total output of each plant be if the external demands for nuclear and coal-produced power are respectively (a) 2020 units and 3030 units? (b) 12,625 units and 13,837 units?

TEST A

1. Find $(I - R)^{-1}$ if $R = \begin{pmatrix} .20 & .50 \\ .40 & .30 \end{pmatrix}$.
2. Using R just above as the input-output matrix of a simple two-sector economic system and with $D = \begin{pmatrix} 12 \\ 20 \end{pmatrix}$ as the external demand vector, calculate the total production levels required from each sector.
3. A company has two departments, A and B. For each dollar of its

operating costs A charges B \$0.15; for each dollar of B's operating costs, A pays \$0.20 to B. Find the total operating costs of the two departments if A's additional expenses are \$9700 and B's additional expenses are \$12,125.

TEST B

1. Find $(I - R)^{-1}$ if $R = \begin{pmatrix} .10 & .60 \\ .40 & .20 \end{pmatrix}$.
2. Using R just above as the input-output matrix of a simple two-sector economic system and with $D = \begin{pmatrix} 15 \\ 8 \end{pmatrix}$ as the external demand vector, calculate the total production levels required for each sector.
3. A university has two branches, endowed (E) and state-supported (S). For each dollar of its operating costs E charges S \$0.30. For each dollar of S's operating costs, E pays S \$0.40. In addition to these costs, E has monthly expenses of \$1,500,000 and S has monthly expenses of \$3,000,000. Determine the total monthly operating costs of each branch of the university.

SECTION 25. MATRICES AND MARKOV PROCESSES

As noted in chapter 2 (page 93), many probability problems involve sequences of identical experiments. It is often of interest in these to speculate about long-term trends: that is, to predict the outcome if the experiment is repeated many times. In the simplest case the outcome of an experiment in a sequence of identical ones does not depend on the outcome of any previous experiment. In chapter 2 we considered just such sequences of *independent* events. If a fair coin, for example, is flipped repeatedly, the probability of a head on any one toss is one half, regardless of the results of previous tosses.

In this section we consider the more interesting situation in which the outcome of an experiment depends on the outcome of the *immediately preceding* experiment, but not on any outcomes earlier than that. A sequence of such experiments is called a *Markov process* or *chain*, in honor of Andrei Andreevich Markov (1856-1922), who developed the mathematical theory of such sequences. We begin with examples.

Example 25-1

Suppose the following pattern has been observed in a certain region: during any calendar year, 80% of those who live in cities continue to do so, while 20% move to suburbs; also, during any calendar year, 90% of the suburban residents stay there but 10% move into cities. Here it is clear that the distribution of population at the end of any calendar year depends on the distribution at the end of the preceding year. The given data can be displayed in the following matrix:

$$\begin{array}{llcc} & & \text{TO} & \\ & & \text{CITY} & \text{SUBURB} \\ \text{FROM} & \text{CITY} & 0.8 & 0.2 \\ & \text{SUBURB} & 0.1 & 0.9 \end{array}$$

Recall (from page 163) that this is a transition matrix: a square matrix in which each row is a probability vector. Every element of a transition matrix is nonnegative and the elements in each row add up to 1. Transition matrices are also called *Markov matrices*, from their use in the theory of Markov chains. The elements in such matrices are often called *transition probabilities* because they are the probabilities of passing from one situation or *state* to another.

In this example there are two states: living in a city, and living in a suburb. Each probability in the Markov matrix is a *conditional* one. If C and S denote respectively living in a city and living in a suburb, then we can let $p(C|C)$ represent the probability, if a person lives in a city at the beginning of a year, that the person will continue to do so, and $p(S|C)$ represent the probability that the person will move to the suburbs. We see from the first row in the matrix above that $p(C|C) = 0.8$ and $p(S|C) = 0.2$. The meanings of $p(C|S)$ and $p(S|S)$ are similar, and from the second row of the matrix we see that $p(C|S) = 0.1$, $p(S|S) = 0.9$.

Suppose we know that this year 60% of the population of our observed region lived in cities and the rest (40%) in suburbs. How will the population be distributed after one year?

Of the 60% who now live in the city, 80% will remain; so $0.6 \times 0.8 = 0.48$ or 48% of the population will remain in the city. Of the 40% now in the suburbs, 10% will move into cities; so $0.4 \times 0.1 = 0.04$ or 4% will move from suburbs to cities. Thus, after one year the cities will have altogether 48% + 4%, or 52% of the population.

That leaves 48% for the suburbs. But we can also compute this directly. Since we know that of the 60% now in cities 20% will move to suburbs, and that of the 40% now in suburbs 90% will remain there, therefore after one year the population in suburbs will be

$$(0.6 \times 0.2) + (0.4 \times 0.9) = 0.12 + 0.36 = 0.48$$

or 48%.

The distribution of population after one year can therefore be given by the probability vector

$$(0.52 \quad 0.48)$$

This result can also be obtained from the following matrix multiplication:

$$(0.6 \quad 0.4)\begin{pmatrix} 0.8 & 0.2 \\ 0.1 & 0.9 \end{pmatrix} = \big((0.6\times0.8) + (0.4\times0.1) \qquad (0.6\times0.2) + (0.4\times0.9)\big)$$
$$= (0.52 \quad 0.48)$$

How will the population be distributed after two years? Of the 52% in cities after one year, 80% will remain; of the 48% in the suburbs after one year, 10% will move to cities. Therefore the fraction of the population in cities after two years will be

$$(0.52 \times 0.80) + (0.48 \times 0.10) = 0.416 + 0.048 = 0.464$$

Also, of the 52% in cities after one year, 20% will move to suburbs during the second year; of the 48% in suburbs, 90% will remain. Therefore the fraction of the region's population in suburbs after two years will be

$$(0.52 \times 0.2) + (0.48 \times 0.9) = 0.104 + 0.432 = 0.536$$

The probability vector for the population distribution after two years is therefore

$$\begin{pmatrix}0.464 & 0.536\end{pmatrix}$$

whose elements also (of course) sum to 1.

Again, we note that we can obtain these results by multiplying appropriate matrices:

$$\begin{aligned}\begin{pmatrix}0.52 & 0.48\end{pmatrix}\begin{pmatrix}0.8 & 0.2\\0.1 & 0.9\end{pmatrix} &= \begin{pmatrix}(0.52\times 0.8)+(0.48\times 0.1) & (0.52\times 0.2)+(0.48\times 0.9)\end{pmatrix}\\ &= \begin{pmatrix}0.464 & 0.536\end{pmatrix}\end{aligned}$$

Since the probability vector (0.52 0.48) for the distribution after one year was the product

$$\begin{pmatrix}0.6 & 0.4\end{pmatrix}\begin{pmatrix}0.8 & 0.2\\0.1 & 0.9\end{pmatrix}$$

we see that the probability vector for the population distribution after two years can be written

$$\left[\begin{pmatrix}0.6 & 0.4\end{pmatrix}\cdot\begin{pmatrix}0.8 & 0.2\\0.1 & 0.9\end{pmatrix}\right]\cdot\begin{pmatrix}0.8 & 0.2\\0.1 & 0.9\end{pmatrix}$$

or, using the associativity of matrix multiplication,

$$\begin{pmatrix}0.6 & 0.4\end{pmatrix}\cdot\left[\begin{pmatrix}0.8 & 0.2\\0.1 & 0.9\end{pmatrix}\cdot\begin{pmatrix}0.8 & 0.2\\0.1 & 0.9\end{pmatrix}\right] = \begin{pmatrix}0.6 & 0.4\end{pmatrix}\cdot\begin{pmatrix}0.8 & 0.2\\0.1 & 0.9\end{pmatrix}^2$$

Continuing in this fashion, we see that the population distribution after three years will be

$$\begin{pmatrix}0.6 & 0.4\end{pmatrix}\cdot\begin{pmatrix}0.8 & 0.2\\0.1 & 0.9\end{pmatrix}^3$$

And if the trend continues, the expected distribution of population after n years will be

$$\begin{pmatrix}0.6 & 0.4\end{pmatrix}\cdot\begin{pmatrix}0.8 & 0.2\\0.1 & 0.9\end{pmatrix}^n$$

We next list the results of these calculations for several increasing values of n (rounding when necessary to three decimal places). The row for $n = 0$ of course gives the initial distribution:

n	DISTRIBUTION (city	suburb)	n	DISTRIBUTION (city	suburb)
0	0.6	0.4	10	0.341	0.659
1	0.52	0.48	15	0.335	0.665
2	0.464	0.536	20	0.334	0.666
3	0.425	0.575	25	0.333	0.667
4	0.397	0.603	30	0.333	0.667
5	0.378	0.622			

In the long run the distribution of population (represented by the probability vectors above) seems to stabilize. Eventually, 0.333 or $33\frac{1}{3}\%$ of the population will live in cities, 0.667 or $66\frac{2}{3}\%$ in suburbs.

Example 25-2

Suppose we consider a different initial distribution, say 95% now living in cities and only 5% in suburbs. To find probability vectors for the population distribution after 1, 2, ..., n years, we now require these calculations:

$$\begin{pmatrix}0.95 & 0.05\end{pmatrix}\cdot\begin{pmatrix}0.08 & 0.2\\ 0.01 & 0.9\end{pmatrix}, \quad \begin{pmatrix}0.95 & 0.05\end{pmatrix}\cdot\begin{pmatrix}0.8 & 0.2\\ 0.1 & 0.9\end{pmatrix}^2,$$

$$\ldots, \begin{pmatrix}0.95 & 0.05\end{pmatrix}\cdot\begin{pmatrix}0.8 & 0.2\\ 0.1 & 0.9\end{pmatrix}^n$$

Here is a table of the results for various values of n:

n	DISTRIBUTION (city	suburb)	n	DISTRIBUTION (city	suburb)
0	0.95	0.05	10	0.351	0.649
1	0.765	0.235	15	0.336	0.664
2	0.635	0.365	20	0.334	0.666
3	0.545	0.455	25	0.333	0.667
4	0.481	0.519	30	0.333	0.667
5	0.437	0.563			

Example 25-3

Suppose the initial distribution of the population is only 10% in cities, 90% in suburbs. The relevant computations now yield these results:

n	DISTRIBUTION (city	suburb)	n	DISTRIBUTION (city	suburb)
0	0.1	0.9	10	0.327	0.673
1	0.170	0.830	15	0.332	0.668
2	0.219	0.781	20	0.333	0.667
3	0.253	0.747	25	0.333	0.667
4	0.277	0.723	30	0.333	0.667
5	0.294	0.706			

CONJECTURES

(1) On the basis of the preceding three examples, we are tempted to conclude that the population distribution will get closer and closer to $\frac{1}{3}$ in cities and $\frac{2}{3}$ in suburbs no matter what the initial distribution is.

Now consider powers of the transition matrix, which form an integral part of the above calculations:

$$T = \begin{pmatrix} 0.8 & 0.2 \\ 0.1 & 0.9 \end{pmatrix} \qquad T^{10} = \begin{pmatrix} 0.352 & 0.648 \\ 0.324 & 0.676 \end{pmatrix}$$

$$T^2 = \begin{pmatrix} 0.66 & 0.34 \\ 0.17 & 0.83 \end{pmatrix} \qquad T^{15} = \begin{pmatrix} 0.336 & 0.664 \\ 0.332 & 0.668 \end{pmatrix}$$

$$T^3 = \begin{pmatrix} 0.562 & 0.438 \\ 0.219 & 0.781 \end{pmatrix} \qquad T^{20} = \begin{pmatrix} 0.334 & 0.666 \\ 0.333 & 0.667 \end{pmatrix}$$

$$T^4 = \begin{pmatrix} 0.493 & 0.507 \\ 0.253 & 0.747 \end{pmatrix} \qquad T^{25} = \begin{pmatrix} 0.333 & 0.667 \\ 0.333 & 0.667 \end{pmatrix}$$

$$T^5 = \begin{pmatrix} 0.445 & 0.555 \\ 0.277 & 0.723 \end{pmatrix} \qquad T^{30} = \begin{pmatrix} 0.333 & 0.667 \\ 0.333 & 0.667 \end{pmatrix}$$

(2) We observe that the more times we multiply the transition matrix by itself the more closely does each row of the product tend to the same limiting population distribution ($\frac{1}{3}$ in cities, $\frac{2}{3}$ in suburbs) obtained in examples 25-1, -2, and -3.

(3) We note, further, that if the probability vector for the limiting population distribution is multiplied on the right by the transition matrix, the product is just the probability vector:

$$\begin{pmatrix} \frac{1}{3} & \frac{2}{3} \end{pmatrix} \cdot \begin{pmatrix} 0.8 & 0.2 \\ 0.1 & 0.9 \end{pmatrix} = \begin{pmatrix} \frac{1}{3} & \frac{2}{3} \end{pmatrix}$$

SUMMARY OF CONCLUSIONS

If T is the transition matrix of examples 25-1, -2, and -3 and we let Q be the probability vector for the initial distribution of population, then there is a unique probability vector, say P, such that

(1) QT^n approaches P as n increases. This is true regardless of the particular Q; that is, no matter what the initial distribution of population is.

(2) T^n approaches a matrix each of whose rows is P. We call this matrix the *stable* matrix of T.

(3) $PT = P$. We say that P is a *fixed* probability vector for T; it is also called a *steady-state* or *stable* vector.

NOTE. Although these conclusions are true for a large class of transition matrices, they do *not* hold for all such matrices. We will return to this important point after considering another set of examples.

Example 25-4

Suppose data gathered in a survey of the readers of three magazines show that, each year: Dime magazine keeps 75% of its readers but loses 15% to Viewsweek and 10% to U.S. Cues and Word Retort; Viewsweek keeps 80% of its readers, losing 10% to Dime and 10% to U.S.Cues; and U.S. Cues keeps 60% of its readers but loses 25% to Viewsweek and 15% to Dime. Denoting the three magazines by D, V, and U, we can display the yearly shifts in readership by the following Markov matrix:

$$\begin{array}{ll} & \quad\text{TO} \\ & \begin{array}{ccc} \text{D} & \text{V} & \text{U} \end{array} \\ \begin{array}{r} \text{FROM D} \\ \text{V} \\ \text{U} \end{array} & \begin{pmatrix} 0.75 & 0.15 & 0.10 \\ 0.10 & 0.80 & 0.10 \\ 0.15 & 0.25 & 0.60 \end{pmatrix} \end{array}$$

Suppose that the distribution of readers at present (the initial distribution) is as follows: 35% read D, 35% read V, and 30% read U. What will the distribution of readers be after one year?

Of the 35% who now read D, 75% will continue to do so; therefore after one year the fraction who still read D is $0.35 \times 0.75 = 0.2625$.

Of the 35% who now read V, 10% will switch to D, and of the 30% who now read U, 15% will switch to D; so altogether, after one year, the fraction of these readers who switch to D is

$$(0.35 \times 0.10) + (0.30 \times 0.15) = 0.35 + 0.045$$

The fraction of all readers, therefore, who will read D after one year is

$$0.2625 + 0.035 + 0.045 = 0.3425$$

So after one year 34.25% of the entire population of readers will be reading D.

Similarly we can calculate the readership of V after one year. It is

$$(0.35 \times 0.15) + (0.35 \times 0.80) + (0.30 \times 0.25) = 0.4075$$

or 40.75% of the population. The fraction of the population reading U after one year is

$$(0.35 \times 0.10) + (0.35 \times 0.10) + (0.30 \times 0.60) = 0.25$$

or 25% of the total readership.

So after one year we can represent the distribution (for D, V, and U in that order) by the probability vector

$$\begin{pmatrix} 0.3425 & 0.4075 & 0.25 \end{pmatrix}$$

with elements summing to 1, as expected.

As in Example 25-1, we observe that we could have used matrix multiplication to obtain the new distribution:

$$\begin{pmatrix} 0.35 & 0.35 & 0.30 \end{pmatrix} \cdot \begin{pmatrix} 0.75 & 0.15 & 0.10 \\ 0.10 & 0.80 & 0.10 \\ 0.15 & 0.25 & 0.60 \end{pmatrix} = \begin{pmatrix} 0.3425 & 0.4075 & 0.25 \end{pmatrix}$$

After two years the distribution of readers will be

$$\begin{pmatrix}0.3425 & 0.4075 & 0.25\end{pmatrix}\cdot\begin{pmatrix}0.75 & 0.15 & 0.10\\0.10 & 0.80 & 0.10\\0.15 & 0.25 & 0.60\end{pmatrix}$$

$$\text{or } \begin{pmatrix}0.35 & 0.35 & 0.30\end{pmatrix}\cdot\begin{pmatrix}0.75 & 0.15 & 0.10\\0.10 & 0.80 & 0.10\\0.15 & 0.25 & 0.60\end{pmatrix}^2$$

which yields (in either case) the distribution vector

$$\begin{pmatrix}0.335 & 0.440 & 0.225\end{pmatrix}$$

with computations rounded to three decimal places.

If the trend continues, we expect the distribution of the readership after n years to equal the product

$$\begin{pmatrix}0.35 & 0.35 & 0.30\end{pmatrix}\cdot\begin{pmatrix}0.75 & 0.15 & 0.10\\0.10 & 0.80 & 0.10\\0.15 & 0.25 & 0.60\end{pmatrix}^n$$

Here are the results for a number of values of n. The initial conditions are those for $n = 0$. We round to three decimal places:

n	D	V	U	n	D	V	U
0	0.35	0.35	0.30	5	0.322	0.475	0.203
1	0.343	0.408	0.250	10	0.314	0.486	0.200
2	0.335	0.440	0.225	20	0.314	0.486	0.200
3	0.329	0.458	0.213	30	0.314	0.486	0.200
4	0.325	0.469	0.206				

We conclude that after many years the distribution of the readership will stabilize with 31.4% reading D, 48.6% reading V, and 20.0% reading U.

Example 25-5

Suppose the initial distribution in example 25-4 were $(1\quad 0\quad 0)$; that is, suppose that to start with everyone reads D. Our calculations would then yield the following table:

n	D	V	U	n	D	V	U
0	1	0	0	5	0.388	0.418	0.194
1	0.75	0.15	0.10	10	0.323	0.477	0.200
2	0.593	0.257	0.150	20	0.314	0.486	0.200
3	0.493	0.332	0.175	30	0.314	0.486	0.200
4	0.429	0.383	0.188				

The limiting distribution, we see, remains unchanged.

Conjectures

(1) We conclude that the limiting distribution in the Markov

process just investigated (examples 25-4 and -5) is the same no matter what the initial distribution is.

(2) Now, as for the earlier set of examples, we compute powers of the transition matrix (rounding to three decimals):

$$T = \begin{pmatrix} 0.75 & 0.15 & 0.10 \\ 0.10 & 0.80 & 0.10 \\ 0.15 & 0.25 & 0.60 \end{pmatrix} \qquad T^5 = \begin{pmatrix} 0.388 & 0.418 & 0.194 \\ 0.272 & 0.534 & 0.194 \\ 0.300 & 0.475 & 0.225 \end{pmatrix}$$

$$T^2 = \begin{pmatrix} 0.592 & 0.258 & 0.150 \\ 0.170 & 0.680 & 0.150 \\ 0.228 & 0.372 & 0.400 \end{pmatrix} \qquad T^{10} = \begin{pmatrix} 0.323 & 0.477 & 0.200 \\ 0.308 & 0.491 & 0.200 \\ 0.314 & 0.486 & 0.200 \end{pmatrix}$$

$$T^3 = \begin{pmatrix} 0.492 & 0.333 & 0.175 \\ 0.218 & 0.607 & 0.175 \\ 0.268 & 0.432 & 0.300 \end{pmatrix} \qquad T^{20} = \begin{pmatrix} 0.314 & 0.486 & 0.200 \\ 0.314 & 0.486 & 0.200 \\ 0.314 & 0.486 & 0.200 \end{pmatrix}$$

$$T^4 = \begin{pmatrix} 0.428 & 0.384 & 0.188 \\ 0.250 & 0.562 & 0.288 \\ 0.289 & 0.461 & 0.250 \end{pmatrix} \qquad T^{30} = \begin{pmatrix} 0.314 & 0.486 & 0.200 \\ 0.314 & 0.486 & 0.200 \\ 0.314 & 0.486 & 0.200 \end{pmatrix}$$

We observe that the more times we multiply the transition matrix by itself the more closely does each row of the product tend to the stable readership distribution--0.314 for D, 0.486 for V, and 0.200 for U, as obtained in examples 25-4 and -5.

(3) Finally, we note here as in the previous case that if the probability vector for the limiting (readership) distribution is multiplied on the right by the transition matrix, the result is precisely the same probability vector:

$$\begin{pmatrix} 0.314 & 0.486 & 0.280 \end{pmatrix} \cdot \begin{pmatrix} 0.75 & 0.15 & 0.10 \\ 0.10 & 0.80 & 0.10 \\ 0.15 & 0.25 & 0.60 \end{pmatrix} = \begin{pmatrix} 0.314 & 0.486 & 0.200 \end{pmatrix}$$

REGULAR MARKOV MATRICES

The conclusions reached in the preceding examples hold whenever the transition matrix for a Markov process is *regular*. A transition matrix T is regular if, for some n, T^n contains no zero. Since, in examples 25-4 and -5, the transition matrices themselves contained no zeros, each was in fact regular: that is, T^1, which is T itself, contains no zero.

Example 25-6

Determine the regularity or irregularity of (a) $\begin{pmatrix} 1 & 0 \\ \frac{1}{2} & \frac{1}{2} \end{pmatrix}$ and (b) $\begin{pmatrix} 0 & 1 \\ \frac{1}{2} & \frac{1}{2} \end{pmatrix}$.

(a) Note that

$$\begin{pmatrix} 1 & 0 \\ \frac{1}{2} & \frac{1}{2} \end{pmatrix}^2 = \begin{pmatrix} 1 & 0 \\ \frac{3}{4} & \frac{1}{4} \end{pmatrix}; \qquad \begin{pmatrix} 1 & 0 \\ \frac{1}{2} & \frac{1}{2} \end{pmatrix}^3 = \begin{pmatrix} 1 & 0 \\ \frac{7}{8} & \frac{1}{8} \end{pmatrix}; \qquad \begin{pmatrix} 1 & 0 \\ \frac{1}{2} & \frac{1}{2} \end{pmatrix}^4 = \begin{pmatrix} 1 & 0 \\ \frac{15}{16} & \frac{1}{16} \end{pmatrix}$$

Here it is apparent that every power of the matrix will have a zero in the upper right-hand corner. The matrix is therefore not regular.

(b) Since

$$\begin{pmatrix} 0 & 1 \\ \frac{1}{2} & \frac{1}{2} \end{pmatrix}^2 = \begin{pmatrix} \frac{1}{2} & \frac{1}{2} \\ \frac{1}{4} & \frac{3}{4} \end{pmatrix}$$

and the latter contains no zero, this matrix is regular.

Theorem on Markov Chains

If T is a regular transition matrix, then there is a unique probability vector P such that

(1) as n becomes large, given any probability vector Q of the right dimension, QT^n approaches P;
(2) as n becomes large, T^n approaches a matrix every row of which is P;
(3) $PT = P$.

Finding the Fixed Vector

We can calculate the fixed (limiting, stable) vector P more readily and more accurately than through the tedious process used in the examples above. We simply use the fact that P is fixed; that is, that $PT = P$.

Example 25-7

Find the fixed vector P for the regular transition matrix $\begin{pmatrix} 0.8 & 0.2 \\ 0.1 & 0.9 \end{pmatrix}$ of example 25-1.

Let $P = (x\ y)$ be the desired fixed probability vector. Then $x + y = 1$ (because P is a probability vector) and $(x\ y)\begin{pmatrix} 0.8 & 0.2 \\ 0.1 & 0.9 \end{pmatrix} = (x\ y)$ (because P is fixed). This matrix equation yields two linear equations which, adjoined to the one already at hand, gives us the system

$$\begin{aligned} x + \quad y &= 1 \\ 0.8x + 0.1y &= x \\ 0.2x + 0.9y &= y \end{aligned}$$

(Note that the second and third of these equations are identical.) We may solve simultaneously to obtain $x = \frac{1}{3}$ and $y = \frac{2}{3}$, the same answer we obtained rather laboriously in example 25-1.

Example 25-8

Find the fixed vector P for the regular transition matrix

$$\begin{pmatrix} 0.75 & 0.15 & 0.10 \\ 0.10 & 0.80 & 0.10 \\ 0.15 & 0.25 & 0.60 \end{pmatrix}$$

of example 25-4.

Let $P = (x\ y\ z)$ be the desired fixed probability vector. Then $x+y+z = 1$ and

$$(x\ y\ z)\begin{pmatrix} 0.75 & 0.15 & 0.10 \\ 0.10 & 0.80 & 0.10 \\ 0.15 & 0.25 & 0.60 \end{pmatrix} = (x\ y\ z)$$

This gives the linear system

$$\begin{aligned} x + y + z &= 1 \\ 0.75x + 0.10y + 0.15z &= x \\ 0.15x + 0.80y + 0.25z &= y \\ 0.10x + 0.10y + 0.60z &= z \end{aligned}$$

in which any one of the last three equations can be discarded as superfluous. Solving this system gives $x = \frac{11}{35}$, $y = \frac{17}{35}$, $z = \frac{1}{5}$, which when rounded to three decimal places is the answer obtained earlier.

SUMMARY; OTHER APPLICATIONS

If a set of events or situations occur at regular intervals and if the set can be modeled by a sequence of experiments in which the outcome of each depends only on the outcome of the immediately preceding one, then we have a Markov process. In most applications we are interested in the long-run trend.

In addition to finding a limiting distribution from observed shifts, say in population or in magazine readership, the methods of this section have been used in sociology to predict class motility, in meteorology to forecast weather, to predict shifts in smoking or drinking patterns, to estimate receipts from fund-raising drives, to project the occurrence of genetic traits, to predict the number of accidents that may occur in the home, to forecast the future of a particular stock, and so on; related mathematical techniques appear in the theory of information and communication.

Whenever it is valid to assume that an observed trend will continue, methods involving Markov matrices can be useful. A variety of applications is illustrated in the exercises below.

EXERCISES

1. Which of the following matrices are regular?

(a) $\begin{pmatrix} \frac{1}{3} & \frac{2}{3} \\ 1 & 0 \end{pmatrix}$ (b) $\begin{pmatrix} \frac{1}{3} & \frac{2}{3} \\ 0 & 1 \end{pmatrix}$ (c) $\begin{pmatrix} 0 & 1 \\ \frac{4}{5} & \frac{1}{5} \end{pmatrix}$ (d) $\begin{pmatrix} 1 & 0 \\ \frac{4}{5} & \frac{1}{5} \end{pmatrix}$

(e) $\begin{pmatrix} 0 & 1 & 0 \\ 0.7 & 0.2 & 0.1 \\ 0 & 0.6 & 0.4 \end{pmatrix}$ (f) $\begin{pmatrix} \frac{1}{2} & 0 & \frac{1}{2} \\ \frac{1}{3} & \frac{1}{3} & \frac{1}{3} \\ 0 & 1 & 0 \end{pmatrix}$

2. Suppose that census-bureau surveys of a particular region show that each year 75% of those living in single-family dwellings (S) continue to do so, while 25% move into multiple-family dwellings (M); also, 60% of those living in multiple-family dwellings stay there while 40% move into single-family dwellings. The transition matrix for these data is

		TO S	TO M
FROM	S	0.75	0.25
	M	0.40	0.60

(a) If the present population is distributed evenly between single- and multiple-family dwellings, what will be the distribution one year from now? 2 years from now? 3? 4? 5? 10? 15? 20? 30?

(b) What is the result of multiplying together two copies of the transition matrix? 3 copies? 4? 5? 10? 15? 20? 30?

(c) Set up and solve an appropriate system of equations to determine the unique fixed probability vector associated with the transition matrix. Compare the answer with the results obtained in (a) and (b).

3. A ball team's performance is influenced by the results of the preceding game as follows: After winning a game, the probability of winning the next is 0.6 and of losing it is 0.4. After losing a game there is an equal probability of winning or losing the next game. The relevant transition matrix is

		NEXT GAME: WIN	NEXT GAME: LOSE
PRECEDING GAME:	WIN	0.6	0.4
	LOSE	0.5	0.5

(a) Suppose the team has just won a game so that the initial probability vector is (1 0). What is the probability of winning or losing the next game? the second game from now? the third? fourth? fifth? tenth? fifteenth? twentieth? What percentage of games can the team be expected to win for the season?

(b) What is the result of multiplying two copies of the transition matrix? three copies? 4? 5? 10? 15? 20?

(c) Find the unique fixed probability vector associated with the transition matrix. Compare the answer with the results of (a) and (b).

4. According to a government labor economist, 95% of those employed

at any time will still be employed one month later while 15% of those who are unemployed will find jobs during the next month.

(a) What is the transition matrix that describes this situation?

(b) Solve an appropriate set of linear equations to determine the long-term distribution of employed and unemployed, assuming the economist's analysis continues to portray the labor market accurately.

5. A credit-manager at a department store determines from her records of charge accounts that the probability that a customer charges something at the store during a particular month depends on whether he charged a purchase during the preceding month. She concludes that 60% of the customers who use their charge accounts one month will charge again the next, but only 20% of those who do not charge one month will charge during the next.

(a) What is the relevant transition matrix?

(b) If the observed trend continues, what is the probability that eventually a customer with a charge account will use it every month?

6. A rat is placed in a maze with three compartments, painted red, yellow, and blue. Records are kept of the rat's location at the end of each minute. The transition probabilities are given by the following matrix:

$$\begin{array}{cc} & \text{TO} \\ & \begin{array}{ccc} \text{R} & \text{Y} & \text{B} \end{array} \\ \begin{array}{r} \text{FROM R} \\ \text{Y} \\ \text{B} \end{array} & \begin{pmatrix} 0.4 & 0.4 & 0.2 \\ 0.4 & 0.6 & 0 \\ 0.4 & 0.5 & 0.1 \end{pmatrix} \end{array}$$

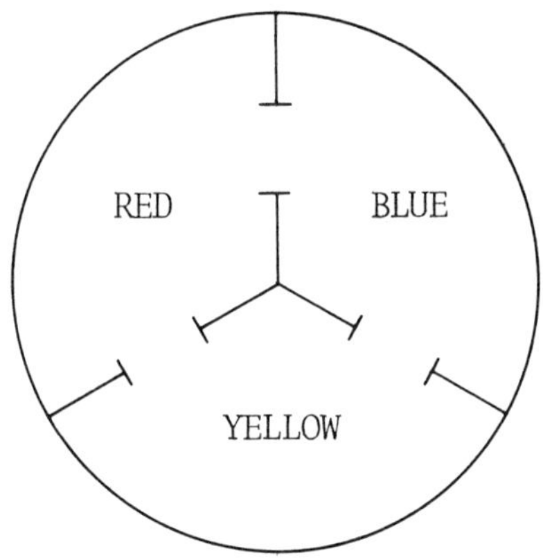

If this pattern of movements between compartments continues, what is the probability in the long run that at a moment chosen at random the rat will be in the red compartment? the yellow? the blue?

7. A small town has two department stores, Missy's and Gamble's. A marketing consultant determines that of the customers who primarily shop at Missy's one year 70% will remain loyal the following year, with 30% switching to Gamble's; of those who do

their major shopping at Gamble's, 60% will continue this pattern the next year, with 40% switching to Missy's. In the long run, what percentage of customers will do their major shopping at Missy's? at Gamble's?

8. The fund-raiser for a professional orchestra, with a fixed mailing list, asserts that 75% of those who make a donation one year also contribute the following year, while only 10% of those who do not contribute one year decide to donate the next year. If this trend holds up, what percentage of those solicited can be counted on in the long run to make yearly contributions?

9. A rumor about the success of an assassination attempt is spreading through a city. Suppose that 90% of the people pass on what they are told, whether or not it is correct, while the remaining 10% tell the opposite of what they hear. In the long run, what is the probability that a person chosen at random will pass on correct information? Answer this question if the percentages are changed to 60% and 40% respectively.

10. A sociological study yields the data in the following matrix on the careers of children of professionals:

PROFESSION OF PARENT:	PROFESSION OF CHILD: LAWYER	DOCTOR	OTHER
LAWYER	0.10	0.05	0.85
DOCTOR	0.05	0.15	0.50
OTHER	0.01	0.02	0.97

If the trend continues, what is the probability that a person chosen at random will become a doctor? a lawyer?

11. A government survey of land use shows the following: If land is used for agricultural purposes one year, then there is a probability of 0.80 that it will be used for agricultural purposes again the following year, a 0.15 probability it will be idle, and a 0.05 probability it will be used for urban development. Land idle one year will remain idle, be used for agriculture, or be used for urban development the following year with the probabilities 0.5, 0.4, and 0.1 respectively. And land used for urban development will be used the same way or will be idle the following year with probabilities 0.99 and 0.01 respectively. In the long run what percentage of the land will be used for agriculture? for urban development?

12. A physicist studies the motion of a particle bouncing around the three chambers A, B, and C (as shown in the figure at top of the facing page). A particle in A is in A, B, or C one minute later with probabilities 0.4, 0.3, and 0.3 respectively. A particle in B is in A, B, or C one minute later with probabilities 0.2, 0.6, and 0.2 respectively. And a particle in C is in A, B, or C one minute later with probabilities 0.4, 0.4, and

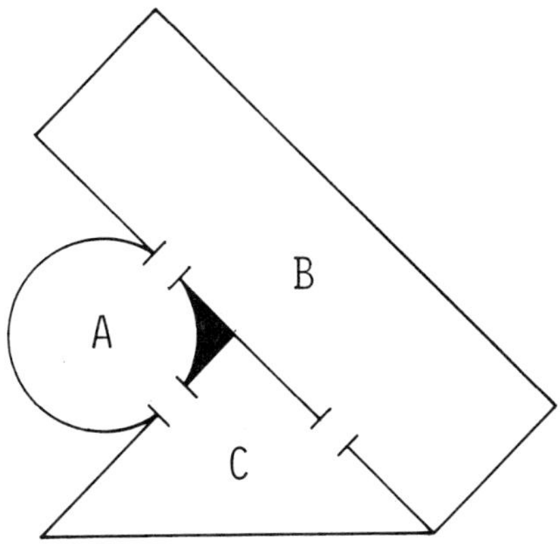

0.2 respectively. If the observed trend continues, what fraction of the time will the particle eventually spend in chamber A? in chamber B? in chamber C?

TEST A

1. Which matrices are regular?

(a) $\begin{pmatrix} 0.9 & 0.1 \\ 1 & 0 \end{pmatrix}$ (b) $\begin{pmatrix} 1 & 0 \\ 0.2 & 0.8 \end{pmatrix}$ (c) $\begin{pmatrix} 0.5 & 0.3 & 0.2 \\ 0.4 & 0.2 & 0.4 \\ 0.1 & 0.6 & 0.3 \end{pmatrix}$

2. A meteorologist in Suva, Fiji notes that if it rains one day the probability is 80% that it will rain again the next day; if there is no rain one day there is still a probability of 50% for rain the following day. If this pattern continues, in the long run what percentage of the days is it expected to rain?
3. A scientist studying social motility in an upstate county in New York notes the following: children of upper-class parents belong to the upper, middle, or lower classes with probabilities 0.85, 0.15, and 0 respectively; children of middle-class parents belong to upper, middle, or lower class with probabilities 0.10, 0.80, and 0.10 respectively; children of lower-class parents belong to upper, middle, or lower class with probabilities 0.05, 0.20, and 0.75 respectively. If these patterns endure, what percentage of the population will eventually be upper class? middle class? lower class?

TEST B

1. Which matrices are regular?

(a) $\begin{pmatrix} 0.6 & 0.4 \\ 0 & 1 \end{pmatrix}$ (b) $\begin{pmatrix} 0 & 1 \\ \frac{2}{3} & \frac{1}{3} \end{pmatrix}$ (c) $\begin{pmatrix} \frac{1}{2} & \frac{1}{2} & 0 \\ \frac{1}{4} & 0 & \frac{3}{4} \\ 0 & 1 & 0 \end{pmatrix}$

2. The records of a mail-order book club show that 60% of their customers who order a book one month order again the following month but only 10% of those who do not order one month do send in an order the next. What percentage of customers will eventually order books each month, if the observed trend continues?

3. An animal is called *dominant* (D) with regard to a particular gene if it has AA genes, *recessive* (R) if it has aa genes, and *hybrid* (H) if it has Aa (or aA) genes. These three classes are called *genotypes*. A sequence of experiments is performed in which successive generations of dominants are bred with other dominants, with hybrids, and with recessives.

(a) Show that the transition probabilities for these experiments are given by the following matrix (remembering that an offspring is equally likely to receive, from each parent, either of the paired genes of that parent):

	GENOTYPE OF OFFSPRING: D	H	R
GENOTYPE OF DOMINANT'S MATE: D	1	0	0
H	.5	.5	0
R	0	1	0

(b) In the long run how will the genotypes be distributed?

SUMMARY OF §§19-25

You should now be able to

(1) specify the dimension of a matrix;
(2) recognize a probability vector and a transition, permutation, diagonal, or symmetric matrix;
(3) understand matrix notation and definitions;
(4) use a matrix to display data;
(5) know and use the rules for operating on matrices involving addition and scalar multiplication;
(6) multiply matrices (when possible);
(7) know the properties of matrix multiplication;
(8) apply the Gauss-Jordan row operations to augmented matrices to solve a system of linear equations;
(9) use the Gauss-Jordan row operations to determine if a system is either inconsistent or redundant;
(10) find the inverse of an invertible matrix by applying row operations;
(11) know the formula for the inverse of a 2×2 matrix;
(12) use inverses to solve a system of linear equations;
(13) write an input-output matrix for given data;
(14) solve matrix equations in applications of input-output analysis;
(15) recognize regular matrices;
(16) write a transition matrix that describes a Markov process;
(17) find the fixed probability vector of a Markov matrix or (equivalently) find the long-term trend in a Markov process.

REVIEW EXERCISES

1. What is the dimension of each matrix?

$$A = \begin{pmatrix} 3 & 0 \\ -1 & 2 \\ 0 & 1 \end{pmatrix} \qquad B = \begin{pmatrix} \frac{1}{2} & 0 & \frac{1}{4} & 0 & \frac{1}{4} \end{pmatrix}$$

$$C = \begin{pmatrix} 0 & 1 & 0 & 0 \\ 1 & 0 & 0 & 0 \\ 0 & 0 & 0 & 1 \\ 0 & 0 & 1 & 0 \end{pmatrix} \qquad D = \begin{pmatrix} -\frac{1}{5} \\ \frac{4}{5} \\ 0 \\ -\frac{2}{5} \\ \frac{4}{5} \end{pmatrix}$$

2. In exercise 1, is B or D a probability vector? Why or why not?
3. Are any of the matrices in exercise 1 transition matrices? permutation? diagonal? symmetric?
4. Data gathered in a survey of the readers of three daily newspapers in a big city show that each year: newspaper A keeps 70% of its readers but loses 10% to B and 20% to C; B keeps 60%, losing 20% to A and 20% to C; C keeps 75%, losing 15% to A and 10% to B. Display the data in matrix form. Is this a transition matrix? Explain.
5. Let

$$A = \begin{pmatrix} 0 & -2 & 3 \\ 1 & 4 & 2 \end{pmatrix}, \quad B = \begin{pmatrix} -1 & -3 & -2 \\ 4 & -6 & 0 \end{pmatrix}, \quad C = \begin{pmatrix} 1 & \frac{1}{2} & -2 \\ -2 & 0 & 1 \\ -1 & 1 & -3 \end{pmatrix}, \quad D = \begin{pmatrix} x \\ y \\ z \end{pmatrix},$$

$$E = \begin{pmatrix} -1 & 5 & 3 \end{pmatrix}, \quad F = \begin{pmatrix} 0 & 1 \\ 1 & 0 \end{pmatrix}$$

Perform the operations indicated below whenever they are defined:

(a) $A + B$ (b) $A - 2B$ (c) AB (d) $D + E$ (e) F^2

(f) CD (g) DC (h) ED (i) AD (j) FA

6. Four stores sold paperback, hard-cover, and deluxe illustrated editions of a certain book one week, as shown below:

NUMBER OF COPIES SOLD:

STORE	PAPER	HARD	DELUXE
A	16	7	2
B	10	4	5
C	20	9	3
D	12	3	4

PRICE IN DOLLARS:

PAPER	4
HARD	9
DELUXE	12

(a) Construct a matrix that shows the total revenue at each store from the week's sale of the books. (b) Which store took in the most? the least?

7. Using Gauss-Jordan row operations, find the unique solution of each of the following, if it exists; if it does not exist, tell whether the system has no solution or many solutions.

(a) $\begin{aligned} x + y + z &= -1 \\ 4x + z &= 2 \\ 2x + 2y - z &= 4 \end{aligned}$ (b) $\begin{aligned} x + y &= 1 \\ y + z &= 3 \\ -x + z &= 2 \end{aligned}$ (c) $\begin{aligned} x + z &= -1 \\ 2y + z &= -5 \\ 2x - 6y + z &= 1 \end{aligned}$

8. Find the inverse, if it exists, of

(a) $\begin{pmatrix} 4 & -2 \\ 3 & 1 \end{pmatrix}$ (b) $\begin{pmatrix} 1 & 3 \\ -2 & -6 \end{pmatrix}$ (c) $\begin{pmatrix} 1 & -1 & 2 \\ 0 & 1 & 3 \\ 2 & -1 & 1 \end{pmatrix}$ (d) $\begin{pmatrix} 1 & 0 & 1 \\ 0 & 2 & 1 \\ 2 & -2 & 1 \end{pmatrix}$

9. Using inverses, solve each system of linear equations in each set.

(a) $\begin{aligned} 2x + y &= 4 \\ x + 3y &= -3 \end{aligned}$ $\begin{aligned} 2x + y &= 8 \\ x + 3y &= 5 \end{aligned}$ $\begin{aligned} 2x + y &= -9 \\ x + 3y &= 1 \end{aligned}$

(b) $\begin{aligned} x + z &= 1 \\ -x + y &= -1 \\ y - 2z &= 0 \end{aligned}$ $\begin{aligned} x + z &= 3 \\ -x + y &= -6 \\ y - 2z &= -3 \end{aligned}$

10. Find $(I - R)^{-1}$ if $R = \begin{pmatrix} .50 & .40 \\ .20 & .30 \end{pmatrix}$

11. Using R in exercise 10 as the input-output matrix of a simple two-sector economic system and with $D = \begin{pmatrix} 8 \\ 12 \end{pmatrix}$, with entries in thousands of dollars, as the external demand vector, find the total production levels required for each sector.

12. A simple economy consists of three industries, A, B, and C, each with a single product. Internal demands for each dollar of output are as follows:

A uses \$0.30 worth of its own product,
none of B's product,
\$0.10 worth of C's product;
B uses \$0.20 worth of A's product,
\$0.10 worth of its own product,
none of C's product;
C uses \$0.40 worth of A's product,
\$0.20 worth of B's product,
none of its own product.

External demands for the products are \$7000 for A's, \$4000 for B's, and \$5000 for C's.

(a) Write the input-output matrix R and the external-demand matrix D.

(b) Find $(I - R)^{-1}$.

(c) Find the total production levels required of the three industries.

13. Which of the following matrices is regular?

(a) $\begin{pmatrix} 0 & 1 \\ 0.6 & 0.4 \end{pmatrix}$ (b) $\begin{pmatrix} 1 & 0 \\ 0.7 & 0.3 \end{pmatrix}$ (c) $\begin{pmatrix} 1 & 0 & 0.8 \\ 0 & 0 & 1 \\ 0.2 & 0.1 & 0.7 \end{pmatrix}$

14. Use the data given in exercise 4 above to determine the long-run distribution of the readership of the three newspapers.

15. A certain gene found in hamsters mutates back and forth between two forms. From the A-form there is a probability of 0.95 that an offspring will also have the A-form (and therefore a probability of .05 that the offspring will have the B-form). From the B-form there is a probability of 0.80 that an offspring will also have the B-form (with probability 0.20 that it will have the A-form). What is the probability that a hamster chosen at random will have (a) the A-form of the gene? (b) the B-form?

CHAPTER TEST

1. What is the dimension of each of the following matrices?

$$A = \begin{pmatrix} 1 & 0 & 4 & 2 \\ -2 & 1 & 3 & -2 \end{pmatrix} \qquad B = \begin{pmatrix} \frac{1}{2} & 0 & -\frac{1}{2} & 0 & 1 \end{pmatrix} \qquad C = \begin{pmatrix} a_{11} & a_{12} \\ a_{21} & a_{22} \\ a_{31} & a_{32} \\ a_{41} & a_{42} \end{pmatrix}$$

2. Define: (a) probability vector; (b) transition matrix.
3. Let

$$A = \begin{pmatrix} 1 & 0 \\ -2 & 3 \\ 0 & -1 \end{pmatrix}, \qquad B = \begin{pmatrix} x & y & z \end{pmatrix}, \qquad C = \begin{pmatrix} -4 & 0 & -2 \\ 3 & 1 & 2 \end{pmatrix},$$

$$D = \begin{pmatrix} -1 & 0 \\ 0 & -1 \end{pmatrix}, \qquad E = \begin{pmatrix} -1 & 2 & 0 & -3 \\ 0 & -4 & 1 & 5 \end{pmatrix}$$

Find each of the following, if defined:

(a) $A + C$ (b) AC (c) ED (d) D^2 (e) DA

(f) AD (g) EC

4. A manufacturer of refrigerators has plants in three cities: Pittsburgh, New Orleans, and San Diego. Sales last year of four models (portable, apartment-size, standard, and deluxe) were as follows:

	PT	APT	ST	DEL
PITTSBURGH:	500	450	460	300
NEW ORLEANS:	460	340	450	290
SAN DIEGO:	520	480	400	340

The selling prices were: portable \$65; apartment-size \$175; standard \$300; deluxe \$500.

(a) Construct a matrix for the sales per model per city.
(b) Construct a matrix for the selling price per model.

5. Use the data given in exercise 4 to find a matrix displaying the total revenue on refrigerators last year for each plant. Which plant had the largest revenue?
6. Using Gauss-Jordan row operations, find the unique solution if it exists; if not, tell if the system has no solution or many:

(a) $\begin{aligned} x + y + z &= -1 \\ 2x - y &= 3 \\ 3x - 3y - z &= 2 \end{aligned}$ (b) $\begin{aligned} x + 2y + z &= 3 \\ 3x - 2z &= 0 \\ 2x + 2y - z &= -1 \end{aligned}$

7. Find the inverse, if it exists, of

(a) $\begin{pmatrix} 5 & -3 \\ 2 & 4 \end{pmatrix}$ (b) $\begin{pmatrix} -2 & 3 \\ 4 & -6 \end{pmatrix}$

8. Solve each of these linear systems using inverses:

(a) $\begin{aligned} x + 2y + z &= 3 \\ 3x - 2z &= 0 \\ 2x + 2y - z &= -1 \end{aligned}$ (b) $\begin{aligned} x + 2y + z &= -5 \\ 3x - 2z &= 3 \\ 2x + 2y - z &= -1 \end{aligned}$

9. A simple economy consists of two sectors, I and II, each with a single product. Internal demands for each dollar's worth of output are as follows:

I uses \$0.30 of its own product and \$0.10 of II's;
II uses \$0.20 of its own product and \$0.40 of I's.

Find the total outputs required of each sector of the external demands on I and II are valued respectively at (a) \$6000 and \$2000; (b) \$8000 and \$3000.

10. Which of the following matrices are regular?

(a) $\begin{pmatrix} 0 & 1 \\ \frac{3}{4} & \frac{1}{4} \end{pmatrix}$ (b) $\begin{pmatrix} \frac{1}{3} & 0 & \frac{2}{3} \\ \frac{1}{4} & \frac{3}{4} & 0 \\ \frac{1}{3} & \frac{1}{3} & \frac{1}{3} \end{pmatrix}$

11. Suppose that the following voting pattern has been observed in a particular region of the country: Of those who vote for the Democratic ticket in one election, 70% do so again at the next election, while 20% of this group switches to Republican and the rest to Independent. Also, of those who vote the Republican ticket in one election, 60% continue to do so in the next, while 20% of them switch to Democratic and the rest to Independent. Of those who vote for Independent candidates at one election, 50% do the same at the next election, while 30% change to Democratic and 20% to Republican. What percentage of this region will eventually vote Democratic if the observed trend continues?

Chapter 5

Linear programming

In this chapter we solve problems in a branch of mathematics known as *linear programming* (LP). For each problem we construct a model--a linear program--which consists of a set of linear inequalities, called *constraints*, together with a linear function called the *objective function*. Our goal is to maximize (or, in some cases, to minimize) the objective function subject to the constraints. We begin the chapter by showing how to solve linear-programming problems geometrically, then turn to an algebraic technique known as the simplex method. Since the latter is readily adaptable to computer use, it is especially suitable when there are many variables and constraints.

The techniques presented in this chapter have been widely and effectively applied, in business, economics, biology, ecology, and other fields involving decision-making. We will consider a variety of these applications here.

SECTION 26. GEOMETRIC SOLUTION OF LP PROBLEMS

In this section we show how geometry is used to solve LP problems in two variables. Since this approach involves graphing systems of linear inequalities, you should review §18, which covers this technique, if necessary, before proceeding.

We begin with examples.

Example 26-1

Find the maximum value of $f(x,y) = x + y$ subject to

$$\begin{aligned} x \geqslant 0, \quad y &\geqslant 0 \\ 2x + 5y &\leqslant 20 \\ 2x + y &\leqslant 12 \end{aligned}$$

Here f is the objective function and the centered inequalities are the constraints.

We start by graphing the *feasible set*: that is, the collection of points satisfying all the constraints, which is the same as the set of all solutions of the system of inequalities. In Figure 26.1 we show the steps used in graphing the feasible set for the given LP problem. Remember that we graph a linear inequality in two variables (pp. 144-147) by graphing the line for the corresponding equality first. We then determine which half-plane (that is, which side of the line) satisfies the inequality, by testing a point not on the line in the inequality. We shade this half-plane. Following this procedure for each inequality yields the feasible set. In this example it is the shaded four-sided region and its boundary shown in Figure 26.1(d).

Next we label the vertices or *corner points* of the feasible set (Figure 26.2). This usually involves solving pairs of linear equations. In this case, we solve the pair

$$\begin{aligned} 2x + 5y &= 20 \\ 2x + y &= 12 \end{aligned}$$

The solution is $x = 5$, $y = 2$. This point and the other corner points, (0,0), (0,4), and (6,0) are labeled in Figure 26.2.

Now we want to find the point *within* the feasible set at which $f(x,y) = x + y$ takes on its maximum value. For example, is there a point in the feasible set at which $f(x,y)$ is equal to ten? To answer this query, we graph the line $f(x,y) = 10$, that is, $x + y = 10$, to see if any point in the feasible set lies on the line.

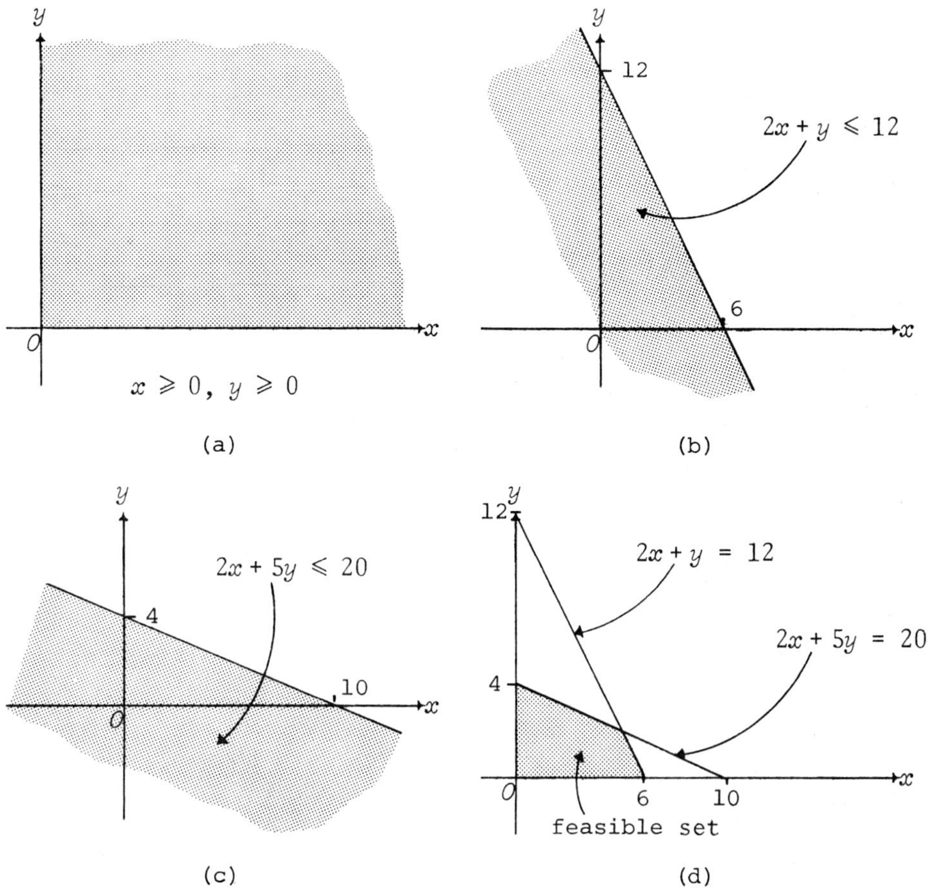

Figure 26.1

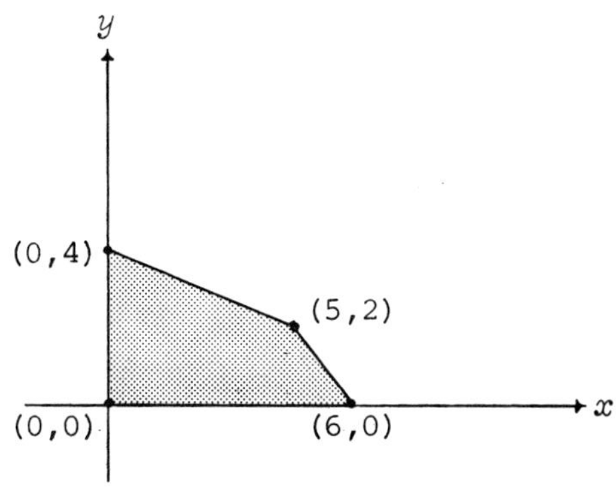

Figure 26.2

We see from Figure 26.3(a) that $f(x,y)$ is never equal to 10 in the feasible set.

How about the value eight?

Figure 26.3(b) shows that $f(x,y)$ is never equal to eight in the feasible set.

How about the value six?

We see from Figure 26.3(c) that there are many points in the feasible set at which $f(x,y)$ has the value six.

What is the largest value of $f(x,y)$ in the feasible set? Based on the above graphs, we conclude that f takes on its largest value at the point (5,2), the corner point of the feasible set that lies on the line parallel to all of the three dotted lines sketched in Figures 26.3 (a), (b), and (c). When $x = 5$ and $y = 2$, then $f(x,y) = x + y = 5 + 2 = 7$. Therefore the maximum value of $f(x,y)$ in the feasible set is 7 and f takes on this maximum value at the corner point (5,2).

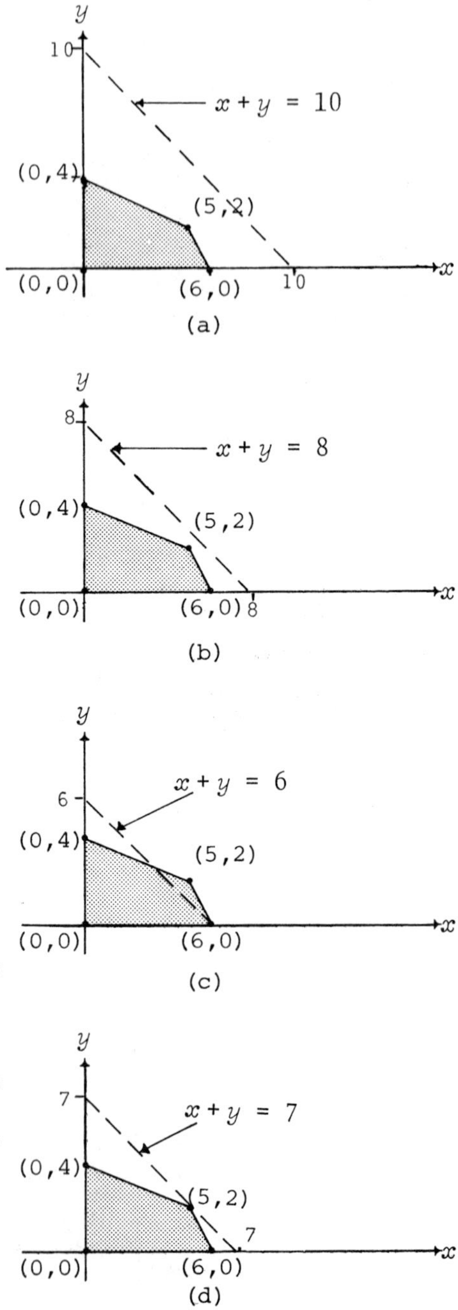

Figure 26.3

Example 26-2

Consider the same feasible set as in the preceding example, but now find the maximum value in this set of the objective function $g(x,y) = y + 3x$.

From the figures on the left below we reach the conclusions given on the right.

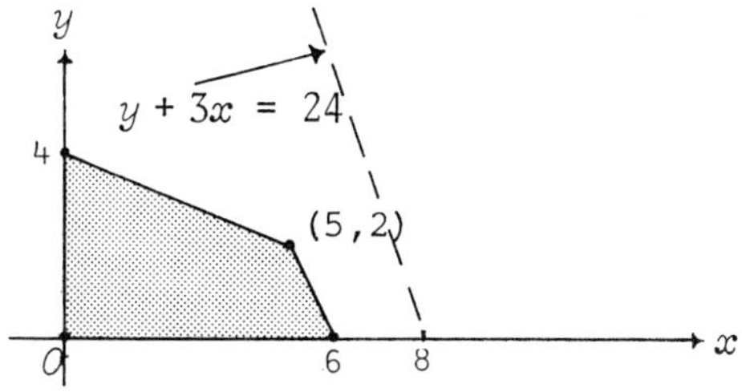

(a)

$g(x,y)$ never has the value 24 in the feasible set.

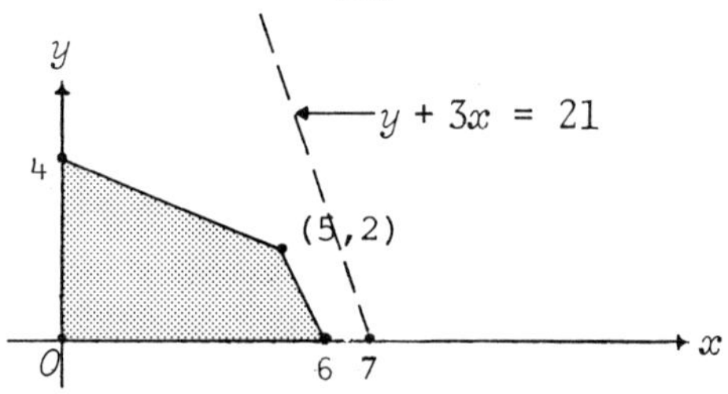

(b)

$g(x,y)$ never has the value 21 in the feasible set.

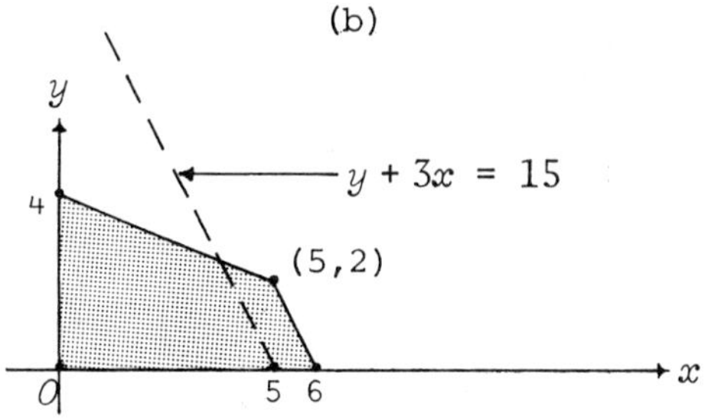

(c)

There are many points in the feasible set at which $g(x,y)$ takes on the value 15.

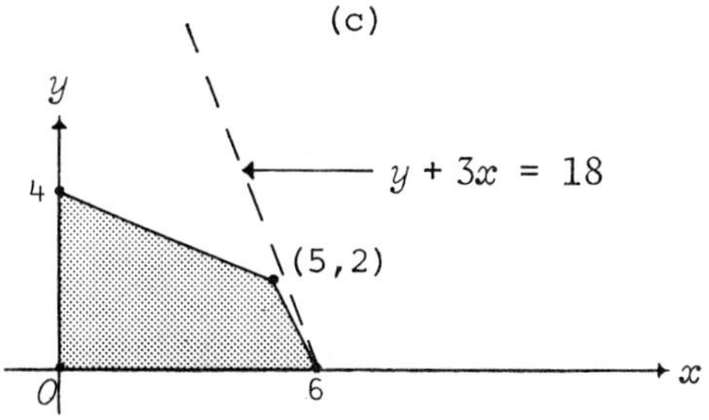

(d)

The function takes on its maximum value in the feasible set at the corner point (6,0) and this maximum value is 18.

Figure 26.4

Example 26-3

Consider the same feasible set once again, but this time find the maximum value of the objective function $h(x,y) = x + 4y$.

We see that $h(x,y)$ is never equal to 20 in the feasible set.

There are many points in the feasible set at which $h(x,y)$ has the value 13.

The maximum value of $h(x,y)$ in the feasible set is 16, which occurs at the corner point (0,4).

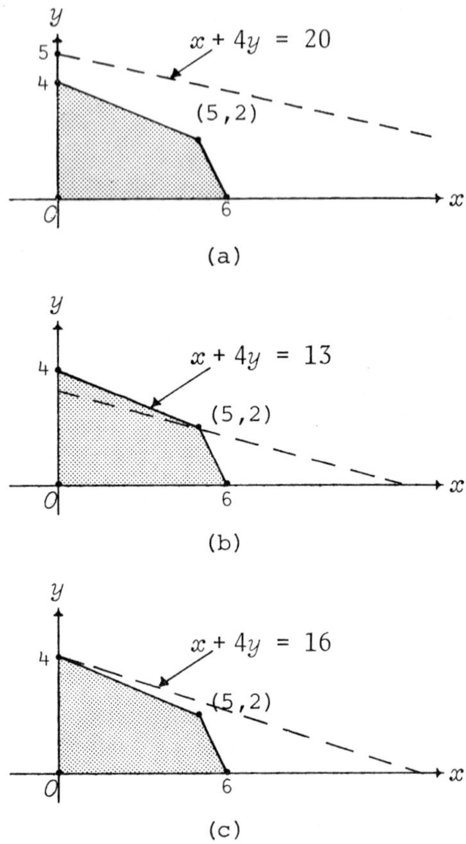

Figure 26.5

PROCEDURE FOR GEOMETRIC SOLUTION OF LP PROBLEMS

The examples given above lead to the following procedure for solving an LP problem geometrically:

(1) Graph the feasible set (as described on page 238).
(2) Find all the corner points of the feasible set, solving pairs of linear equations simultaneously when necessary.
(3) Find the value of the objective function at each corner point determined in the second step.
(4) The largest value obtained in the third step is the

maximum of the function in the feasible set. (Also, the smallest value obtained in the third step is the minimum of the function in the feasible set.)

Example 26-4

Find the maximum and the minimum values of the objective function $f(x,y) = x + 3y$ subject to the constraints

$$x \geqslant 0, \quad y \geqslant 0$$
$$2x + y \geqslant 2$$
$$3x + 4y \leqslant 12$$

STEP (1). Graph the feasible set. The procedures are shown in Figure 26.6.

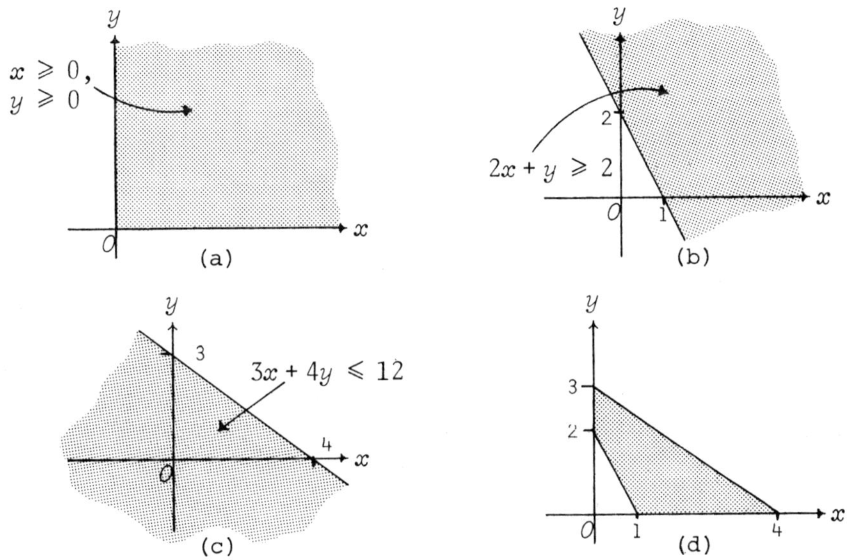

Figure 26.6

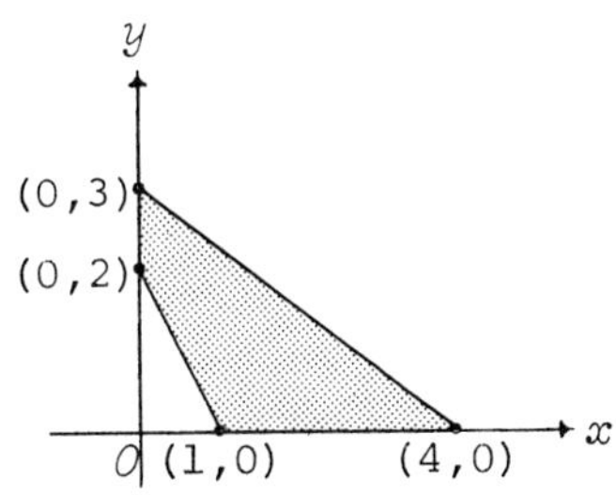

Figure 26.7

STEP (2). Find all the corner points. The vertices of the four-sided region are labeled in Figure 26.7.

STEP (3). Evaluate the objective function at each corner point:

(x,y)	$f(x,y) = x + 3y$
(0,2)	6
(0,3)	9
(1,0)	1
(4,0)	4

STEP (4). The maximum value of f in the feasible set is 9, when $x = 0$ and $y = 3$; the minimum value of f in the feasible set is 1, when $x = 1$ and $y = 0$.

MULTIPLE SOLUTIONS

In some problems the slope of the linear objective function is the same as that of one of the boundaries of the feasible set. In this case these lines are parallel and the maximum (or minimum) value of the objective function may occur at two corner points (the endpoints of the relevant boundary). Then every point on the line segment between these corner points is also a solution.

Example 26-5

Find the maximum and minimum values of the function $g(x,y) = 3x + 4y$ for the feasible region of Example 26-4.

The first two steps are identical with those for Example 26-4; see Figures 26.6 and 26.7.

STEP (3). Evaluate g at each corner point:

(x,y)	$g(x,y) = 3x + 4y$
(0,2)	8
(0,3)	12
(1,0)	3
(4,0)	12

STEP (4). The corner points (0,3) and (4,0) both yield a maximum. Noting that these points lie on the line $3x + 4y = 12$, we verify that g is also equal to 12 at other points on the line between these corner points, for example at $(2,1\frac{1}{2})$ and at $(1,\frac{9}{4})$:

$$g(2,1\tfrac{1}{2}) = 3(2) + 4(1\tfrac{1}{2}) = 12$$

$$g(1,\tfrac{9}{4}) = 3(1) + 4(\tfrac{9}{4}) = 12$$

In this case an infinite number of points yield the maximum value of the objective function.

The minimum value of g is 3 and occurs at (1,0).

UNBOUNDED FEASIBLE SETS

The feasible sets in the preceding examples of this section have all been bounded. Sometimes we encounter unbounded feasible sets (see examples 18-4, 18-5, 18-6, 18-8, and 18-9 on pp. 148-54).

Example 26-6

Find the minimum value of $f(x,y) = 2x + 3y$ subject to

$$x \geqslant 0, \quad y \geqslant 0$$
$$3x + y \geqslant 3$$
$$2x + 5y \geqslant 10$$

STEP (1). We graph the feasible set. It is the four-sided unbounded region shaded in Figure 26.8(d).

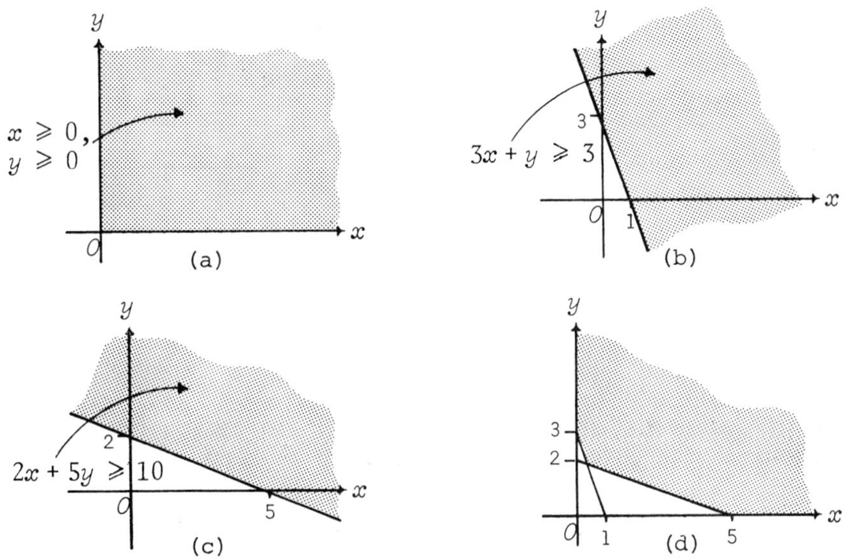

Figure 26.8

STEP (2). We find the corner points. The simultaneous solution of the pair of equations $3x + y = 3$ and $2x + 5y = 10$ is $x = \frac{5}{13}$, $y = \frac{24}{13}$. This and the other corner points are labeled in Figure 26.9.

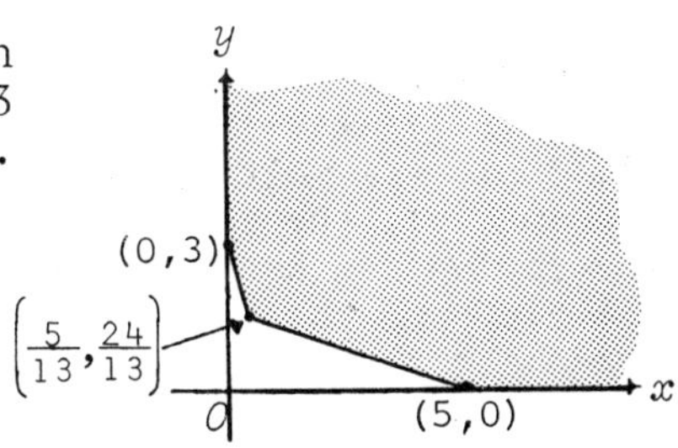

Figure 26.9

As can be seen from Figure 26.10(a), a minimum does exist.

STEP (3). We evaluate f at each corner point:

(x,y)	$f(x,y) = 2x + 3y$
$(0,3)$	9
$(\frac{5}{13},\frac{24}{13})$	$6\frac{4}{13}$
$(5,0)$	10

STEP (4). We note that the minimum of f in the feasible region is $6\frac{4}{13}$, at $x = \frac{5}{13}$, $y = \frac{24}{13}$. But as Figure 26.10(b) shows, f can be made as large as we please in the feasible set. Therefore it has no maximum value.

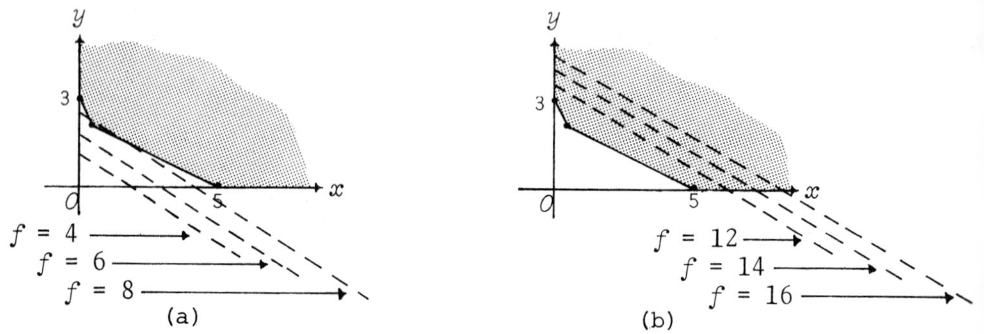

Figure 26.10

THEOREM ON LP PROBLEMS

In the preceding examples we have used the following fundamental theorem about linear-programming problems:

> If the objective function in a linear-programming problem attains a maximum (or minimum) value, that value occurs at a corner point of the feasible set.

This theorem describes the situation fully if the problem has a solution; that is, if the function takes on a maximum (or minimum) value. If there are multiple solutions, at least one must occur at a corner point. If the feasible set of a problem is bounded (that is, if the region can be enclosed in a circle) then the problem does have a solution and the conclusion of the theorem is valid.

WORD PROBLEMS

Linear programming has been used to solve problems of the following types:

Production-scheduling problems: to schedule production of two or more products so as to maximize profit, with constraints on the availability of raw-materials, of machine or other processing time, of labor, of storage costs, and so on. (A variant of this is site-selection: for example, finding the best location for bank branches, supermarkets, factory outlets, and the like.)

Transportation problems: (1) to minimize the costs of shipping a product from different factories to several outlets with constraints on production capacities, product demands, varying delivery costs; (2) to schedule air flights and determine routes so as to maximize service subject to restrictions on available money, equipment, and labor.

Diet or product-mixing problems: (1) to produce at least cost a product that meets certain nutritional constraints (minimum vitamins and minerals, maximum calories and protein, etc.); (2) to determine how to combine ingredients or to mix products in obtaining another product (a food, animal feed, fertilizer, drug, chemical) so as to minimize cost, with limitations on amounts of raw-materials available and with specified nutritional or medical requirements.

Investment problems: to maximize yield, say, in a pension, mutual, or retirement fund from diversified investments where there are constraints on how the amounts are to be allocated, on the risk factors, and on minimum expected yields.

Linear programming is especially useful in the management sciences, but it has also been used in numerous other industries and areas: in advertising, to select from among suitable media so as to reach the largest number of potential consumers for a product; in fire-department operations for the most efficient deployment (allocation) of firefighting resources; in wildlife management, to determine "ideal" populations of different animal species so as to assure survival; in pollution control, to minimize the cost of, say, air-pollution control subject to restrictions on production and specified pollution emission reductions; in personnel management, to find the assignment of individuals to jobs which will maximize the "yield" (where the yield or "worth" of a particular person in a particular job is assumed to be known).

Most real-life applications of linear programming similar to those above involve complicated models and numerous variables and constraints--so many as to render the use of a computer essential. However, the examples that follow can all be solved with only two variables. They are therefore, of necessity, simplified (contrived) applications of the methods presented earlier. They are nonetheless instructive to the novice in LP.

Example 26-7

A company manufactures both black-and-white and color television sets, with profits of \$50 and \$150 per set respectively. Each black-and-white set requires three hours of assembly time, two hours of finishing time, and one hour of testing time; each color set requires respectively ten, three, and two hours of these three processes. Each week, 300 man-hours are available for assembly, 126 hours for finishing, and 68 hours for testing. How many of each kind of TV set should be produced weekly to maximize profit?

The question "How many of each kind?" suggests how to assign the variables: let x and y be the number of black-and-white and of color sets respectively. We wish to maximize the profit

$$P(x,y) = 50x + 150y$$

We note that $x \geqslant 0$ and $y \geqslant 0$ because the number of sets manufactured cannot be negative. There are three other constraints because of the limited number of hours available for the various manufacturing processes:

$$\begin{aligned} 3x + 10y &\leqslant 300 \\ 2x + 3y &\leqslant 126 \\ x + 2y &\leqslant 68 \end{aligned}$$

We now apply the techniques presented earlier. We graph the five inequalities to get the feasible set (the five-sided shaded region in Figure 26.11(a)).

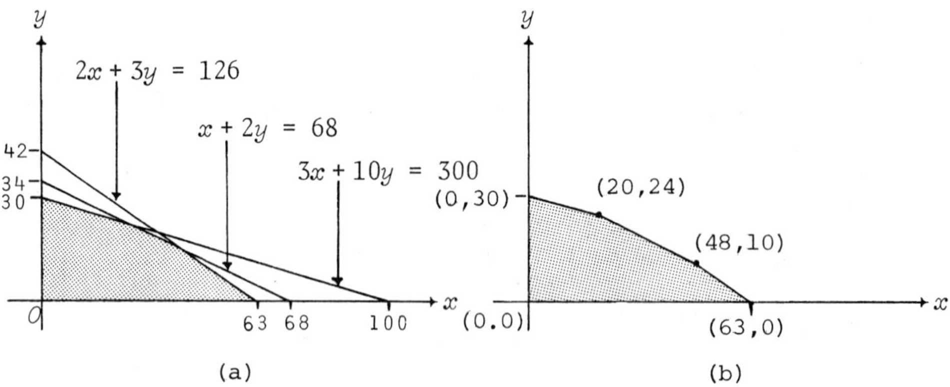

Figure 26.11

Next we solve the appropriate pairs of equations. The pair

$$\begin{aligned} 3x + 10y &= 300 \\ x + 2y &= 68 \end{aligned}$$

yields the corner point $x = 20$, $y = 24$. The pair

$$x + 2y = 68$$
$$2x + 3y = 126$$

yields the corner point $x = 48$, $y = 10$. Note from Figure 26.11(a) that

$$3x + 10y = 300$$
$$2x + 3y = 126$$

intersect *outside* the feasible set. In Figure 26.11(b) we see the feasible set and its five corner points.

Next we evaluate the profit function $P(x,y) = 50x + 150y$ at each of the corner points:

(x,y)	$P(x,y) = 50x + 150y$
(0,0)	0
(0,30)	4500
(20,24)	4600
(48,10)	3900
(63,0)	3150

We see that the maximum weekly profit is \$4600; it is achieved when the production-level is set at 20 black-and-white and 24 color sets.

Example 26-8

A farmer wants to provide each of his beef cattle daily with at least 600 units of vitamin I, 400 of vitamin II, and 550 of III. Two food supplements, A and B, are available. The first costs \$0.02 per ounce and supplies 30, 10, and 10 units respectively of vitamins I, II, and III. The second costs \$0.03 per ounce and supplies 10, 10, and 20 units respectively of the three vitamins. How many ounces of each supplement should be used to minimize cost yet meet the daily nutritional needs of the cattle?

We let x and y be the number of ounces of supplements A and B respectively. The objective function to be minimized is the cost $C(x,y) = 2x + 3y$, where C is in cents. The constraints are

$$x \geqslant 0, \quad y \geqslant 0$$
$$30x + 10y \geqslant 600$$
$$10x + 10y \geqslant 400$$
$$10x + 20y \geqslant 550$$

The feasible set is the unbounded region shaded in Figure 26.12 (a) (next page). Solving the pair of equations

$$30x + 10y = 600$$
$$10x + 10y = 400$$

gives $x = 10$ and $y = 30$. Solving the pair

$$10x + 10y = 400$$
$$10x + 20y = 550$$

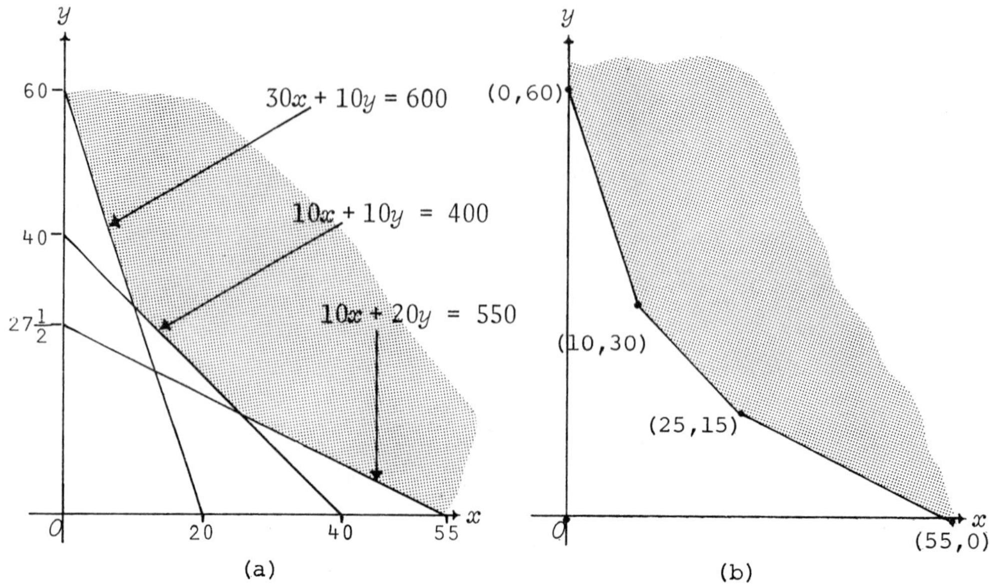

Figure 26.12

gives $x = 25$ and $y = 15$. We note that the pair

$$\begin{aligned} 30x + 10y &= 600 \\ 10x + 20y &= 550 \end{aligned}$$

intersect outside the feasible set. Figure 26.12(b) shows the feasible set and its corner points.

Now evaluate $C(x,y)$ at the corner points:

(x,y)	$C(x,y) = 2x + 3y$
(0,60)	180
(10,30)	110
(25,15)	95
(55,0)	110

Since the cost at the right is in cents, we see that the minimum cost is \$0.95, obtained by using 25 ounces of supplement A and 15 ounces of B.

Example 26-9

Environmentalists plan to stock a small mountain lake with two species of fish. There are 1000 of the first species and 1500 of the second available for stocking purposes. Fish of the first species need two units of plant life daily while fish of the second species need three units. Each day 5900 units of plant life grow in the lake. If each fish of the first species weighs $\frac{1}{2}$ pound and each of the second one pound, how many of each should be stocked to maximize the initial poundage of fish, yet insure that they have a sufficient food supply?

Let x and y be the number of fish of the first and second species respectively. The objective function to be maximized is the total weight, $W(x,y) = \frac{1}{2}x + y$. The constraints imposed by availability of fish and of food are

$$x \geqslant 0, \quad y \geqslant 0$$
$$x \leqslant 1000, \quad y \leqslant 1500$$
$$2x + 3y \leqslant 5900$$

Figure 26.13 shows the feasible set and its corner points.

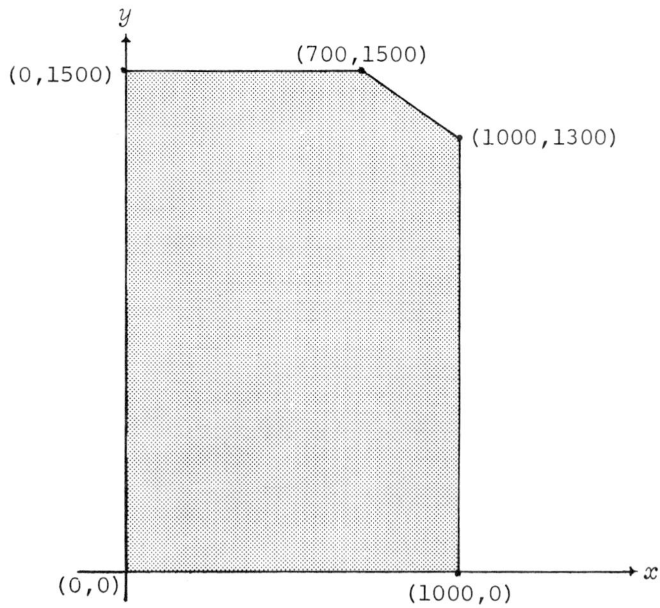

Figure 26.13

We evaluate $W(x,y)$ at each of the corner points:

(x,y)	$W(x,y) = \frac{1}{2}x + y$
(0,0)	0
(0,1500)	1500
(700,1500)	1850
(1000,1300)	1800
(1000,0)	500

Subject to the availability of fish and food, we see that the maximum initial weight of fish is 1850 pounds, obtained by stocking the lake with 700 fish of the first species and 1500 of the second.

EXERCISES

In exercises 1 through 8, find both the maximum and the minimum values of the given function subject to the given constraints.

1. $f(x,y) = 2x + 3y$

   $x \geqslant 0,\ y \geqslant 0$
   $3x + 5y \leqslant 30$
   $x + y \leqslant 8$

2. $f(x,y) = 4x + y$

   $x \geqslant 0,\ y \geqslant 0$
   $x + 2y \leqslant 12$
   $2x + y \leqslant 18$

3. $f(x,y) = 2x + 3y$

   $x \geqslant 0,\ y \geqslant 0$
   $x + 3y \geqslant 3$
   $3x + 5y \leqslant 15$

4. $f(x,y) = 2x + y$

   $x \geqslant 0,\ y \geqslant 0$
   $x + 2y \geqslant 4$
   $x + y \leqslant 3$

5. $f(x,y) = 2x + 3y$

   $x \geqslant 0,\ y \geqslant 0$
   $x + 5y \leqslant 15$
   $x + 2y \leqslant 9$
   $x + y \leqslant 8$

6. $f(x,y) = x + 2y$

   $x \geqslant 0,\ y \geqslant 0$
   $3x + 4y \leqslant 24$
   $x + y \leqslant 7$
   $2x + y \leqslant 12$

7. $f(x,y) = 5x + 4y$

   $y \leqslant 4$
   $3x + y \leqslant 16$
   $x + y \geqslant 6$

8(a). $f(x,y) = x + y$

   $y \geqslant 2,\ y \leqslant 6$
   $2x - y \geqslant 0$
   $4x + y \leqslant 30$

8(b). $g(x,y) = 4x + y$

   $y \geqslant 2,\ y \leqslant 6$
   $2x - y \geqslant 0$
   $4x + y \leqslant 30$

In exercises 9 through 12, find the minimum value of the given function subject to the given constraints.

9. $f(x,y) = 2x + y$

   $x \geqslant 0,\ y \geqslant 0$
   $4x + y \geqslant 8$
   $x + y \geqslant 5$

10. $f(x,y) = 6x + y$

   $x \geqslant 0,\ y \geqslant 0$
   $5x + y \geqslant 15$
   $5x + 8y \geqslant 50$

11. $f(x,y) = 4x + 3y$

   $x \geqslant 0,\ y \geqslant 0$
   $3x + y \geqslant 14$
   $2x + y \geqslant 13$
   $x + y \geqslant 10$

12. $f(x,y) = x + y$

   $x \geqslant 0,\ y \geqslant 0$
   $5x + 2y \geqslant 24$
   $3x + 2y \geqslant 20$
   $4x + 3y \geqslant 28$

13. An artist sells finger paintings and oil paintings with profits of \$75 and \$500 per painting respectively, The materials for each finger painting cost \$10 and each takes three hours to complete. An oil painting requires materials costing \$30 and requires 25 hours. If the artist can spend up to \$400 each month for supplies and wishes to work no more than 200 hours monthly, how many of each kind should he paint if his only goal is to maximize his profit?

14. A city plans to build parks of two types. The first type costs \$100,000 and has five baseball fields and three tennis courts; a park of the second type costs \$120,000 and has two baseball fields and eight tennis courts. If the city wishes to have at least 16 baseball fields and at least 30 tennis courts, what is the least amount the city can spend to achieve its goal?

15. An animal sanctuary is to be stocked with two species of animals. Each individual of the first species needs three units of plant type A and five units of type B daily; each member of the second species needs six units of A and three units of B. There is enough growth in the sanctuary to provide daily up to 900 units of A and 1080 units of B. What is the maximum number of animals combined, from the two species, that can survive under the above food restrictions?
16. A dietician working for the manufacturer of a popular beer has been asked to determine how much protein an adult male can obtain daily from drinking beer. She makes the following assumptions: the typical consumer can drink either regular or light beer; a can of regular beer provides two grams of protein and 300 calories; a can of light beer provides one gram of protein and 200 calories; a person's daily caloric intake from beer should not exceed 1800 calories; no one should consume more than eight cans of beer daily. What is the maximum amount of protein our typical male can derive daily from beer subject to the constraints assumed by the dietician?
17. A manufacturer of restaurant tables offers two types, standard and small booth-size. Each model requires one unit of wood per table; also, the standard table needs two units of metal, the booth-size only one. It takes four man-hours to manufacture a standard table, only one hour for a booth-size table. Because of limited storage space at least 19 units of wood and 24 of metal must be used daily. Furthermore, at most 40 hours of labor are available daily. How many tables of each type should be made daily for maximum profit if the profit is \$12 on the standard model and \$8 on the booth-size?

TEST A

1. Maximize and minimize $f(x,y) = x + 2y$ subject to

$$\begin{aligned} x \geqslant 0, \quad y &\geqslant 0 \\ x + y &\leqslant 4 \\ 3x + 5y &\leqslant 15 \end{aligned}$$

2. Find the minimum value of $f(x,y) = 4x + 5y$ subject to

$$\begin{aligned} x \geqslant 0, \quad y &\geqslant 0 \\ 6x + 2y &\geqslant 12 \\ 3x + 4y &\geqslant 15 \end{aligned}$$

3. A factory produces jeeps that require five hours in the assembly shop and three hours in the finishing shop, and pick-up trucks that require seven hours in the assembly and four hours in finishing. Each week 350 hours are available in the assembly shop and 204 in the finishing shop. If the profit per jeep is \$410 and per pick-up truck \$560, how many of each should be produced weekly to maximize profit?

TEST B

1. Maximize and minimize $f(x,y) = 2x + 3y$ subject to

$$x \geqslant 0, \quad y \geqslant 0$$
$$x + y \geqslant 7$$
$$10x + 7y \leqslant 70$$

2. Find the maximum value of $f(x,y) = x + y$ subject to

$$x \geqslant 0, \quad y \geqslant 0$$
$$x + 5y \geqslant 5$$
$$3x + 2y \geqslant 6$$

3. A dealer packages one-pound cans of cashews and one-pound cans of mixed nuts containing equal amounts of peanuts and cashews. Each can of cashews sells for \$5, each can of mixed nuts for \$3. If the dealer has 1250 pounds of cashews and 1000 pounds of peanuts, how many cans of each type should he pack to maximize revenue?

SECTION 27. SOLVING MAXIMIZATION PROBLEMS BY THE SIMPLEX METHOD

The geometric solution of an LP problem, described in §26, is straightforward and simple--as long as only two variables are involved. Even with as few as three variables the graphical technique is impractical. The simplex method is an algebraic *algorithm*--that is, a step-by-step procedure--that can be applied to LP problems involving any number of variables and constraints. The method was developed after World War II by the mathematician George B. Dantzig. As noted earlier, its adaptability to computers renders it of great value.

In this section we show how to set up and solve LP maximization problems by the simplex method. Setting up the problem is the important human input required even when computer programs are available to carry out the arithmetic manipulations. We will provide enough examples and exercises here to enable you to develop facility with this technique. And, when practical, we will compare solutions obtained by the simplex method with those obtained geometrically. Later in the section we will discuss briefly why the simplex method works; the explanation is an optional topic.

The individual steps of the method are all easy and straightforward; the method seems complicated only because there are quite a number of steps. Here is an outline of the steps, which can serve to keep you oriented as you go through the detailed explanation that follows:

I. Putting the problem into standard form.

II. Preparing to construct a simplex tableau:

- IIA. Converting inequalities to equations; slack variables;
- IIB. Standardizing the objective function;

III. Constructing the initial simplex tableau;

IV. Obtaining a new tableau: the pivot operation (may be iterated);

V. Reading a solution from a tableau.

I. THE STANDARD FORM

An LP problem is said to be in *standard form* if it has the following characteristics:

(1) The objective function is to be maximized;
(2) The variables are all nonnegative;
(3) All other constraints are of the form

$$a_1x_1 + a_2x_2 + \dots + a_nx_n = b$$

where the constant b is also nonnegative.* The sense of the inequality ("$\leqslant$", never "$\geqslant$") is a necessary characteristic of a problem in standard form.

In the rest of this chapter (§§27 and 28), for convenience, the term "constraint" will refer *only to the constraints covered by point* (3) *above*.

The following LP problem is in standard form: Maximize the objective function $f(x,y,z) = 2x + y + 3z$ subject to $x,y,z \geqslant 0$ and to the constraints

$$\begin{aligned} 2x + 3y - 3z &\leqslant 6 \\ 4x - y + z &\leqslant 2 \end{aligned}$$

Example 27-1

Convert each of the following to standard form:

(a) Maximize $f(x,y) = 3x + y$ subject to

$$\begin{aligned} x \geqslant 0, \quad y &\leqslant 0 \\ x + y &\leqslant 4 \\ 3x - 2y &\leqslant 2 \end{aligned}$$

(b) Maximize $f(x,y) = 2x + 5y$ subject to

$$\begin{aligned} x,y &\geqslant 0 \\ 2x - 3y &\geqslant -2 \\ x + y &\leqslant 4 \end{aligned}$$

For (a), let $z = -y$. Then we have: Maximize $f = 3x - z$ subject to

$$\begin{aligned} x,z &\geqslant 0 \\ x - z &\leqslant 4 \\ 3x + 2z &\leqslant 2 \end{aligned}$$

For (b), to obtain the standard form we merely multiply the aberrant inequality by -1, getting: Maximize $f(x,y) = 2x + 3y$ subject to

$$\begin{aligned} x,y &\geqslant 0 \\ -2x + 3y &\leqslant 2 \\ x + y &\leqslant 4 \end{aligned}$$

II. PREPARING FOR THE SIMPLEX TABLEAU

In order to apply the simplex method to an LP problem we must

*The simplex method can be extended to cover cases in which b in (3) is negative, but we do not consider such cases in this text.

represent the problem by a type of matrix called a *simplex tableau.* Two modifications must be made in any LP problem before we can represent it in this form.

IIA. First Modification

We must first transform the inequalities given in the constraints into equations. We achieve this by introducing *slack variables.*

For example, the inequality

$$4x + y \leqslant 12$$

implies that there is a number s that is either positive or zero and is such that

$$4x + y + s = 12$$

In this equation s is the difference (if any) between $4x + y$ and 12; in other words, s "takes up the slack" between $4x + y$ and 12. We must introduce a different nonnegative slack variable for each constraint.

Example 27-2

Each system with inequalities on the left below is equivalent to the system with equations and slack variables on the right:

(a)
$$\begin{aligned} &x,y \geqslant 0 \\ &x + 2y \leqslant 3 \\ &5x - y \leqslant 18 \end{aligned} \qquad \begin{aligned} &x,y,s,t \geqslant 0 \\ &x + 2y + s = 3 \\ &5x - y + t = 18 \end{aligned}$$

(b)
$$\begin{aligned} &x,y,z \geqslant 0 \\ &3x - y + z \leqslant 10 \\ &x + y \leqslant 8 \end{aligned} \qquad \begin{aligned} &x,y,z,s,t \geqslant 0 \\ &3x - y + z + s = 10 \\ &x + y + t = 8 \end{aligned}$$

(c)
$$\begin{aligned} &x,y \geqslant 0 \\ &x - 2y \leqslant 15 \\ &3x + 4y \leqslant 19 \\ &5x - y \leqslant 2 \\ &4y \leqslant 9 \end{aligned} \qquad \begin{aligned} &x,y,s,t,u,v \geqslant 0 \\ &x - 2y + s = 15 \\ &3x + 4y + t = 19 \\ &5x - y + u = 2 \\ &4y + v = 9 \end{aligned}$$

IIB. Second Modification

The second modification required in preparing an LP problem for the construction of a simplex tableau is to rewrite the equation for the objective function so that the right side equals zero. For example:

$$f(x,y) = 2x + 3y \quad \text{becomes} \quad -2x - 3y + f = 0$$
$$f(x,y,z) = x + 2y - z \quad \text{becomes} \quad -x - 2y + z + f = 0$$

III. THE INITIAL TABLEAU

We are now ready to display the information in an LP problem, modified as indicated above, in matrix form--to construct a simplex tableau. The following three examples show how the *initial* simplex tableau is assembled.

Example 27-3

To maximize $f(x,y) = 2x + 3y$, subject to $x,y \geqslant 0$ and to

$$\begin{aligned} 3x + 4y &\leqslant 5 \\ 4x - y &\leqslant 8 \end{aligned}$$

we rewrite as shown below on the left; the initial tableau is on the right (the shading will be explained in a moment):

$$\begin{aligned} &x,y,s,t \geqslant 0 \\ &3x + 4y + s = 5 \\ &4x - y + t = 8 \\ &-2x - 3y + f = 0 \end{aligned}$$

x	y	s	t	f	
3	4	1	0	0	5
4	-1	0	1	0	8
-2	-3	0	0	1	0

Example 27-4

To maximize $f(x,y,z) = 4x - y + z$ subject to $x,y,z \geqslant 0$ and to

$$\begin{aligned} 4x + y &\leqslant 7 \\ x + 2z &\leqslant 10 \end{aligned}$$

We rewrite as shown on the left and extract the initial tableau as shown on the right:

$$\begin{aligned} &x,y,z,s,t \geqslant 0 \\ &4x + y + s = 7 \\ &x + 2z + t = 10 \\ &-4x + y - z + f = 0 \end{aligned}$$

x	y	z	s	t	f	
4	1	0	1	0	0	7
1	0	2	0	1	0	10
-4	1	-1	0	0	1	0

Example 27-5

To maximize $f(x,y) = 3x + y$, subject to $x,y \geqslant 0$ and to

$$\begin{aligned} 2x - y &\leqslant 4 \\ 6x + y &\leqslant 2 \\ 3x - 5y &\leqslant 9 \\ 4y &\leqslant 10 \end{aligned}$$

We rewrite and extract thus:

$$\begin{aligned} &x,y,s,t,u,v \geqslant 0 \\ &2x - y + s = 4 \\ &6x + y + t = 2 \\ &3x - 5y + u = 9 \\ &4y + v = 10 \\ &-3x - y + f = 0 \end{aligned}$$

x	y	s	t	u	v	f	
2	-1	1	0	0	0	0	4
6	1	0	1	0	0	0	2
3	-5	0	0	1	0	0	9
0	4	0	0	0	1	0	10
-3	-1	0	0	0	0	1	0

NOTES ON THE INITIAL TABLEAU

In assembling the initial tableau,

(1) we list all the variables, including the slack variables and the objective function f to be maximized, across the top;
(2) we draw a vertical line to separate the coefficients of the variables from the rightmost column of constants;
(3) we draw a horizontal line to separate the rows derived from the constraints from the row (the bottom one) derived from the objective function.

We now note that

(4) all the numbers in the rightmost column are nonnegative;
(5) when there are n slack variables, the tableau contains an identity matrix I_{n+1} (shaded in the above examples).

IV. OBTAINING A NEW TABLEAU: THE PIVOT OPERATION

We now come to the heart of the method.

The simplex method is closely related to the geometric approach described in the preceding section. With the graphical solution, we test all the corner points of the feasible region to find the one at which the objective function attains its maximum; with the simplex method, we test only a select few of these corner points. We start always at the origin and travel from one corner point to another along a path (a series of line segments) along which the objective function increases most dramatically. We stop when the objective function cannot be further increased--that is, when it has achieved its maximum value.

If an LP problem involves only two variables, the path-tracing is easy and yields, of course, the same result as that obtained graphically. With three variables, the path traced is along the edges of a three-dimensional figure. The figure is constructible, and working a problem out graphically with the aid of such a figure is very informative, but usually highly impractical.

In the simplex method, the act of moving from one corner point to another takes the algebraic form of converting one tableau into another. The initial tableau corresponds to the origin, at which the value of the objective function (given by the entry in the lower right-hand corner of the tableau) is zero. To obtain a new tableau we engage in a *pivot operation*.

In describing the pivot operation we illustrate using the following LP problem:*

---

*For the moment we ignore certain special situations that may arise with some problems, requiring special handling; these are listed and dealt with later in this section.

Example 27-6

Maximize $f(x,y) = 4x + 3y$ subject to $x,y \geqslant 0$, $x + y \leqslant 4$, $3x + y \leqslant 6$. The equivalent system of equations with slack variables is at the left below, and the corresponding initial tableau is at the right:

$$\begin{aligned} &x,y,s,t \geqslant 0 \\ &x + y + s = 4 \\ &3x + y + t = 6 \\ &-4x - 3y + f = 0 \end{aligned}$$

INITIAL TABLEAU

x	y	s	t	f		
1	1	1	0	0	4	
3	1	0	1	0	6	
-4	-3	0	0	1	0	($\leftarrow f$)

To *pivot* is to use Gauss-Jordan row operations Ma2 and Ma3 (pp. 190-197) to obtain a column with one element equal to 1 and all others equal to 0.

INSTRUCTIONS	EXAMPLE
(1) Find the smallest negative number in the bottom row. The column containing this element is the *pivot column*.	The smallest negative element in the bottom row of the example is -4. The first column (headed by x) is therefore the pivot column.
(2) Divide each *positive* element in the pivot column *into* the corresponding element in the rightmost column. The row in which this yields the smallest quotient is the *pivot row*. The pivot column and pivot row intersect in the *pivot element*.	In column one, both 1 and 3 are positive. In row one $4 \div 1 = 4$; in row two, $6 \div 3 = 2$. Since row two yields the smaller quotient, it is the pivot row. The pivot element is in row two, column one. We shade it below:

x	y	s	t	f	
1	1	1	0	0	4
3	1	0	1	0	6
-4	-3	0	0	1	0

(3) Divide each element in the pivot row by the pivot element (Ma2). The pivot element is now equal to 1.

x	y	s	t	f	
1	1	1	0	0	4
1	$\frac{1}{3}$	0	$\frac{1}{3}$	0	2
-4	-3	0	0	1	0

(4) Apply Ma3 in turn to each other row, adding to it an appropriate multiple of the pivot row, to obtain 0 in every other position in the pivot column.

Multiply row two by -1 and add to row one:

x	y	s	t	f	
0	$\frac{2}{3}$	1	$-\frac{1}{3}$	0	2
1	$\frac{1}{3}$	0	$\frac{1}{3}$	0	2
-4	-3	0	0	1	0

(4) *(continued)*

Multiply row two by 4 and add to row three:

x	y	s	t	f			
0	$\frac{2}{3}$	1	$-\frac{1}{3}$	0	2		SECOND
1	$\frac{1}{3}$	0	$\frac{1}{3}$	0	2	$(\leftarrow x)$	TABLEAU
0	$-\frac{5}{3}$	0	$\frac{4}{3}$	1	8	$(\leftarrow f)$	

The tableau obtained by steps (1) through (4) is the desired replacement for the initial tableau; it corresponds to a new corner point. In our example, the new tableau corresponds to the corner point $x = 2$, $y = 0$, at which point the value of f is 8 (we will explain shortly how this is read from the display).

But whether the new tableau is the *final* simplex tableau, yielding the desired maximum value of f, depends on the bottom row. If any element in the bottom row is negative, then steps (1) through (4) must be repeated.

In our example the new tableau still contains a negative element in the bottom row; namely, $-\frac{5}{3}$. We must therefore repeat the pivot procedure:

(1) The pivot column is column two.

(2) $2 \div \frac{1}{3} = 6$; $2 \div \frac{2}{3} = 3$. The quotient is smaller in row one, so that is the pivot row. The pivot element is in row one, column two:

x	y	s	t	f	
0	$\frac{2}{3}$	1	$-\frac{1}{3}$	0	2
1	$\frac{1}{3}$	0	$\frac{1}{3}$	0	2
0	$-\frac{5}{3}$	0	$\frac{4}{3}$	1	8

(3) Divide each element in row one by $\frac{2}{3}$:

x	y	s	t	f	
0	1	$\frac{3}{2}$	$-\frac{1}{2}$	0	3
1	$\frac{1}{3}$	0	$\frac{1}{3}$	0	2
0	$-\frac{5}{3}$	0	$\frac{4}{3}$	1	8

(4) Multiply row one by $-\frac{1}{3}$ and add to row two:

x	y	s	t	f	
0	1	$\frac{3}{2}$	$-\frac{1}{2}$	0	3
1	0	$-\frac{1}{2}$	$\frac{1}{2}$	0	1
0	$-\frac{5}{3}$	0	$\frac{4}{3}$	1	8

Multiply row one by $\frac{5}{3}$ and add to row three:

x	y	s	t	f			
0	1	$\frac{3}{2}$	$-\frac{1}{2}$	0	3	$(\leftarrow y)$	THIRD
1	0	$-\frac{1}{2}$	$\frac{1}{2}$	0	1	$(\leftarrow x)$	TABLEAU
0	0	$\frac{5}{2}$	$\frac{1}{2}$	1	13	$(\leftarrow f)$	

We now have the third tableau It yields the corner point $x = 1$, $y = 3$ and the value 13 for the objective function f. Since there are no negative elements left in the bottom row, the simplex routine ends. The tableau just obtained is the *final* tableau. The maximum value of f is 13 and it is achieved as $x = 1$, $y = 3$.

READING A SOLUTION FROM A TABLEAU

As noted above each tableau corresponds to a corner point. The initial tableau corresponds to the origin, with all variables equal to zero, and at that point the objective function f is also zero.

As each new tableau is obtained, it yields a new solution (a new set of values for the original variables) and a corresponding value for the objective function, as follows:

(1) If a column headed by one of the original variables (x and y in our example) contains a single 1, with all other entries 0, then the entry in the rightmost column in the same row as that 1 is the value of that variable.

(2) Any original variable whose column does not satisfy that test is assigned the value zero.

(3) The value of the objective function corresponding to the solution thus obtained for the original variables appears in the lower right-hand corner of the tableau.

These instructions should be verified for the tableaus in example 27-6.

ADDITIONAL COMMENTS ON THE PIVOT OPERATION

In the course of application of the pivot operation certain situations can arise which were not provided for in our earlier presentation. The following comments tell how some of these are to be handled.

COMMENT 1. If in step (1) of the procedure (page 260) the smallest negative entry in the bottom row appears twice, we arbitrarily choose either column as the pivot column. For example, for the tableau

x	y	s	t	f	
1	-1	1	0	0	3
-2	1	0	1	0	2
-1	-1	0	0	1	0

we can select either the first or the second column as pivot column.

COMMENT 2. If in step (2) of the pivot procedure we obtain two equal quotients, then we can use either row yielding the least ratio as pivot row. For example, for the tableau

x	y	s	t	f	
3	2	1	0	0	4
1	1	0	1	0	2
-2	-3	0	0	1	0

the second column is the pivot column; but since, in row one, $4 \div 2 = 2$ and, in row two, $2 \div 1 = 2$, we can choose either row one or row two as pivot row.

COMMENT 3. If, in a *final* tableau, two variables, say x and y, head identical columns each with a single 1 in the *same* row and with zeros elsewhere, then arbitrarily assign the value in the rightmost column to either variable, letting the other equal zero.

COMMENT 4. If no element in the pivot column is positive, then the pivot procedure is aborted after step (1). This situation tells us that there is no maximum value for the objective function subject to the given constraints. An example is

x	y	s	t	f	
-3	0	1	2	0	6
-2	1	0	1	0	2
-5	0	0	2	1	2

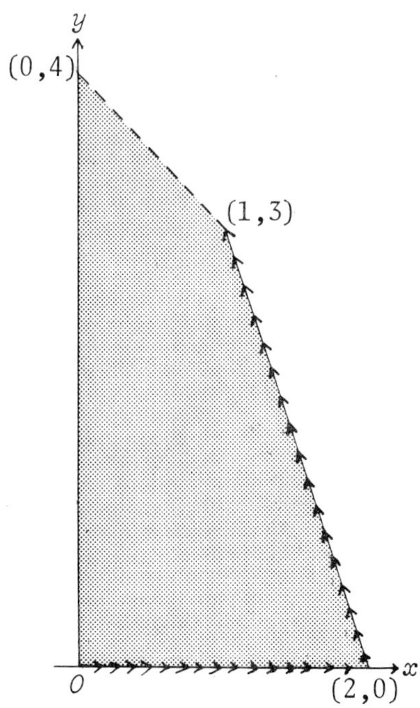

The first column is the pivot column, but since no element in it is positive the procedure grinds to a halt. The LP maximization problem leading to this tableau has no solution.

GEOMETRIC INTERPRETATION

The procedure described for example 27-6 can be interpreted geometrically in terms of the Figure. The initial tableau corresponds to corner point (0,0) of the feasible region (shaded). Proceeding from the initial tableau to the second tableau is equivalent to moving along the x-axis from (0,0) to corner point (2,0). Along *this boundary* of the feasible region, the corner point (2,0) yields the maximum attainable value of the objective function. Obtaining the third tableau from the second is equivalent to moving from corner point (2,0) to corner point (1,3) along the boundary of the feasible region that contains those two points. The latter point yields the maximum value of the objective function that is attainable along *this* boundary. The fact that the third tableau is the final tableau (with no negative elements in the bottom row) shows that f attains its maximum value in the whole feasible region at (1,3). The path-tracing therefore stops at this point.

HOW THE SIMPLEX METHOD WORKS*

We now take still another look at example 27-6, this time with the goal of understanding why we do what we do--the goal of explaining the steps in the simplex method.

To maximize $f(x,y) = 4x + 3y$, subject to the constraints $x,y \geqslant 0$, $x + y \leqslant 4$, $3x + y \leqslant 6$, we first obtained the system of linear equations at the left below (where s and t are also nonnegative), then constructed the initial tableau shown at the right:

$$\begin{array}{lr} \text{(Eq. 1)} & x + y + s = 4 \\ \text{(Eq. 2)} & 3x + y + t = 6 \\ \text{(Eq. 3)} & -4x - 3y + f = 0 \end{array}$$

x	y	s	t	f		
1	1	1	0	0	4	INITIAL
3	1	0	1	0	6	TABLEAU
-4	-3	0	0	0	0	

We want the solution of the linear system (x,y,s,t)--that is, the values of the variables x, y, s, and t--for which f is largest. In our final answer, we give the values for x and for y and the maximum value of f yielded by those values.

We begin by letting x and y both equal zero. From the linear system at the left above,

it follows from	that
Equation 1	$s = 4$
Equation 2	$t = 6$
Equation 3	$f = 0$

and these can be read from the initial tableau (rightmost column):

$$x = 0, \quad y = 0; \quad s = 4, \quad t = 6, \quad f = 0$$

Is f as large as possible? No, because the bottom row of this tableau contains a negative entry--in fact, two of them. To see that f can be increased, we rewrite it in the form

$$f = 4x + 3y$$

Since x and y are nonnegative, f is increased if either x or y is increased. We choose that variable which will increase f more--in this case, x (since its coefficient is greater than the coefficient of y). This choice is equivalent to step (1) of the pivot proce-

*Optional topic.

dure, in which we choose as pivot column the one in the tableau whose bottom-line element is the smallest *negative* one (recall that we chose column one, the x-column, because $-4 < -3$). For convenience we set y equal to zero.

Of course, during the process of increasing x we continue to be bound by the given constraints. If we let $y = 0$ in the top two equations of the linear system, then we have

$$x + s = 4 \quad \text{and} \quad 3x + t = 6$$

Since s and t are both nonnegative, it follows from these, respectively, that

$$x \leqslant \frac{4}{1} = 4 \quad \text{and} \quad x \leqslant \frac{6}{3} = 2$$

But *both* conditions must hold. We therefore conclude that $x \leqslant 2$. Thus the largest x can be is 2.

The reasoning used in reaching this conclusion is the basis of step (2) of the pivot procedure, in which we chose the *smaller* of the two quotients, $\frac{4}{1}$ (from row one) and $\frac{6}{3}$ (from row two), to determine the pivot row and thus, also, the pivot element. So the pivot element is in column one, row two.

In steps (3) and (4) we obtained the second tableau by Gauss-Jordan row operations. Since these operations on a linear system yield an equivalent linear system, any solution of the system given by the new tableau is also a solution of the original system.

x	p	s	t	f			
0	$\frac{2}{3}$	1	$-\frac{1}{2}$	0	2		SECOND
1	$\frac{1}{3}$	0	$\frac{1}{3}$	0	2	$(\leftarrow x)$	TABLEAU
0	$-\frac{5}{3}$	0	$\frac{4}{3}$	1	8	$(\leftarrow f)$	

Note that the modified (but equivalent) linear system given by the second tableau is

$$\begin{aligned} \tfrac{2}{3}y + s - \tfrac{1}{3}t \qquad &= 2 \\ x + \tfrac{1}{3}y \qquad + \tfrac{1}{3}t \qquad &= 2 \\ -\tfrac{5}{3}y \qquad + \tfrac{4}{3}t + f &= 8 \end{aligned}$$

The second of these equations shows that, with $x = 2$, both y and t equal zero. From the first equation it then follows that $s = 2$, and from the third we see that $f = 8$. We therefore have the solution

$$x = 2, \quad y = 0; \quad s = 2, \quad t = 0, \quad f = 8$$

Is f now as large as it can be? No, because there is still a negative element in the bottom row of the second tableau. From the bottom row, we get the equation

$$f = 8 + \tfrac{5}{3}y - \tfrac{4}{3}t$$

An increase in y will increase f. (Thus column two, the y-column, is our new pivot column.) We observe also that any increase in t will diminish f and that t currently is equal to zero. How large can y be? The second tableau shows us the following:

$$\text{from row one: } \tfrac{2}{3}y \leqslant 2, \quad y \leqslant \frac{2}{\frac{2}{3}} \quad \text{so} \quad y \leqslant 3$$

$$\text{from row two: } \tfrac{1}{3}y \leqslant 2, \quad y \leqslant \frac{2}{\frac{1}{3}} \quad \text{so} \quad y \leqslant 6$$

To satisfy both of these inequalities, we conclude that y cannot exceed 3. (The pivot element is thus in column two, row one.) Incidentally, from the second equation of the modified linear system we also see that any increase in y, from its current value of 0, would necessarily produce a decrease in x from its current value of 2.

We now pivot on the shaded element in the second tableau, to get

x	y	s	t	f		
0	1	$\frac{3}{2}$	$-\frac{1}{2}$	0	3	$(\leftarrow y)$
1	0	$-\frac{1}{2}$	$\frac{1}{2}$	0	1	$(\leftarrow x)$
0	0	$\frac{5}{2}$	$\frac{1}{2}$	1	13	$(\leftarrow f)$

THIRD (AND FINAL) TABLEAU

From row one, with $y = 3$, we see that both s and t are 0; from row two, since $s = t = 0$, we see that $x = 1$; from row three, since $s = t = 0$, it follows that $f = 13$.

Can we make f any bigger? There are no negative elements left in the bottom row. From the bottom row we can write the following equation for f:

$$f = 13 - \tfrac{5}{2}s - \tfrac{1}{2}t$$

To increase f we would have to decrease s or t. But both s and t are now equal to zero, and they cannot be negative. Therefore f is now a maximum; its maximum value is 13, which is achieved when $x = 1$ and $y = 3$.

FURTHER EXAMPLES

Example 27-7

Maximize $f(x,y,z) = 2x + 3y + z$ subject to the constraints $x,y,z \geqslant 0$, $2y + z \leqslant 11$, $-x + 7y \leqslant 28$, $5x - y \leqslant 30$.

The modified linear system of equations is

$$\begin{aligned} 2y + z + s \quad &= 11 \\ -x + 7y \quad + t \quad &= 28 \\ 5x - y \quad + u \quad &= 30 \\ -2x - 3y - z \quad + f &= 0 \end{aligned}$$

The initial tableau is

x	y	z	s	t	u	f		
0	2	1	1	0	0	0	11	
-1	7	0	0	1	0	0	28	
5	-1	0	0	0	1	0	30	
-2	-3	-1	0	0	0	1	0	($\leftarrow f$)

INITIAL TABLEAU

This tableau represents the solution or corner point (0,0,0), where the objective function f has value 0. We now use the simplex method to obtain a new tableau.

Since the smallest negative number in the bottom row is -3, the pivot column is column two, headed by y.

In row one, $11 \div 2 = 5\frac{1}{2}$; in row two, $28 \div 7 = 4$. The latter quotient is smaller, so the pivot element is 7, in row two, column two, shaded above.

Divide the pivot row by 7:

x	y	z	s	t	u	f	
0	2	1	1	0	0	0	11
$-\frac{1}{7}$	1	0	0	$\frac{1}{7}$	0	0	4
5	-1	0	0	0	1	0	30
-2	-3	-1	0	0	0	1	0

Add -2 times row two to row one; add row two to row three; add 3 times row two to the bottom row. This yields the second tableau:

x	y	z	s	t	u	f		
$\frac{2}{7}$	0	1	1	$-\frac{2}{7}$	0	0	3	
$-\frac{1}{7}$	1	0	0	$\frac{1}{7}$	0	0	4	($\leftarrow y$)
$\frac{34}{7}$	0	0	0	$\frac{1}{7}$	1	0	34	
$-\frac{17}{7}$	0	-1	0	$\frac{3}{7}$	0	1	12	($\leftarrow f$)

SECOND TABLEAU

This tableau corresponds to the solution (0,4,0), where the value of the objective function is 12.

The smallest negative entry in the bottom row is $-\frac{17}{7}$; so the pivot column is the first, headed by x.

In row one, $3 \div \frac{2}{7} = 10.5$; in row three, $34 \div \frac{34}{7} = 7$. So the third row is the pivot row. We pivot on $\frac{34}{7}$, shaded above.

Multiply the pivot row by $\frac{7}{34}$:

x	y	z	s	t	u	f	
$\frac{2}{7}$	0	1	1	$-\frac{2}{7}$	0	0	3
$-\frac{1}{7}$	1	0	0	$\frac{1}{7}$	0	0	4
1	0	0	0	$\frac{1}{34}$	$\frac{7}{34}$	0	7
$-\frac{17}{7}$	0	-1	0	$\frac{3}{7}$	0	1	12

Add $-\frac{2}{7}$ times the pivot row to row one; add $\frac{1}{7}$ times the pivot row to row two; and add $\frac{17}{7}$ times the pivot row to the bottom row. We thus get the third tableau:

x	y	z	s	t	u	f			
0	0	1	1	$-\frac{5}{17}$	$-\frac{1}{17}$	0	1		THIRD TABLEAU
0	1	0	0	$\frac{5}{34}$	$\frac{1}{34}$	0	5	$(\leftarrow y)$	
1	0	0	0	$\frac{1}{34}$	$\frac{7}{34}$	0	7	$(\leftarrow x)$	
0	0	-1	0	$\frac{1}{2}$	$\frac{1}{2}$	1	29	$(\leftarrow f)$	

This tableau yields the solution (7,5,0) for (x,y,z) and at that point f has the value 29.

Since there is still a negative number in the bottom row of the tableau, we must again go through the pivot procedure. The pivot column is now the third, headed by z; the pivot row is the first (the only row that has a positive element in column three).

We now pivot on 1, shaded above.

Since the pivot element is already 1 and there are already zeros in rows two and three, we have only to get a zero in the bottom row of column three. Just add the pivot row to the bottom row and we will have the fourth tableau:

x	y	z	s	t	u	f			
0	0	1	1	$-\frac{5}{17}$	$-\frac{1}{17}$	0	1	$(\leftarrow z)$	FOURTH (FINAL) TABLEAU
0	1	0	0	$\frac{5}{34}$	$\frac{1}{34}$	0	5	$(\leftarrow y)$	
1	0	0	0	$\frac{1}{34}$	$\frac{7}{34}$	0	7	$(\leftarrow x)$	
0	0	0	1	$\frac{7}{34}$	$\frac{15}{34}$	1	30	$(\leftarrow f)$	

This is the final tableau since there are no negative numbers in the bottom row. The objective function is maximized at $x = 7$, $y = 5$, $z = 1$, and has the value 30 at that point.

GEOMETRIC INTERPRETATION. The geometric path-tracing described earlier for example 27-6 is applicable to example 27-7 too, as shown in the figure at the top of the facing page (p. 269). Since three variables are involved, the path traced is along the edges of a (three-dimensional) solid.

Example 27-8

Using the simplex method, solve the TV-manufacturing problem of example 26-7 (page 248).

If we let f be the profit, then we want to maximize $f = 50x + 150y$, where x and y are the number respectively of black-and-white and of color sets. The constraints are $3x + 10y \leqslant 300$, $2x + 3y \leqslant 126$, and

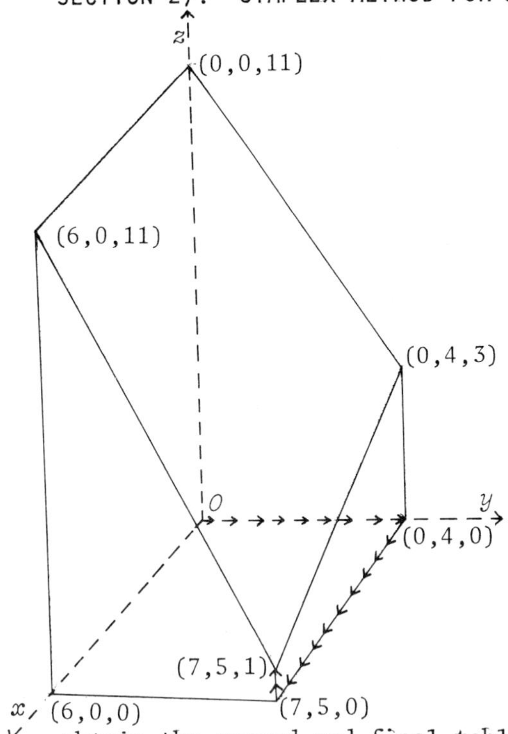

$x + 2y \leqslant 68$. The relevant system of linear equations is

$$\begin{aligned} 3x + 10y + s \qquad\qquad &= 300 \\ 2x + 3y \quad + t \qquad &= 126 \\ x + 2y \qquad + u \quad &= 68 \\ -50x - 150y \qquad\qquad + f &= 0 \end{aligned}$$

The initial tableau, using the simplex method, is

x	y	s	t	u	f	
3	**10**	1	0	0	0	300
2	3	0	1	0	0	126
1	2	0	0	1	0	68
-50	-150	0	0	0	1	0

Calculations to find the pivot element:

ROW 1: $300 \div 10 = 30$
ROW 2: $126 \div 3 = 42$
ROW 3: $68 \div 2 = 34$

As before we have shaded the pivot element. Follow the step-by-step procedure outline on page 260 to obtain the second and final tableaus given below:

SECOND TABLEAU

x	y	s	t	u	f	
$\frac{3}{10}$	1	$\frac{1}{10}$	0	0	0	30
$\frac{11}{10}$	0	$-\frac{3}{10}$	1	0	0	36
$\frac{2}{5}$	0	$-\frac{2}{10}$	0	1	0	8
-5	0	15	0	0	1	4500

$30 \div \frac{3}{10} = 100$

$36 \div \frac{11}{10} = \frac{360}{11} \approx 32.7$

$8 \div \frac{2}{5} = 20$ *(minimum)*

FINAL TABLEAU

x	y	s	t	u	f		
0	1	$\frac{1}{4}$	0	$-\frac{3}{4}$	0	24	$(\leftarrow y)$
0	0	$\frac{1}{4}$	1	$-\frac{11}{4}$	0	14	
1	0	$-\frac{1}{2}$	0	$\frac{5}{2}$	0	20	$(\leftarrow x)$
0	0	$\frac{25}{2}$	0	$\frac{25}{2}$	1	4600	$(\leftarrow f)$

From the final tableau we see that f is maximized at $x = 20$, $y = 24$, which corresponds to a maximum profit of \$4600 at a production level of 20 black-and-white sets and 24 color sets. This agrees with the solution obtained geometrically in example 26-7.

SUMMARY OF THE SIMPLEX METHOD FOR MAXIMIZATION

Given an LP maximization problem, the steps are:

put the problem into standard form (p. 255);

convert the constraints into equations with slack variables (p. 257);

standardize the objective function (p. 257);

find the initial tableau (p. 258);

find the pivot element; pivot to obtain a new tableau (pp. 260-261);

repeat the preceding step as necessary until a final tableau is obtained;

read the answer from the final tableau (p. 262).

EXERCISES

1. Convert each of the following into standard form, if necessary:
   (a) Maximize $f(x,y) = 2x + 3y$ subject to $x \leqslant 0$, $y \geqslant 0$, $3x - y \leqslant 4$, $5x + 2y \leqslant 8$.
   (b) Maximize $f(x,y) = 8x - y$ subject to $x \geqslant 0$, $y \geqslant 0$, $9x + y \leqslant 2$, $4x - y \leqslant 1$.
   (c) Maximize $f(x,y) = x + y$ subject to $x \geqslant 0$, $y \geqslant 0$, $3x - y \geqslant -4$, $x + y \leqslant 4$.
2. Rewrite each of the following use slack variables:
   (a) $x,y \geqslant 0$, $5x - y \leqslant 3$, $x + 2y \leqslant 7$.
   (b) $x,y,z \geqslant 0$, $x + 3y - z \leqslant 4$, $2x - y + z \leqslant 5$.
   (c) $x,y \geqslant 0$, $4x + y \leqslant 1$, $x - 5y \leqslant 2$, $3x + 2y \leqslant 5$, $-x + 4y \leqslant 7$.

NOTE. In the remaining exercises and tests of this section it is to be understood, without further mention, that the original variables (x, y, z, etc.) and the slack variables (s, t, u, etc.) are always nonnegative.

3. Construct the initial tableau for each of the following LP problems:
   (a) Maximize $f(x,y) = 2x - y$ subject to $4x + y \leqslant 2$, $3x - 9y \leqslant 5$.
   (b) Maximize $f(x,y,z) = -x + y - 2z$ subject to $2x - z \leqslant 11$, $4y + 2z \leqslant 7$.
   (c) Maximize $f(x,y) = x + 7y$ subject to $x + 3y \leqslant 5$, $2x - y \leqslant 6$, $4x + y \leqslant 7$, $-x + 2y \leqslant 8$.

Use the simplex method to solve the LP problems in 4 through 16:

4. Maximize $f(x,y) = 4x + 3y$ subject to $3x + 2y \leqslant 12$, $x + y \leqslant 5$.
5. Maximize $f(x,y) = x + 2y$ subject to $3x + 2y \leqslant 12$, $x + y \leqslant 5$.
6. Maximize $f(x,y) = 2x + y$ subject to $3x + 2y \leqslant 12$, $x + y \leqslant 5$.
7. Maximize $f(x,y) = 4x + 5y$ subject to $2x + 3y \leqslant 30$, $x + y \leqslant 11$, $5x + 3y \leqslant 45$.

8. Maximize $f(x,y) = 4x + 3y$ subject to $2x + 3y \leqslant 30$, $x + y \leqslant 11$, $5x + 3y \leqslant 45$.
9. Maximize $f(x,y,z) = x + 2y + z$ subject to $2x + y + 3z \leqslant 3$, $x + 2y + z \leqslant 10$.
10. Maximize $f(x,y,z) = 4x + 3y + z$ subject to $2x + y + 3z \leqslant 3$, $x + 2y + z \leqslant 10$.
11. Maximize $f(x,y,z) = x + 3y + 4z$ subject to $x + 5y \leqslant 20$, $2x + 3z \leqslant 10$, $4y + z \leqslant 3$.
12. Maximize $f(x,y,z) = -x + 2y + 3z$ subject to $4x + z \leqslant 8$, $-x + 2y \leqslant 7$, $x - y + z \leqslant 10$.
13. Maximize $f(x,y) = x + y$ subject to $x + 2y \leqslant 30$, $6x + 5y \leqslant 91$, $5x + 2y \leqslant 65$, $3x + 4y \leqslant 62$.
14. Maximize $f(x,y) = 3x + 2y$ subject to $x + y \leqslant 15$, $6x + y \leqslant 24$, $3x + y \leqslant 17$, $5x + y \leqslant 21$.
15. Maximize $f(x,y,z,w) = x + 2y + z + w$ subject to $4x - y + 2z \leqslant 2$, $x + 3y + z + 2w \leqslant 3$.
16. Maximize $f(x,y,z,w) = x + y + z + 2w$ subject to $x - z \leqslant 10$, $2y + 3z \leqslant 12$, $4x + y + w \leqslant 8$.
17. Solve exercise 13, page 252, by the simplex method.
18. Use the simplex method to solve the problem on stocking a lake with fish, in example 26-9, page 250.

TEST A

1. Find the initial tableau for this problem: maximize $f(x,y,z) = 3x - 5y + z$ subject to $4x + y \leqslant 5$, $3x - z \leqslant 8$, $2y + z \leqslant 9$.
2. Maximize $f(x,y) = 3x + 2y$ subject to $5x + 2y \leqslant 20$, $5x + y \leqslant 15$.
3. Maximize $f(x,y,z,w) = x + y + z + w$ subhect to $2x + y + 3z + 2w \leqslant 7$, $3x + 4y + z + 5w \leqslant 2$, $x + 2y + 4z + 3w \leqslant 9$.
4. A dressmaker has 20 sq yd of polyester, 18 sq yd of silk, and 15 sq yd of cotton. He makes dresses and fancy long gowns. A dress calls for 2 sq yd of polyester, 1 of silk, and 1 of cotton. A fancy gown needs 2 sq yd of polyester, 3 of silk, and 2 of cotton. A dress sells for \$50, a gown for \$80. How many of each should be made for maximum revenue? (Assume all made are sold.)

TEST B

1. Find the initial tableau for the following: maximize $f(x,y) = 3x - y$ subject to $5x - y \leqslant 1$, $3x + y \leqslant 2$, $-x + y \leqslant 7$.
2. Using the simplex method, maximize $f(x,y) = 2x + y$ subject to $x + 2y \leqslant 10$, $3x + y \leqslant 12$, $x + y \leqslant 6$.
3. Using the simplex method, maximize $f(x,y,z) = x + 2y + 3y$ subject to $2x + z \leqslant 5$, $y - z \leqslant 6$, $x + 2y \leqslant 7$.
4. Use the simplex method to solve exercise 15 on page 253.

SECTION 28. SOLVING MINIMIZATION PROBLEMS BY THE SIMPLEX METHOD

To solve an LP minimization problem we construct a maximization problem that corresponds to it, called its *dual*. We solve the latter by the methods of the preceding section. The solution of the original minimization problem can then be read off directly from the final tableau obtained for the dual maximization problem.

Every LP problem is actually a pair of problems, each the *dual* of the other. According to a fundamental theorem, if one of such a pair of LP problems has a solution, then so does the other. Furthermore, the maximum value of the objective function in an LP maximization problem is equal to the minimum value of the objective function of the associated dual minimization problem.

In this section we describe how to construct the dual of an LP problem, then how to use that in solving minimization problems.

DEFINITION OF MINIMIZATION PROBLEM

A *minimization* problem is an LP problem in which

(1) The objective function is to be minimized;

(2) The constraints are all of the form

$$c_1w_1 + c_2w_2 + \dots + c_nw_n \geqslant d$$

where each of the variables w_i and the constant d are all nonnegative. The sense of the inequality in each constraint is "$\geqslant$".

The following is a typical minimization problem: Minimize $g(p,q,r) = 3p - 2q + 4r$ subject to

$$p \geqslant 0,\ q \geqslant 0,\ r \geqslant 0 \qquad \text{(i)}$$
$$4p + 5q - 9r \geqslant 4 \qquad \text{(ii)}$$
$$p - q + r \geqslant 5 \qquad \text{(iii)}$$

NOTE. Recall, from §27 (p. 256) that we have limited the word "constraint" to refer only to those of the sort shown in lines (ii) and (iii) above. In maximization problems these constraints are inequalities in which the sense is "$\leqslant$"; in minimization problems the sense is "$\geqslant$". The requirement specified in line (i), that all variables be nonnegative, holds for minimization and maximization problems alike.

THE TRANSPOSE OF A MATRIX

To apply the procedure about to be described for finding the dual of an LP problem, we must know what is meant by the *transpose* of a matrix. Given any matrix A, its transpose A^T is obtained merely by interchanging rows and columns. That is, the rows of A^T are the columns of A, and the columns of A^T are the rows of A:

if A is	then A^T is
$\begin{pmatrix} 1 & 2 & 3 \\ 4 & 5 & 6 \end{pmatrix}$	$\begin{pmatrix} 1 & 4 \\ 2 & 5 \\ 3 & 6 \end{pmatrix}$
$\begin{pmatrix} 1 & 4 \\ 2 & 5 \\ 3 & 6 \end{pmatrix}$	$\begin{pmatrix} 1 & 2 & 3 \\ 4 & 5 & 6 \end{pmatrix}$
$\begin{pmatrix} 1 & 2 & 3 \\ 4 & 5 & 6 \\ 7 & 8 & 9 \end{pmatrix}$	$\begin{pmatrix} 1 & 4 & 7 \\ 2 & 5 & 8 \\ 3 & 6 & 9 \end{pmatrix}$
$\begin{pmatrix} 1 & 0 & 0 \\ 0 & 5 & 0 \\ 0 & 0 & 9 \end{pmatrix}$	$\begin{pmatrix} 1 & 0 & 0 \\ 0 & 5 & 0 \\ 0 & 0 & 9 \end{pmatrix}$

We see that

(1) If A is $m \times n$ then its transpose A^T is $n \times m$.
(2) If $B = A^T$, then $A = B^T$: the transpose of the transpose of A is A itself.
(3) The transpose of a square matrix is square.
(4) If A is a diagonal matrix, then $A^T = A$: a diagonal matrix is its own transpose.

FINDING THE DUAL PROBLEM

To find the dual of a given minimization problem,

(1) construct the matrix for the given problem;
(2) find the transpose of the matrix;
(3) write the maximization problem described by the transposed matrix.

Example 28-1

Given the following minimization problem, find its dual: Minimize $g(p,q) = 2p + 3q$ subject to

$$5p + q \geqslant 7$$
$$4p - q \geqslant 6$$

Following the steps described above, we have:

Step (1):
$$\begin{array}{cc|c} 5 & 1 & 7 \\ 4 & -1 & 6 \\ \hline 2 & 3 & 0 \end{array}$$

Step (2):
$$\begin{array}{cc|c} 5 & 4 & 2 \\ 1 & -1 & 3 \\ \hline 7 & 6 & 0 \end{array}$$

Step (3): Maximize $f(x,y) = 7x + 6y$ subject to

$$\begin{aligned} 5x + 4y &\leqslant 2 \\ x - y &\leqslant 3 \end{aligned}$$

Example 28-2

Given the following problem, find its dual: Minimize $g(p,q) = 6p + q$ subject to

$$\begin{aligned} 3p - 2q &\geqslant 1 \\ 4p &\geqslant 3 \\ 7p + 5q &\geqslant 4 \end{aligned}$$

We have:

Step (1):
$$\begin{array}{cc|c} 3 & -2 & 1 \\ 4 & 0 & 3 \\ 7 & 5 & 4 \\ \hline 6 & 1 & 0 \end{array}$$

Step (2):
$$\begin{array}{ccc|c} 3 & 4 & 7 & 6 \\ -2 & 0 & 5 & 1 \\ \hline 1 & 3 & 4 & 0 \end{array}$$

Step (3): Maximize $f(x,y,z) = x + 3y + 4z$ subject to

$$\begin{aligned} 3x + 4y + 7z &\leqslant 6 \\ -2x \qquad + 5z &\leqslant 1 \end{aligned}$$

Example 28-3

Find the dual of the following: Minimize $g(p,q,r,m) = 3p + q + 2r + 4m$ subject to

$$\begin{aligned} 2p \qquad + r + 6m &\geqslant 3 \\ 4p + q \qquad + 7m &\geqslant 5 \end{aligned}$$

Step (1):
$$\begin{array}{cccc|c} 2 & 0 & 1 & 6 & 3 \\ 4 & 1 & 0 & 7 & 5 \\ \hline 3 & 1 & 2 & 4 & 0 \end{array}$$

Step (2):
$$\begin{array}{cc|c} 2 & 4 & 3 \\ 0 & 1 & 1 \\ 1 & 0 & 2 \\ 6 & 7 & 4 \\ \hline 3 & 5 & 0 \end{array}$$

Step (3): Maximize $f(x,y) = 3x + 5y$ subject to

$$\begin{aligned} 2x + 4y &\leqslant 3 \\ y &\leqslant 1 \\ x &\leqslant 2 \\ 6x + 7y &\leqslant 4 \end{aligned}$$

Notes on Finding the Dual

When constructing the matrix for a given minimization problem, we do *not* change the signs of the coefficients of the objective function.

When writing down the dual maximization problem from the transpose in step (3) above, we do *not* change the signs of the coefficients in the bottom row.

There are as many constraints in the dual maximization problem as there are variables in the given minimization problem, and as many variables in the former as constraints in the latter.

GETTING THE ANSWER TO A MINIMIZATION PROBLEM

When one has found the dual maximization problem for a given minimization problem, and has solved the maximization problem by the simplex method, the answer to the minimization problem can be read from the final tableau for the dual, as follows: The values of the variables that yield the minimum value for the objective function are those that appear as the bottom entries in the columns headed by slack variables. The minimum value of the objective function is the lower right-hand entry.

That minimum value, as noted earlier, is equal to the maximum value of the objective function of the dual maximization problem.

We illustrate this procedure.

Example 28-4

Minimize $g(p,q) = 7p + 8q$, subject to $p + q \geqslant 5$, $3p + 4q \geqslant 18$.

First we find the dual problem.

Matrix:

$$\begin{array}{cc|c} 1 & 1 & 5 \\ 3 & 4 & 18 \\ \hline 7 & 8 & 0 \end{array}$$

Transpose:

$$\begin{array}{cc|c} 1 & 3 & 7 \\ 1 & 4 & 8 \\ \hline 5 & 18 & 0 \end{array}$$

Dual problem: Maximize $f(x,y) = 5x + 18y$ subject to $x + 3y \leqslant 7$, $x + 4y \leqslant 8$.

Applying the simplex method, we have

$$\begin{array}{ccccc|c} x & y & s & t & f & \\ 1 & 3 & 1 & 0 & 0 & 7 \\ 1 & \boxed{4} & 0 & 1 & 0 & 8 \\ \hline -5 & -18 & 0 & 0 & 1 & 0 \end{array} \qquad \begin{array}{l} 7 \div 3 = 2\tfrac{1}{3} \\ 8 \div 4 = 2 \ \checkmark \end{array}$$

$$\begin{array}{ccccc|c} x & y & s & t & f & \\ \boxed{\tfrac{1}{4}} & 0 & 1 & -\tfrac{3}{4} & 0 & 1 \\ \tfrac{1}{4} & 1 & 0 & \tfrac{1}{4} & 0 & 2 \\ \hline -\tfrac{1}{2} & 0 & 0 & \tfrac{9}{2} & 1 & 36 \end{array} \qquad \begin{array}{l} 1 \div \tfrac{1}{4} = 4 \ \checkmark \\ 2 \div \tfrac{1}{4} = 8 \end{array}$$

FINAL TABLEAU

x	y	s	t	f		
1	0	4	-3	0	4	($\leftarrow x$)
0	1	-1	1	0	1	($\leftarrow y$)
0	0	2	3	1	38	($\leftarrow f = g$)
		$\uparrow p$	$\uparrow q$			

We can now read off the answers to both problems, the original minimization problem and its maximization dual:

(max.:) $f(x,y) = 5x + 18y$ has maximum value 38 at $x = 4$, $y = 1$;
(min.:) $g(p,q) = 7p + 8q$ has minimum value 38 at $p = 2$, $q = 3$.

Example 28-5

Minimize $g(p,q,r,m) = 31p + 13q + 33r + 40m$ subject to

$$3p + q + r + 5m \geqslant 3$$
$$2p + q + 3r + m \geqslant 1$$

Find the dual maximization problem:

Matrix:

3	1	1	5	3
2	1	3	1	1
31	13	33	40	0

Transpose:

3	2	31
1	1	13
1	3	33
5	1	40
3	1	0

Dual problem: Maximize $f(x,y) = 3x + y$ subject to

$$3x + 2y \leqslant 31$$
$$x + y \leqslant 13$$
$$x + 3y \leqslant 33$$
$$5x + y \leqslant 40$$

Applying the simplex method:

x	y	s	t	u	v	f		
3	2	1	0	0	0	0	31	$31 \div 3 = 10\frac{1}{3}$
1	1	0	1	0	0	0	13	$13 \div 1 = 13$
1	3	0	0	1	0	0	33	$33 \div 1 = 33$
5	1	0	0	0	1	0	40	$40 \div 5 = 8$ ✓
-3	-1	0	0	0	0	1	0	

x	y	s	t	u	v	f		
0	$\frac{7}{5}$	1	0	0	$-\frac{3}{5}$	0	7	$7 \div \frac{7}{5} = 5$ ✓
0	$\frac{4}{5}$	0	1	0	$-\frac{1}{5}$	0	5	$5 \div \frac{4}{5} = 6.2$
0	$\frac{14}{5}$	0	0	1	$-\frac{1}{5}$	0	25	$25 \div \frac{14}{5} = 8.9$
1	$\frac{1}{5}$	0	0	0	$\frac{1}{5}$	0	8	$8 \div \frac{1}{5} = 40$
0	$-\frac{2}{5}$	0	0	0	$\frac{3}{5}$	1	24	

x	y	s	t	u	v	f		
0	1	$\frac{5}{7}$	0	0	$-\frac{3}{7}$	0	5	($\leftarrow y$)
0	0	$-\frac{4}{7}$	1	0	$\frac{1}{7}$	0	1	
0	0	-2	0	1	1	0	11	
1	0	$-\frac{1}{7}$	0	0	$\frac{2}{7}$	0	7	($\leftarrow x$)
0	0	$\frac{2}{7}$	0	0	$\frac{3}{7}$	1	26	($\leftarrow f = g$)
		$\uparrow p$	$\uparrow q$	$\uparrow r$	$\uparrow m$			

FINAL TABLEAU

The answers to the pair of problems are:

$f(x,y) = 3x + y$ has maximum value 26 at $x = 7$, $y = 5$;

$g(p,q,r,m) = 31p + 13q + 33r + 40m$ has minimum value 26 at $p = \frac{2}{7}$, $q = 0$, $r = 0$, $m = \frac{3}{7}$.

Example 28-6

Solve example 26-8 (p. 249) by the methods of this section. For convenience we repeat the problem here:

A farmer wants to provide each of his beef cattle daily with at least 600 units of vitamin I, 400 units of vitamin II, and 550 units of vitamin III. Two food supplements, A and B, are available. The first costs \$0.02 per ounce and supplies 30, 10, and 10 units of vitamins I, II, and III respectively. The second costs \$0.03 per ounce and supplies 10, 10, and 20 units respectively of vitamins I, II, and III. How many ounces of each supplement should be used to minimize cost yet meet the daily nutritional needs of the cattle?

We again let x and y be the number of ounces of supplements A and B respectively, again denote the objective cost function by $C(x,y)$. Then the problem is to minimize $C(x,y) = 2x + 3y$ subject to $x,y \geqslant 0$, $30x + 10y \geqslant 600$, $10x + 10y \geqslant 400$, $10x + 20y \geqslant 550$.

Find the dual maximization problem:

Matrix:

30	10	600
10	10	400
10	20	550
2	3	0

Transpose:

30	10	10	2
10	10	20	3
600	400	550	0

Dual problem: Maximize $f(p,q,r) = 600p + 400q + 550r$ subject to $30p + 10q + 10r \leqslant 2$, $10p + 10q + 20r \leqslant 3$.

Solve the dual problem by the simplex method (note that in the tableaus s and t are the slack variables):

p	q	r	s	t	f	
30	10	10	1	0	0	2
10	10	20	0	1	0	3
-600	-400	-550	0	0	1	0

$2 \div 30 = \frac{1}{5}$ ✓

$3 \div 10 = \frac{3}{10}$

p	q	r	s	t	f	
1	$\frac{1}{3}$	$\frac{1}{3}$	$\frac{1}{30}$	0	0	$\frac{1}{15}$
0	$\frac{20}{3}$	$\frac{50}{3}$	$-\frac{1}{3}$	1	0	$\frac{7}{3}$
0	-200	-350	20	0	1	40

$\frac{1}{15} \div \frac{1}{3} = \frac{1}{5} = 0.2$

$\frac{7}{3} \div \frac{50}{3} = \frac{7}{50} = 0.14$ ✓

p	q	r	s	t	f	
1	$\frac{1}{5}$	0	$\frac{1}{25}$	$-\frac{1}{50}$	0	$\frac{1}{50}$
0	$\frac{2}{5}$	1	$-\frac{1}{50}$	$\frac{3}{50}$	0	$\frac{7}{50}$
0	-60	0	13	21	1	89

$\frac{1}{50} \div \frac{1}{5} = 0.1$ ✓

$\frac{7}{50} \div \frac{2}{5} = 0.35$

p	q	r	s	t	f		
5	1	0	$\frac{1}{5}$	$-\frac{1}{10}$	0	$\frac{1}{10}$	$(\leftarrow q)$
-2	0	1	$-\frac{1}{10}$	$\frac{1}{10}$	0	$\frac{1}{25}$	$(\leftarrow r)$
300	0	0	25	15	1	95	$(\leftarrow f = C)$
			$(\uparrow x)$	$(\uparrow y)$			

FINAL TABLEAU

Thus f attains its maximum value at $p = 0$, $q = \frac{1}{10}$, $r = \frac{1}{25}$.

The answer to the given problem is: the minimum cost is \$0.95, obtained by using 25 ounces of supplement A and 15 ounces of supplement B. This is the result obtained previously with the geometric approach.

EXERCISES

Remember throughout these problems that all variables $(p,q,r,\ldots)$ remain nonnegative.

In 1 through 5, find the dual maximization problem:

1. Minimize $g(p,q) = 5p + 3q$ subject to $4p + q \geq 2$, $7p - 3q \geq 5$.
2. Minimize $g(p,q) = 2p + 9q$ subject to $-3p + q \geq 6$, $14p + 11q \geq 18$.
3. Minimize $g(p,q) = 3p + 4q$ subject to $p + q \geq 2$, $9p - q \geq 4$, $3p + 7q \geq 5$.
4. Minimize $g(p,q,r) = 2p + q + 3r$ subject to $p - q + r \geq 5$, $3p + 2q - 4r \geq 6$.
5. Minimize $g(p,q,r,m) = 4p + 2q + 5r + 3m$ subject to $4p - q + m \geq 5$, $3p + r - m \geq 7$, $p - q + 2r + 3m \geq 6$.

In 6 through 15, solve each problem by the simplex method:

6. Minimize $g(p,q) = p + q$ subject to $6p + 5q \geq 48$, $7p + 3q \geq 39$.
7. Minimize $g(p,q) = 2p + q$ subject to the constraints in problem 6.
8. Minimize $g(p,q) = 3p + q$ subject to the constraints in problem 6.

9. Minimize $g(p,q) = 3p + 2q$ subject to $4p + 3q \geqslant 32$, $2p + q \geqslant 12$, $p + q \geqslant 9$.
10. Minimize $g(p,q) = 2p + 3q$ subject to the constraints in problem 9.
11. Minimize $g(p,q,r) = 5p + 2q + 4r$ subject to $p + q + 2r \geqslant 2$, $2p + q + r \geqslant 6$.
12. Minimize $g(p,q,r) = 30p + 11q + 45r$ subject to $2p + q + 5r \geqslant 4$, $3p + q + 3r \geqslant 5$.
13. Minimize $g(p,q,r) = 20p + 10q + 3r$ subject to $p + 2q \geqslant 1$, $5p + 4r \geqslant 3$, $3q + r \geqslant 4$.
14. Minimize $g(p,q,r,m) = 30p + 91q + 65r + 62m$ subject to $p + 6q + 5r + 3m \geqslant 9$, $2p + 5q + 2r + 4m \geqslant 9$.
15. Minimize $g(p,q,r) = 10p + 12q + 8r$ subject to $p + 4r \geqslant 1$, $2q + r \geqslant 1$, $-p + 3q \geqslant 1$, $r \geqslant 2$.

16. Suppose a housekeeper wants her family's daily meals to contain at least 60 units of protein, 40 units of carbohydrates, and 100 units of fat. A pound of food I has 20 units of protein, 8 of carbohydrates, and 20 of fat and costs \$2.50; a pound of food II has 10 units of protein, 8 of carbohydrates, and 50 of fat and costs \$1.50. How many pounds of each food will satisfy these needs at the least possible cost? What is this minimum daily cost?
17. Solve exercise 14 on page 252 by the simplex method.

TEST A

1. Using the simplex method, minimize $g(p,q) = 3p + 4q$ subject to $p + q \geqslant 2$, $p + 2q \geqslant 3$.
2. Using the simplex method, minimize $g(p,q,r) = 2p + q + 5r$ subject to $2p + q \geqslant 1$, $p - q + 2r \geqslant 2$.
3. A company operates two mines that yield three kinds of coal: anthracite, high-grade bituminous, and low-grade bituminous. Mine I yields 10, 20, and 40 tons daily of the three kinds respectively; mine II yields 30 tons daily of each of the bituminous coals (and no anthracite). The company has agreed to supply at least 300 tons of anthracite, 1800 tons of high-grade bituminous, and 3000 tons of low-grade. If it costs \$800 a day to operate mine I and \$600 a day for mine II, how many days should the mines be worked to minimize the operating cost? What is the minimum cost?

TEST B

1. Using the simplex method, minimize $g(p,q) = 3p + 4q$ subject to $3p + 5q \geqslant 35$, $2p + 5q \geqslant 30$.

2. Using the simplex method, minimize $g(p,q) = p + 3q$ subject to $2p + q \geqslant 4$, $p + q \geqslant 3$, $p + 2q \geqslant 4$.
3. A TV manufacturer has two factories. At each, three models are made: standard, superior, and deluxe. At factory I the daily production is 40 standard, 30 superior, and 20 deluxe; plant II produces daily 30 standard, 20 superior, and 10 deluxe sets. It costs \$1200 a day to operate plant I, \$1000 a day for plant II. If the manufacturer has orders for 1200 standard, 850 superior, and 450 deluxe sets, how many days should each plant be operated to minimize the cost?

SUMMARY OF §§26-28

You should now be able to

(1) graph the feasible set in an LP problem and specify the corner points;
(2) find graphically, if they exist, the maximum and minimum values of an objective function subject to given constraints;
(3) set up mathematically an LP problem presented in words;
(4) solve graphically an LP problem in two variables;
(5) put an LP problem into standard form;
(6) write the initial tableau for a maximization LP problem;
(7) solve a maximization LP problem by the simplex method;
(8) write the dual maximization problem for a minimization LP problem;
(9) solve a minimization LP problem by the simplex method.

REVIEW EXERCISES

In exercises 1 through 3, find graphically the maximum or minimum values, if they exist, of the given function, subject to the given constraints.

1. $f(x,y) = 3x + 5y$
   $x,y \geqslant 0$
   $x + 2y \leqslant 16$
   $2x + y \leqslant 11$

2. $f(x,y) = 4x + 3y$
   $x,y \geqslant 0$
   $x + y \leqslant 4$
   $x \leqslant 2$
   $3x + 2y \geqslant 9$

3. $f(x,y) = 2x + 3y$
   $x,y \geqslant 0$
   $4x + y \geqslant 12$
   $x + y \geqslant 6$

Solve 4 through 6 graphically.

4. A furniture manufacturer makes deluxe and standard coffee tables. The deluxe model needs 4, 1, and 3 units of materials I, II, and III respectively and sells for \$120. The standard model takes 3, 2, and 1 units respectively of materials I, II, and III and sells for \$80. The number of units of I, II, and III available to the manufacturer are respectively 74, 36, and 48. How many of each table should the manufacturer make to maximize income? What is the maximum income?
5. Suppose the deluxe table in the preceding exercise sells for \$150, the standard one for \$50; answer the same questions.
6. A dietician wants to prepare a food that provides at least 32 units of nutrient I and at least 40 of nutrient II. Food R provides 2 units of I and 4 of II, and costs 3¢ per ounce. Food S has 10 units of I and 8 of II, and costs 7¢ per ounce. What amounts of the two foods should be combined to yield the specified nutrients at least cost?

7. Rewrite in standard form: Maximize $f(x,y) = 2x - 5y$ subject to $x,y \geqslant 0$, $4x - 2y \geqslant -1$, $3x - y \leqslant 2$.
8. Write the initial tableau for the following: Maximize $f(x,y,z) = -3x + 9y + 4z$ subject to $x,y,z \geqslant 0$, $3x + 9y - 2z \leqslant 3$, $2x + 3y + 6z \leqslant 12$, $-4x + y + 5z \leqslant 3$.

Solve 9 and 10 by the simplex method.

9. Maximize $f(x,y) = 2x + 3y$ subject to $x,y \geqslant 0$, $2x + 5y \leqslant 20$, $x + 5y \leqslant 15$.

10. Maximize $f(x,y,z) = 5x + 9y + 2z$ subject to $x,y,z \geqslant 0$, $-2x + 3y + z \leqslant 2$, $-5x + 2y + 3z \leqslant 12$, $3x + 2y - z \leqslant 3$.

11. Find the dual problem: Minimize $g(p,q,r) = 5p + 3q + 2r$ subject to $p,q,r \geqslant 0$, $p - q + r \geqslant 8$, $-4p + 2q + 3r \geqslant 3$.

Solve 12 and 13 by the simplex method.

12. Minimize $g(p,q) = 2p + 3q$ subject to $p,q \geqslant 0$, $p + 2q \geqslant 12$, $p + q \geqslant 9$, $3p + 4q \geqslant 32$.
13. Minimize $g(p,q,r) = p + 4q + 5r$ subject to $p,q,r \geqslant 0$, $q + 2r \geqslant 4$, $p + q + r \geqslant 3$, $-p + 2q + 2r \geqslant 2$.

14. Solve exercise 4 above by the simplex method.
15. Solve exercise 6 above by the simplex method.

CHAPTER TEST

1. Use the graphical method to
   (a) maximize $f(x,y) = 2x + 5y$ subject to $x,y \geqslant 0$, $3x + 2y \leqslant 13$, $2x + 3y \leqslant 12$.
   (b) minimize $f(x,y) = 4x + 2y$ subject to $x,y \geqslant 0$, $3x + y \geqslant 6$, $2x + 3y \geqslant 11$.
2. A Hawaiian entrepreneuse makes leis daily for visitors from the Mainland. Each day she gets 250 plumerias and 200 hibiscus flowers. She uses 20 plumerias and 10 hibiscuses on her Lani Lei and sells it for \$2.50. Her Kehena Lei has 10 plumerias and 20 hibiscuses and sells for \$3.50. How many leis of each kind should she make for maximum income?

Solve problems 3 and 4 by the simplex method.

3. Maximize $f(x,y,z) = 3x + 4y + 5z$ subject to $x,y,z \geqslant 0$, $3x + y \leqslant 40$, $x + y + 2z \leqslant 30$.
4. Minimize $g(p,q) = 4p + 3q$ subject to $p,q \geqslant 0$, $4p + q \geqslant 16$, $p + q \geqslant 10$, $4p + 2q \geqslant 26$.
5. Solve number 2 above by the simplex method.

Chapter 6

Mathematics of finance

We consider in this chapter some everyday topics in finance. Since the matters to be discussed are likely to be of concern to most people, you should be familiar with the basic ideas we cover. In particular, you should understand the difference between simple and compound interest, be aware that money does not hold steady as time passes, and be able to evaluate various investment options or liabilities and to compare them on a sound mathematical basis so as to choose wisely from among them.

The applications in this chapter are practical, ranging from finding the amount of interest earned on a deposit in a savings account to calculating how much one must stash away regularly over a period of years to generate a specified sequence of, say, retirement benefits in later life.

SECTION 29. SIMPLE INTEREST

Interest is money paid for the use of money. If you borrow money, you pay the lenders interest for using their money. If you invest (deposit money in a bank, buy bonds, make a loan) then you earn interest for the use of your money. In this section we consider only simple interest.

DEFINITIONS; FORMULA FOR SIMPLE INTEREST

The original amount of money borrowed or lent (or invested) is called the *principal*. *Simple interest* is interest earned only on the principal. The *interest rate*, usually given as a percentage, indicates what part of the principal is earned in a unit of time, ordinarily one year.

If I is the (simple) interest, P the principal, R the yearly interest rate, and t the number of years (sometimes called the *term* of the loan or investment), then

$$I = PRt$$

In calculating I, we always express R in decimal form.

Example 29-1

If you invest \$1200 for four years at 8% simple interest, how much interest will you earn?

Applying the formula we have, immediately,

$$I = 1200(0.08)(4) = \$384$$

Example 29-2

If you borrow \$1000 at 18% per year, how much interest will you owe after six months?

Here $P = 1000$, $R = 0.18$, and $t = \frac{1}{2}$. So

$$I = 1000(0.18)(\tfrac{1}{2}) = \$90$$

FORMULA FOR AMOUNT

The *amount* is the accumulated value of an investment or loan at the end of the term. If A is the amount, then it is equal to the sum of the principal and interest:

$$A = P + I = P + PRt = P(1 + Rt)$$

Example 29-3

If you deposit \$700 in a bank today at 9% simple interest, what amount will it accumulate to after three years?

$$A = 700[1 + (0.09)(3)] = \$889$$

Note how this investment grows, year by year:

$$\begin{aligned} &\text{After 1 year:} && A = 700 + (0.09)(700) = 700 + 63 = \$763 \\ &\text{After 2 years:} && A = 763 + (0.09)(700) = 763 + 63 = \$826 \\ &\text{After 3 years:} && A = 826 + (0.09)(700) = 826 + 63 = \$889 \end{aligned}$$

From this example we see that, since simple interest is computed only on the original principal, the annual interest earned is constant--in this case, \$63 each year.

FINDING ANY ONE VARIABLE GIVEN THE OTHERS

Example 29-4

How much must be invested at 8% simple interest to earn \$100 in interest over a five-year period?

We seek P, given $I = 100$, $R = 0.08$, and $t = 5$:

$$I = PRt$$

$$100 = P(0.08)(5)$$

$$P = \frac{100}{0.40} = \$250$$

Example 29-5

Suppose you borrow \$300 at 10% simple interest. When will the interest you owe on your loan reach \$75?

Again we use the interest formula

$$I = PRt$$

$$75 = 300(0.10)(t)$$

$$t = \frac{75}{300(0.10)} = 2.5 \text{ years}$$

Example 29-6

If you have \$1500 to invest for four years, and want to earn at least \$540 in interest, what is the minimum simple interest rate you must get?

Since

$$I = PRt = 1500(t)\cdot 4$$

and I must equal at least 540, it follows that R must satisfy

$$(1500)(4)R \geqslant 540$$

$$R \geqslant \frac{540}{6000} = 0.09 = 9\%$$

Example 29-7

A businessman is confident he can pay his cousin \$14,000 five years from now. His cousin agrees to charge him only 15% simple interest. How much can the businessman borrow from his cousin today?

Here we use

$$A = P(1 + Rt)$$

with $A = 14{,}000$, $R = 0.15$, and $t = 5$:

$$14000 = P[1 + (0.15)(5)]$$

$$= P(1 + 0.75)$$

$$P = \frac{14000}{1.75} = \$8000$$

Example 29-8

How long does it take an investment to double at 5% simple interest?

We seek t such that $A = 2P$ in the formula for amount:

$$2P = P(1 + 0.05t)$$

$$2 = 1 + 0.05t$$

$$1 = 0.05t$$

$$t = \frac{1}{0.05} = 20 \text{ years}$$

Example 29-9

What simple interest rate is required for an investment to triple in 25 years?

Let $A = 3P$ in the formula for amount:

$$3P = P(1 + 25R)$$

$$3 = 1 + 25R$$

$$2 = 25R$$

$$R = \frac{2}{25} = \frac{8}{100} = 0.08 = 8\%$$

NOTE

We see from the last two examples that the actual value of the principal is irrelevant. *Any* principal doubles in 20 years at 5%; *any* principal triples in 25 years at 8%.

EXERCISES

1. Find the interest you must pay and the final amount if you borrow
   (a) \$500 for seven years at 10% simple interest.
   (b) \$25,000 for one year at 12% simple interest.
   (c) \$900 for $3\frac{1}{2}$ years at 8% simple interest.
   (d) \$1500 for 5 years at $7\frac{1}{2}$% simple interest.
2. How much principal must you invest
   (a) to earn \$1000 interest after four years at 10% simple interest?
   (b) to earn \$350 interest after seven years at $12\frac{1}{2}$% simple interest?
   (c) to obtain a final amount of \$1000 after three years at $13\frac{1}{3}$% simple interest?
   (d) to obtain a final amount of \$1500 after $2\frac{1}{2}$ years at 8% simple interest?
3. How long does it take
   (a) to earn \$100 interest on a \$500 investment at 5% simple interest?
   (b) to earn \$350 interest on a \$1000 investment at 11% simple interest?
   (c) to accumulate an amount of \$2000 on a \$1500 investment at 10% simple interest?
   (d) to accumulate an amount of \$5000 on a \$750 investment at 15% simple interest?
   (e) to double your investment at $7\frac{1}{2}$% simple interest?
   (f) to triple your investment at 10% simple interest?
4. What rate of interest will
   (a) yield \$100 interest on a \$500 investment in four years?
   (b) yield \$630 interest on a \$700 investment in six years?
   (c) result in a final amount of \$290 after five years on a \$200 investment?
   (d) let you accumulate \$2010 on a \$1500 investment in four years?
   (e) result in doubling your investment in $12\frac{1}{2}$ years?
   (f) result in tripling your investment in 16 years?

TEST A

1. A college coed borrows \$2300 for $2\frac{1}{2}$ years at 6% simple interest. How much interest must she pay and what is the final amount due?
2. How much principal must be invested to obtain a final amount of \$1000 after four years at $12\frac{1}{2}$% simple interest?
3. How much time is needed to double one's investment at 8% simple interest?
4. Determine the interest rate if you must pay interest of \$250 in three years on a loan of \$700 today.

TEST B

1. If you can borrow \$18,000 for five years at 12% simple interest, how much interest will you have to pay and what is the final amount?
2. How much principal must be invested to earn \$100 interest after three years at 9% simple interest?
3. How long will it take to repay a \$100 loan at 10% simple interest if the final amount is \$350?
4. What interest rate is required for an investment to triple in ten years?

SECTION 30. COMPOUND INTEREST

As noted in the preceding section, when simple interest is involved only the principal earns interest. The principal and the interest earned during a time unit are both constant. When *compound interest* is involved the interest earned during any time unit is added to the amount accumulated by the end of the preceding time unit. The interest so added thus becomes principal and itself earns interest. An important feature of an investment at compound interest is the *frequency* of compounding: that is, the number of times during a year that interest is converted into principal. This is often used as a factor in the competition between banks for depositors' funds.

ANNUAL COMPOUNDING

Example 30-1

Suppose you plan to invest $1000 for three years at 7%. Compare the amounts obtainable with simple interest and with yearly compounding:

AFTER YEAR NO.	SIMPLE INTEREST	INTEREST COMPOUNDED YEARLY
1	1000 + 0.07(1000) = $1070	1000 + 0.07(1000) = $1070
2	1070 + 0.07(1000) = $1140	1070 + 0.07(1070) = $1144.90
3	1140 + 0.07(1000) = $1210	1144.90 + 0.07(1144.90) = $1225.04

Example 30-2

Compare the amounts, as in the preceding example, if $20,000 is invested for four years at 10%:

AFTER YR NO.	SIMPLE INTEREST	INTEREST COMPOUNDED YEARLY
1	20,000 + 0.10(20,000) = $22,000	20,000 + 0.10(20,000) = $22,000
2	22,000 + 0.10(20,000) = $24,000	22,000 + 0.10(22,000) = $24,200
3	24,000 + 0.10(20,000) = $26,000	24,200 + 0.10(24,200) = $26,620
4	26,000 + 0.10(20,000) = $28,000	26,620 + 0.10(26,620) = $29,282

MORE FREQUENT COMPOUNDING

Examples 30-1 and 30-2 show that an investment grows a good deal faster with compound interest than with simple, even when interest is compounded only once a year. Often the interest is compounded more frequently, as in the following examples.

Example 30-3

Suppose \$500 is invested at 10% per year, compounded semiannually. Find the balance after three years.

Since the stated interest rate of 10% is *per year*, the rate of interest each *half*-year is half of that, or 5%. We have the following amounts in the account at the times indicated after successive half-years:

after $\frac{1}{2}$ year:	500 + 0.05(500)	=	\$525
after 1 year:	525 + 0.05(525)	=	\$551.25
after $1\frac{1}{2}$ years:	551.25 + 0.05(551.25)	=	\$578.81
after 2 years:	578.81 + 0.05(578.81)	=	\$607.75
after $2\frac{1}{2}$ years:	607.75 + 0.05(607.75)	=	\$638.14
after 3 years:	638.14 + 0.05(638.14)	=	\$670.05

Note that with simple interest of 10%, the amount after three years would have been only $500[1 + (0.10)(3)] = \$650$.

Example 30-4

Suppose a deposit of \$2000 is made at an annual interest rate of 12%. How much is in the account after one year if interest is compounded (a) yearly? (b) twice a year? (c) four times a year? (d) six times a year?

(a) If interest is compounded once a year, then after one year the amount is $2000(1 + 0.12) = \$2240$.

(b) With semiannual compounding, the interest rate per *interest period* is $12\% \div 2 = 6\%$. Therefore the amounts are

after $\frac{1}{2}$ year:	2000 + 0.06(2000)	=	\$2120
after 1 year:	2120 + 0.06(2120)	=	\$2247.20

(c) With quarterly compounding, the rate per interest period is $12\% \div 4 = 3\%$. In this case the amounts are

after 1 quarter:	2000 + 0.03(2000)	=	\$2060
after 2 quarters:	2060 + 0.03(2060)	=	\$2121.80
after 3 quarters:	2121.80 + 0.03(2121.80)	=	\$2185.45
after 1 year:	2185.45 + 0.03(2185.45)	=	\$2251.02

(d) Compounding six times a year (or bimonthly) yields an interest rate per period of $12\% \div 6 = 2\%$. Now the amounts are as shown at the top of the facing page.

This example shows that the amount increases with the frequency of compounding. You may wonder whether it increases indefinitely without bound as the frequency increases. We will return shortly to the consequences of more frequent compounding.

(results of compounding six times a year)

after 2 months:	2000	+ 0.02(2000)	= 2040
after 4 months:	2040	+ 0.02(2040)	= 2080.80
after 6 months:	2080.80	+ 0.02(2080.80)	= 2122.42
after 8 months:	2122.42	+ 0.02(2122.42)	= 2164.86
after 10 months:	2164.86	+ 0.02(2164.86)	= 2208.16
after 1 year:	2208.16	+ 0.02(2208.16)	= 2252.32

FORMULA FOR AMOUNT WITH COMPOUND INTEREST

From example 30-4 we note that, if interest is compounded k times a year, then there are k interest periods per year, and the rate r per interest period is obtained by dividing the annual rate by k. For example, if the annual rate is 8% and interest is compounded quarterly, then there are four interest periods and $r = 8\% \div 4 = 2\%$. We will frequently describe this by saying "the interest is 8%, compounded quarterly." Thus, "interest is 12%, compounded monthly" means that r is $12\% \div 12 = 1\%$.

If P is the principal or original investment, r the interest rate per interest period, and n the number of interest periods, then the amounts in the account are as follows:

after 1 period: $P + rP = P(1+r)$

after 2 periods: $P(1+r) + r[P(1+r)] = P(1+r)(1+r) = P(1+r)^2$

after 3 periods: $P(1+r)^2 + r[P(1+r)^2] = P(1+r)^2(1+r) = P(1+r)^3$

•

•

•

after n periods: $P(1+r)^n$

So, if A is the amount (or balance) after n periods, then

$$A = P(1+r)^n$$

With a hand calculator that has an x^y (or y^x) function, it is easy to compute the amount accumulated using this formula. Appendix 2 has a table of values of $(1+r)^n$ for selected values of r and for n from 1 to 100. (Occasionally answers obtained using calculator and tables will differ slightly because of rounding-off errors.)

Example 30-5

If you deposit \$800 in a savings account that pays 8%, compounded quarterly, how much will you have after ten years?

Here $P = 800$, $r = 8\% \div 4 = 2\%$, and $n = 10\text{(years)} \times 4\text{(times per year)} = 40$. Thus, after ten years

$$A = 800(1 + 0.02)^{40} = \$1766.43$$

This result was obtained using a Sharp EL-506H calculator: we keyed in 1.02, then the y^x function key, then 40, and multiplied the answer by 800.

Using the table in Appendix 2, we see that $(1+r)^n$ for $r = 2\%$ and $n = 40$ is 2.20804, so that $A = \$1766.43$, as above.

Example 30-6

Suppose a loan of $2300 is to be paid back with interest in nine years. If the interest rate is 14%, compounded semiannually, what total amount is due nine years from now? How much interest is to be paid?

Here $A = 2300$, $r = 14\% \div 2 = 7\%$, and $n = 9 \times 2 = 18$. Thus

$$A = 2300(1+0.07)^{18} = \$7773.84$$

The interest is equal to the amount minus the principal:

$$I = A - P = 7773.84 - 2300 = \$5473.84$$

EFFECTIVE RATE OF INTEREST

Two annual interest rates with different frequencies of compounding are said to be *equivalent* if they yield the same amount of interest in one year. In particular, given a certain interest rate which is compounded more than once a year, we are often interested in the rate equivalent to it that is compounded only once a year. The latter is called the *effective rate of interest* or just the *effective rate*. (Note that the effective rate of interest is just the equivalent *simple* interest rate.)

Example 30-7

Suppose an investment pays 12%, compounded quarterly. What is the effective rate?

Using the principal $P = \$1$, we note that the amount after one year at 12% compounded quarterly is

$$1(1+0.03)^4 = 1.12551$$

To obtain the interest earned, we subtract the principal from the amount. So the interest at 12%, compounded quarterly, is

$$1.12551 - 1 = 0.12551$$

To calculate the simple interest rate R, we use $I = PRt$ with $P = 1$ and $t = 1$; thus $I = R$. Then

$$R = 0.12551 = 12.55\%$$

So 12.55% is the effective interest rate.

Example 30-8

Find the effective rate equivalent to 18% compounded (a) annually; (b) three times a year; (c) six times a year; (d) monthly.

We have

$$\begin{aligned}
&\text{(a) } R = (1+0.18)^1 - 1 = 18\% \\
&\text{(b) } R = (1+0.06)^3 - 1 = 19.10\% \\
&\text{(c) } R = (1+0.03)^6 - 1 = 19.41\% \\
&\text{(d) } R = (1+0.015)^{12} - 1 = 19.56\%
\end{aligned}$$

Formula for Effective Rate

From the examples above, we see that the effective rate R equivalent to $r\%$ *per interest period* with k periods a year is given by

$$R = (1+r)^k - 1$$

Note that a rate of $r\%$ per interest period is equivalent to an annual rate of $kr\%$ which is compounded k times a year.

PRESENT VALUE

Often we wish to know how much to invest now to obtain a certain amount at a specified time in the future. The *present value* P of an amount A to be accumulated (or paid) after n periods at rate $r\%$ per compounding period is obtained from the formula

$$A = P(1+r)^n$$

by solving for P:

$$P = \frac{A}{(1+r)^n}$$

Tables are available that give $(1+r)^{-n}$ for selected values of r and n. However, it is more convenient, with a hand calculator, to key in A, then divide by $(1+r)^n$. This makes it unnecessary to use pencil and paper. The table in Appendix 2 can be used for $(1+r)^n$ if desired.

Example 30-9

How much should you invest at 6%, compounded twice a year, to accumulate \$1000 in five years?

Here $r = 6\% \div 2 = 3\%$, $n = 5 \times 2 = 10$, and $A = 1000$. So

$$P = \frac{1000}{(1+0.03)^{10}} = \frac{1000}{1.34392} = \$744.09$$

Example 30-10

Find the present value of \$25,000 to be paid in ten years, if the interest rate is 6% compounded quarterly.

Since $r = 6\% \div 4 = 1.5\%$, $n = 10 \times 4 = 40$, and $A = 25{,}000$,

$$A = \frac{25{,}000}{(1+0.015)^{40}} = \frac{25{,}000}{1.81402} = \$13{,}781.55$$

FINDING r OR n FROM COMPOUND INTEREST FORMULA

Logarithms to the base 10 are used when solving the formula

$$A = P(1+r)^n$$

for r or n, as illustrated in the following examples.

Example 30-11

How many years will it take to double an investment at 10% compounded semiannually?

Note that we set $A = 2P$. Since $r = 5\%$, we have

$$\begin{aligned} 2P &= P(1+0.05)^n \\ 2 &= (1.05)^n \\ \log_{10} 2 &= n \log_{10} 1.05 \\ n &= \frac{0.3010}{0.0212} = 14.2 \end{aligned}$$

(Both logarithms were found using a calculator.) Since $n = 14.2$ semiannual periods or half-years, it will take $14.2 \div 2 = 7.1$ years to double an investment at 10% compounded semiannually.

Example 30-12

At what rate of interest, compounded quarterly, will an investment triple in ten years?

Here $A = 3P$ and $n = 10 \times 4 = 40$. We must solve

$$3P = P(1+r)^{40} \quad \text{or} \quad 3 = (1+r)^{40}$$

for r. Using logarithms to base 10, we get

$$\log 3 = 40 \log(1+r)$$

$$\log(1+r) = \frac{\log 3}{40} = \frac{0.47712}{40} = 0.011928$$

To compute $(1+r)$ with a calculator, we find the antilog of 0.011928 using the 10^x function key. This yields

$$1 + r = 1.0278459$$

Then

$$r = 0.0278459$$

Since r is the periodic interest rate, for compounding quarterly, the annual rate is $4r = 0.1114$, which is equivalent to 11.14%. (With a little practice on a calculator, the student will be able to find the final answer in an exercise of this sort with little or no pencil-and-paper work.)

CONTINUOUS COMPOUNDING*

We now investigate the effect of increasing the frequency of compounding. The following unrealistic example is instructive.

Example 30-13

Suppose the Gobroke Bank of Sillyville lets you invest $1 at 100% interest for one year. Find how the amount increases with increasing frequency of compounding.

We use the formula $A = (1+r)^n$ and a calculator to obtain the following table:

RATE PER PERIOD	FREQUENCY OF COMPOUNDING	AMOUNT AFTER ONE YEAR	
1	once a year	$(1+1)^1$	= \$2
$1 \div 2 = \frac{1}{2}$	twice a year	$\left(1+\frac{1}{2}\right)^2$	= \$2.25
$1 \div 4 = \frac{1}{4}$	four times a year	$\left(1+\frac{1}{4}\right)^4$	= \$2.44141
$\frac{1}{12}$	twelve times a year	$\left(1+\frac{1}{12}\right)^{12}$	= \$2.61304
$\frac{1}{365}$	daily (365 times a year)	$\left(1+\frac{1}{365}\right)^{365}$	= \$2.71456
$\frac{1}{8760}$	hourly (8760 times a year)	$\left(1+\frac{1}{8760}\right)^{8760}$	= \$2.71800
$\frac{1}{1\,000\,000}$	1 000 000 times a year	$\left(1+\frac{1}{1\,000\,000}\right)^{1\,000\,000}$	= \$2.71828

Note how small the differences are in the last three rows in the rightmost column. Although the amount does increase with increased frequency of compounding it does *not* increase without bound. In more advanced mathematics courses it is shown that the quantity

$$\left(1+\frac{1}{n}\right)^n$$

as n gets larger and larger, gets closer and closer to an irrational number denoted by "e". The definition of e is

$$e = \lim_{n\to\infty} \left(1+\frac{1}{n}\right)^n$$

*Optional topic.

and the value of e to five decimal places is 2.71828. We say, in the limiting case, that the interest is *compounded continuously*. It can be shown that if the sum of $\$P$ is invested at $r\%$ compounded continuously, the amount after t years is given by

$$A = Pe^{rt}$$

THE POWER OF COMPOUND INTEREST

To show how rapidly money grows under compound interest, we calculate the amount that would have been accumulated as of today if one cent had been invested 2000 years ago at 6%, compounded annually:

$$A = 0.01(1+0.06)^{2000} = 0.01 \times (4.0901 \times 10^{50}) =$$

$$\$4,090,100,000,000,000,000,000,000,000,000,000,000,000,000,000,000,000,000$$

EXERCISES

1. \$5000 is invested for four years at 6% interest. Show the year-by-year calculations both for simple interest and for yearly compounding.
2. \$30,000 is invested for two years at 8% interest. Show the six-month calculations if interest is compounded twice a year. How much would be accumulated if the interest were simple?
3. \$800 is invested for one year at 18% interest. Find the amount in the account if interest is compounded (a) once a year; (b) four times a year; (c) six times a year; (d) monthly.
4. Find the amount in an account after 15 years if \$25,000 is invested at 12% interest, compounded quarterly.
5. Suppose \$100 is invested at 15% compounded every four months. How much is in the account after three years?
6. If the interest rate is 10% compounded twice a year, what is the effective rate?
7. Find the effective rate equivalent to an annual rate of 9% compounded (a) twice a year; (b) three times a year; (c)* monthly.
8. How much should be invested now to yield \$250 after four years if the interest rate is 6%, compounded twice a year?
9. Find the present value of \$1000 to be paid in six years if the interest rate is 12%, (a) compounded twice a year; (b) compounded quarterly; (c) compounded monthly.

†10. How long will it take to double an investment if the interest rate is 8%, compounded quarterly?

†11. If the interest rate is 5%, compounded yearly, how long will it take an investment of \$250 to grow to \$600?

---

*Use of calculator is recommended.

†Exercise involves logarithms.

†12. What interest rate compounded twice a year must you obtain if you want to double your initial investment in five years?

†*13. If the interest is to be compounded monthly, what must the annual rate be if you have $400 to invest and you want it to grow to $1000 in ten years?

14. Suppose you pay $1000 in cash now for a used car and agree to pay an additional lump sum of $800 in two years. If the car dealer tells you his money can earn 6% compounded quarterly, how much should he be willing to settle for now altogether for the car?

15. If a newly married couple wants to have $12,000 as a down payment for a house in ten years how much must they invest now at 6% compounded semiannually?

16. Which of the following investment options yields the most interest after one year?

 (a) 7% compounded annually;
 *(b) 6.9% compounded semiannually;
 *(c) 6.8% compounded quarterly;
 *(d) 6.6% compounded monthly.

TEST A

1. Find the amount in an account if $400 is invested for seven years at 12% compounded quarterly.
2. How much must be invested today to yield $2000 after 20 years if the interest rate is 8% compounded twice a year?
3. What effective rate is equivalent to 6% compounded monthly?

†4. How long will it take to triple an investment if the interest rate is 6% compounded quarterly?

Test B

1. What initial investment must be made to yield $750 after six years if the interest rate is 10% compounded quarterly?
2. If $500 is invested at 6% compounded semiannually, how much is in the account after 25 years?

*3. Which investment option yields the greatest interest after one year, 6.2% compounded annually, 6.1% compounded quarterly, or 6% compounded monthly?

†4. What interest rate compounded quarterly will double an investment in seven years?

---

*Use of calculator is recommended.
†Exercise involves logarithms.

SECTION 31. APPLICATIONS WITH A TIME LINE

What money is worth depends on the time. Money increases as time passes, and is worth less when evaluated at earlier times. As noted in §30, if r is the interest rate per compounding period, then a sum of x dollars today is worth $x(1+r)^n$ dollars n periods from now but was worth only $x(1+r)^{-n}$ dollars n periods ago.

We say that x dollars today has a *future value* of $x(1+r)^n$ dollars n periods from now.

In §30 we defined the *present value* of a sum worth x dollars n periods from now as $x(1+r)^{-n}$ dollars.

In this section we solve problems involving present or future value using a device called a *time line*, which we now describe.

Example 31-1

If \$1200 is invested today at 18% compounded monthly, find the future value of the investment after two years.

We draw a line segment on which we indicate the relevant times (or dates) and the values of various sums of money at those times. Letting x dollars be the future value sought, we have Figure 31.1.

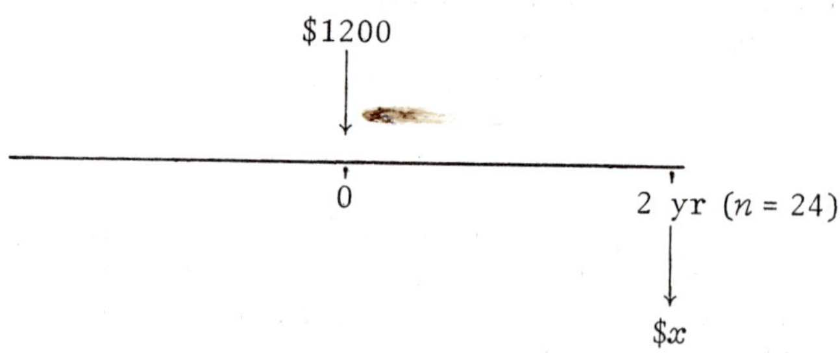

Figure 31.1

Since $r = 18\% \div 12 = 1.5\%$ and $n = 2 \times 12 = 24$, we see that

$$x = 1200(1+0.015)^{24} = \$1715.40$$

Example 31-2

Find the present value of a sum worth \$1500 three years from now if the interest rate is 12%, compounded every two months.

We let x dollars be the desired present value, and draw the time line (Figure 31.2).

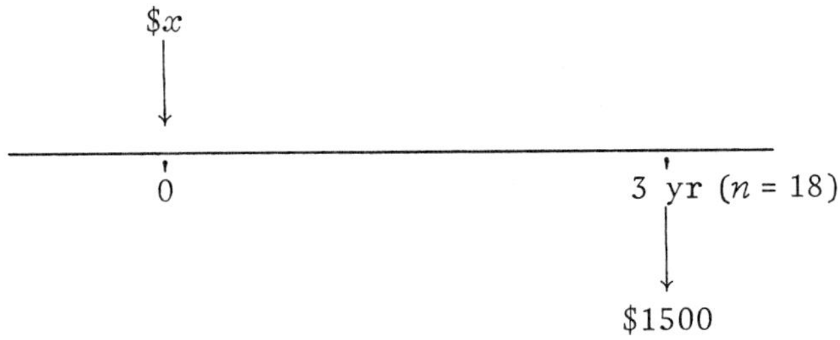

Figure 31.2

Since $r = 12\% \div 6 = 2\%$ and $n = 3 \times 6 = 18$,

$$x = 1500(1 + 0.02)^{-18} = \$1050.24$$

(As in §30, a hand calculator is being used in this section. To avoid pencil-and-paper work, we obtain the above answer by keying in 1500, then *dividing* by 1.02^{18}.)

NOTES ON THE TIME LINE

From the preceding examples and figures, we see that "0" below the line denotes "now" or "today," and that we move to the right on the line as time passes. An arrow above the line, pointing down towards it, generally indicates a deposit, an investment, an amount that is put *in*; an arrow below the line, pointing down away from it, indicates a withdrawal, a payment, an amount that is--or can be--taken *out*.

In example 31-1, the future value, $\$x$, is located 24 periods to the *right* on the time line and the evaluation of x involves raising $(1 + r)$ to the 24th power. In example 31-2, the present value, $\$x$, is located 18 periods to the *left* on the time line; the evaluation of x now involves raising $(1 + r)$ to the power -18.

EQUATION OF VALUE

We now consider some applications in which various amounts of money "enter" or "leave" at different points in time. We will solve what is called an *equation of value* for one or more unknowns. To obtain this equation, we first evaluate *all* the relevant sums of money at the *same* point in time, to be called the *reference point*. We then equate these sums so as to satisfy the given conditions. It is interesting to discover that the answer does *not* depend on the choice of reference point. However, often a wise choice can simplify the calculations.

Example 31-3

Suppose you deposit \$500 in an account now and \$1000 more in two years.* If you earn 10% compounded annually, how much will there be in the account five years from now?

Let x be the amount in the account in five years. We draw a time line (Figure 31.3).

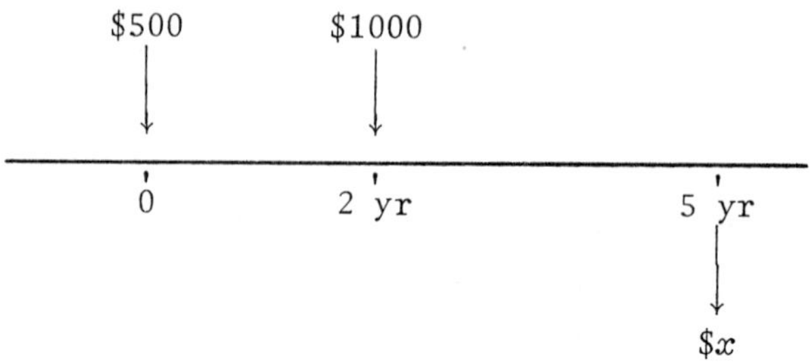

Figure 31.3

Choosing five years from now as the reference point, we get the following equation of value:

$$500(1+0.10)^5 + 1000(1+0.10)^3 = x$$

Figure 31.4 shows how we evaluate each sum of money at the reference point (n denotes, as usual, the number of periods).

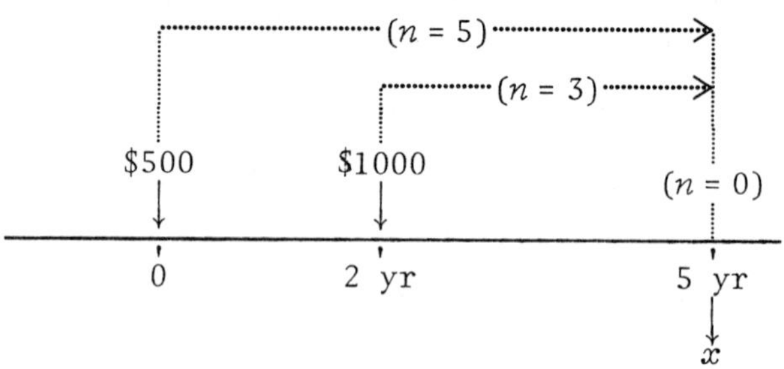

Figure 31.4

A hand calculator yields x = \$2136.26.

*Note that the expressions "two years from now," "at the end of two years," "after two years," and "in two years" *all mean the same thing*.

Suppose we were to choose two years from now, instead of five years from now, as the reference point. The sketch is modified as shown in Figure 31.5.

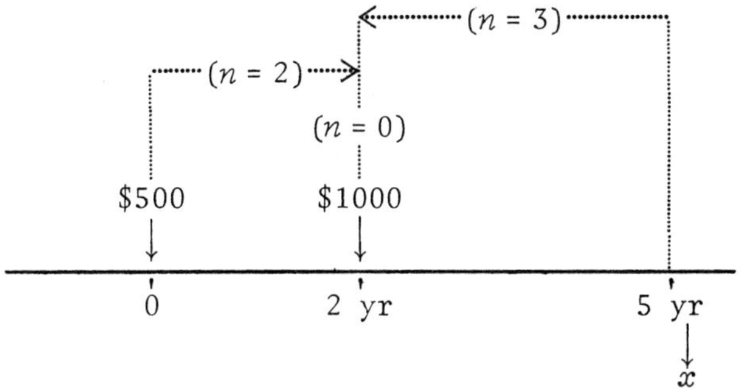

Figure 31.5

Now the equation of value becomes

$$500(1.10)^2 + 1000 = x(1.10)^{-3}$$

We see that "moving" to the right raises $(1+r)$ to a positive power, moving to the left to a negative power. To calculate x from the equation of value above it is most convenient to multiply the equation by $(1.10)^3$. But then we have exactly the equation obtained earlier with five years from now as the reference point (figure 31.4.)

As a final illustration that the choice of reference point is immaterial, let us focus on the time now (0 years). The new sketch appears in Figure 31.6.

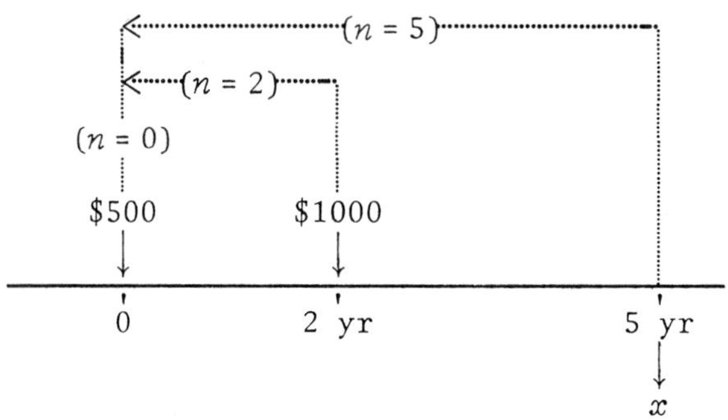

Figure 31.6

We see that

$$500 + 1000(1.10)^{-2} = x(1.10)^{-5}$$

With this choice, multiplying the equation by $(1.10)^5$ also yields exactly the equation of value obtained from Figure 31.4.

Example 31-4

How much should be invested now at 8% compounded semiannually if one withdrawal of $400 is to be made in three years and another, of $600, is to be made in ten years?

Let x dollars be the amount to be invested now. Figure 31.7 shows the time line. We choose now as the reference point. In this

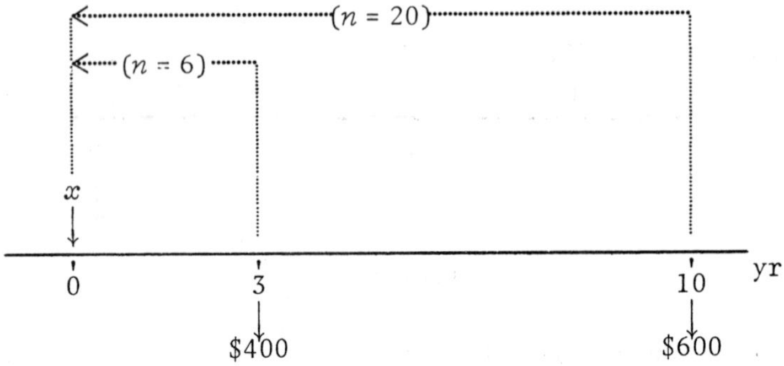

Figure 31.7

and all the sketches to follow, we omit "$n = 0$" for an amount evaluated at the reference point. We have

$$x = 400(1.04)^{-6} + 600(1.04)^{-20} = \$589.96$$

(Using a calculator, we divided 400 by 1.04^6, then added the term 600 divided by 1.04^{20}.)

Example 31-5

A college student wants to borrow $1000. He is considering paying it off in two equal installments, one after three years, the other after seven years. If he is to be charged 15% interest, compounded annually, how much will each installment be?

The money the student borrows can be regarded as going *into* his account; the payments he is to make will come *out of* his account. Let us use seven years from now at the reference point (Figure 31.8). The equation of value is

$$1000(1.15)^7 = x(1.15)^4 + x$$

So

$$1000(1.15)^7 = x[1 + (1.15)^4]$$

$$x = \frac{1000(1.15)^7}{1 + (1.15)^4} \qquad (*)$$

$$= \$967.63$$

[On a calculator we obtained x by dividing the numerator in (*) by the quantity $(1 + (1.15)^4)$.]

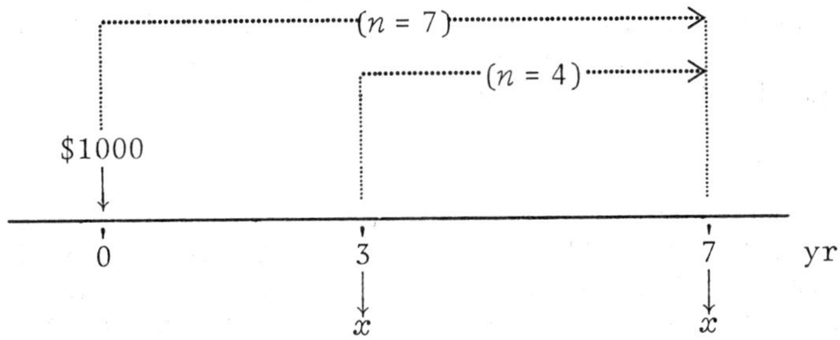

Figure 31.8

Verify that the equation obtained using the three-year point as reference point is

$$1000(1.15)^3 = x + x(1.15)^{-4}$$

How is this latter equation of value related to the one obtained using the seven-year point?

Example 31-6

A young couple wants to have \$1500 three years from now for a nursery. Their bank will give them 8% compounded quarterly. If they decide to make a deposit today and another twice as big one year from now, how much must they deposit today?

We use three years from now as the reference point (Figure 31.9).

Figure 31.9

The equation of value is

$$x(1.02)^{12} + 2x(1.02)^8 = 1500$$

So

$$x(1.02)^8((1.02)^4 + 2) = 1500$$

$$x = \frac{1500}{(1.02)^8((1.02)^4 + 2)} = \$415.33$$

[In solving the equation of value for example 31-6 with a hand calculator, we first divided 1500 by 1.02^8, then divided that quotient by the quantity $((1.02)^4 + 2)$.]

Example 31-7

A father has three daughters who are celebrating their 13th, 15th, and 18th birthdays (all three of them!) today. He wishes each of them to receive \$10,000 on her 21st birthday. If he can earn 10% compounded semiannually, how much must he invest today to fulfill his wish?

The daughters will become 21 in 8, 6, and 3 years respectively. We use today as reference point (Figure 31.10).

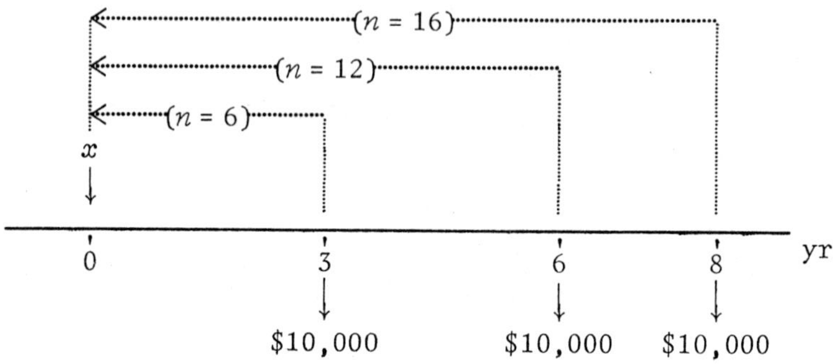

Figure 31.10

We have

$$x = 10{,}000(1.05)^{-6} + 10{,}000(1.05)^{-12} + 10{,}000(1.05)^{-16}$$

$$= \frac{10{,}000}{1.05^6} + \frac{10{,}000}{1.05^{12}} + \frac{10{,}000}{1.05^{16}} = \$17{,}611.64$$

(With a Sharp EL-506H, proceeding as shown above, we found no need for paper and pencil.)

Example 31-8

A company had agreed to discharge a debt by paying \$50,000 two years from now and \$75,000 five years from now. They have now decided instead to settle both obligations by making a single payment three years from now. If money is worth 9% compounded annually, how much will the single payment be?

We choose three years from now as reference point (Figure 31.11). The equation of value is then

$$50{,}000(1.09)^1 + 75{,}000(1.09)^{-2} = x$$

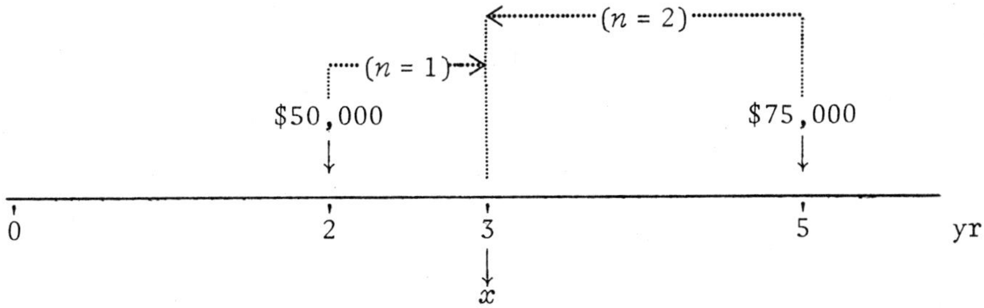

Figure 31.11

from which

$$x = 50{,}000(1.09) + \frac{75{,}000}{1.09^2} = \$117{,}626.00$$

Example 31-9

A year ago, an investor deposited \$5000 in a bank paying 6% compounded quarterly. How much should she invest now in bonds paying 10% compounded annually if she wants to have \$10,000 altogether four years from now?

We choose four years from now as reference point (Figure 31.12).

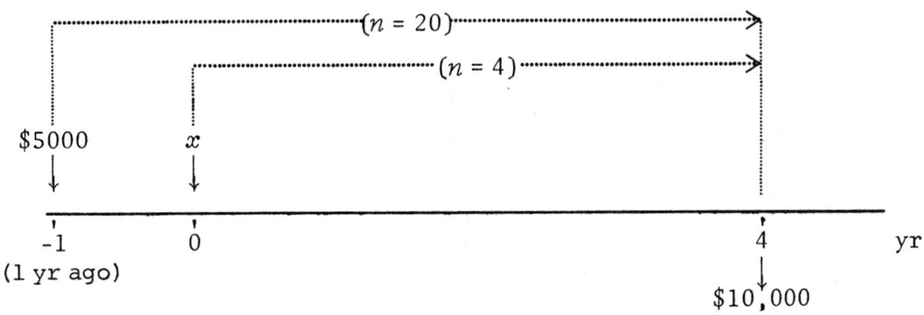

Figure 31.12

The equation of value is

$$5000(1.015)^{20} + x(1.10)^4 = 10{,}000 \qquad (*)$$

So

$$x = \frac{10{,}000 - 5000(1.015)^{20}}{1.10^4} = \$2230.53$$

NOTE. In this example, the amount to be accumulated (\$10,000) is based on two investments (sums of \$5000 and of x dollars) earning interest at different rates. With four years from now as reference point each investment is accumulated separately at its interest rate, as can be seen from the left side of equation (*) above. If we focus on any other time, however, then we do not know how the final amount of \$10,000 is to be divided between the two different investments.

EXERCISES

1. Calculate the future value of each of the following investments:
   (a) \$900 at 10% compounded annually for seven years.
   (b) \$1700 at 12% compounded quarterly for ten years.
   (c) \$2600 at 7% compounded twice a year for six years.
   (d) \$55 at 6% compounded monthly for five years.
2. Calculate the present value of each of the following amounts:
   (a) \$1250 in four years if money is worth 9% compounded monthly.
   (b) \$15,000 in three years if money is worth 8% compounded quarterly.
   (c) \$375 in ten years if the interest rate is 14% compounded twice a year.
   (d) \$1800 in 25 years if the interest rate is 6% compounded annually.
3. If you deposit \$1000 today and \$750 in three years, and the interest rate is 16% compounded quarterly, how much will be in your account seven years from now?
4. Suppose an investor deposits \$100 now, \$300 after two years, and \$200 more after two more years. If the interest rate is 9% compounded every two months, what will the balance be two years after the third deposit?
5. How big a loan can be made today to someone who agrees to make one payment of \$1000 in five years and another of \$1500 in ten years? Assume the interest rate is 11% compounded twice a year.
6. How much should you invest now if you want to withdraw \$600 after one year, \$800 after two years, and \$1000 after three years? Assume the interest rate is 15% compounded annually.
7. A man receives a \$3000 loan which he plans to repay in two equal installments, one after two years and the other after five years. If he is being charged 18% interest compounded monthly, what will be the size of each installment?
8. A man borrows \$7500 for a car. He plans to pay back the loan in three equal installments: at the end of one, two, and three years. If he is paying 13% interest compounded annually, how big will each installment be?
9. An investor plans to make a deposit today and another, three times as large, in two years, so that he can withdraw \$1000 after three years. If the interest rate is 8% compounded quarterly, how much should the first deposit be?

10. An investor plans to make one deposit today, and to deposit an additional \$1000 in two years. He wants to withdraw \$750 three years from now and to withdraw an additional \$750 four years from now, which will close out his account. If the interest rate is 6% compounded every two months, what should be the size of the initial deposit?
11. A company is supposed to make a \$10,000 payment in one year and a \$15,000 payment in four years. They wish to settle both obligations by making a single payment in three years. If money is worth 9% compounded monthly, what should the payment be?
12. (i) Draw a time line and find an equation of value for the following problem using as reference point (a) three years from now; (b) four years from now; (c) five years from now: A man must make a \$3000 payment in one year, a \$4000 payment in three years, and a \$5000 payment in five years. He wishes to settle all three obligations by making two equal payments, one in two years, the other in four years. If money is worth 11% compounded annually, what should be the size of the two payments? (ii) Show that the equations of value obtained in (i) are all equivalent. (iii) Find the numerical answer to the problem.

TEST A

1. If you deposit \$150 today and \$200 in two years and the bank pays interest at 12% compounded quarterly, how much will be in your account five years from now?
2. How much should be invested now if \$5000 is to be withdrawn after one year and an additional \$5000 after three years? Assume the interest rate is 6% compounded every two months.
3. A man borrows \$800 which he plans to repay in two equal installments, one after one year, one after four years. If he is being charged 10% interest compounded annually, what will be the size of the installments?

TEST B

1. Suppose you deposit \$4500 now and \$1000 more after one year. If the interest rate is 12% compounded monthly, how much will the fund be worth four years from now?
2. How much should be invested now if \$300 is to be withdrawn after two years and \$350 more after three years? Assume the interest rate is 9% compounded annually.
3. A company has agreed to make a payment of \$5000 in three years and one of \$5000 in seven years. They now wish to settle both obligations with a single payment in four years. If money is worth 10% compounded twice a year, how much is the single payment?

SECTION 32. THE AMOUNT OF AN ANNUITY

An *annuity* is a sequence of equal transfers of funds (payments, deposits, or withdrawals) made at equal intervals of time. Some examples are monthly mortgage payments, quarterly interest payments on bonds, weekly deposits in a Christmas Club saver's account, semi-annual premiums on an insurance policy, monthly installment payments on a car loan.

In this section and the next we consider only annuities in which payments are made at the *end* of each interval and in which the payment intervals and the interest (or compounding) periods are the same.

The *amount* A of an annuity is the total value accumulated from all the payments together with all the interest earned.

Example 32-1

Suppose you deposit \$100 at the end of each year for the next five years. How much is in the account after five years if the interest rate is 9% compounded annually?

A time line is helpful here (Figure 32.1).

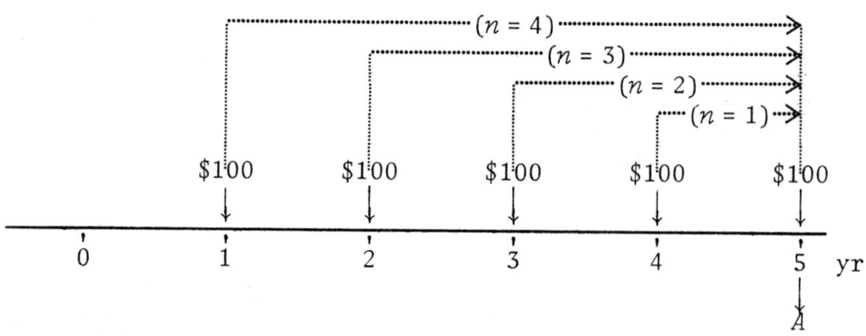

Figure 32.1

We have

$$\begin{aligned} A &= 100(1.09)^4 + 100(1.09)^3 + 100(1.09)^2 + 100(1.09) + 100 \\ &= 100[(1.09)^4 + (1.09)^3 + (1.09)^2 + 1.09 + 1] \quad (*) \\ &= \$598.47 \end{aligned}$$

Example 32-2

Verify that the sum in brackets in (*) above is equal to

$$\frac{(1.09)^5 - 1}{0.09}$$

Let

$$S = (1.09)^4 + (1.09)^3 + (1.09)^2 + 1.09 + 1$$

Multiply both sides of this equation by 1.09; then align the terms of the equation for S under the terms of the new equation; then subtract:

$$\begin{aligned} 1.09S &= (1.09)^5 + (1.09)^4 + (1.09)^3 + (1.09)^2 + 1.09 \\ S &= (1.09)^4 + (1.09)^3 + (1.09)^2 + 1.09 + 1 \\ \hline 0.09S &= (1.09)^5 - 1 \end{aligned}$$

So

$$S = \frac{(1.09)^5 - 1}{0.09}$$

With a calculator we find S equal to 5.9847. When we multiply this by 100, we get the same answer as in example 32-1.

Example 32-3

Verify the following formula:

$$(1+r)^{n-1} + (1+r)^{n-2} + \ldots + (1+r)^1 + 1 = \frac{(1+r)^n - 1}{r}$$

We use the trick of the preceding example. Let S be the sum above to the left of the equals sign; multiply S by $1 + r$; then subtract S from $(1+r)S$ as shown here:

$$\begin{aligned} (1+r)S &= (1+r)^n + (1+r)^{n-1} + \ldots + (1+r)^2 + (1+r) \\ S &= (1+r)^{n-1} + \ldots + (1+r)^2 + (1+r) + 1 \\ \hline rS &= (1+r)^n - 1 \\ S &= \frac{(1+r)^n - 1}{r} \end{aligned}$$

NOTATION FOR THIS FUNCTION

The quantity S defined in example 32-3 is usually denoted as

$$s_{\overline{n}|r}$$

which we read "s angle n at r." It is the amount of an annuity of \$1 per payment for n payments at interest rate r% per interest period. Thus

$$s_{\overline{n}|r} = \frac{(1+r)^n - 1}{r}$$

Note that we can express S in example 32-2 using this notation. It is

$$s_{\overline{5}|0.09}$$

Tables are available for $s_{\overline{n}|r}$ for selected values of n and r. However, we will continue to use a hand calculator for computations in this section.

FORMULA FOR THE AMOUNT OF AN ANNUITY

Let an annuity consist of n payments, each of P dollars, with interest at r% per interest (or payment) period. The amount A of the annuity can be found from Figure 32.2.

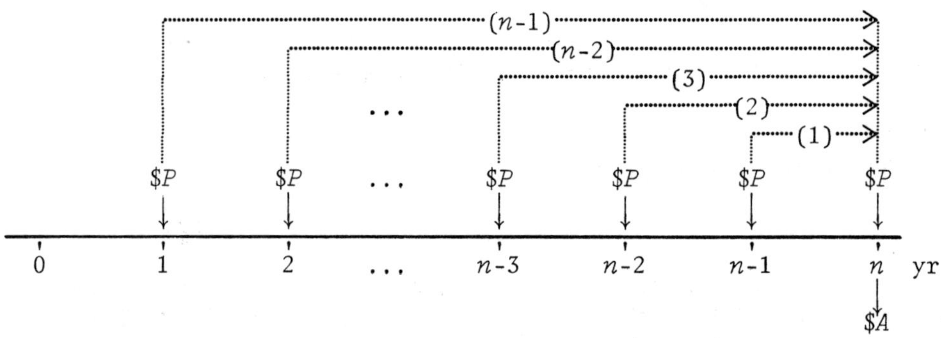

Figure 32.2

We have

$$\begin{aligned} A &= P(1+r)^{n-1} + P(1+r)^{n-2} + \ldots + P(1+r) + P \\ &= P[(1+r)^{n-1} + (1+r)^{n-2} + \ldots + (1+r) + 1] \end{aligned}$$

We now use the formula verified in Example 32-3:

$$A = P\left(\frac{(1+r)^n - 1}{r}\right) = Ps_{\overline{n}|r}$$

Example 32-4

Suppose $50 is deposited monthly in a bank that pays 12% compounded monthly. Find the amount accrued after five years.

Here $P = 50$, $r = 12\% \div 12 = 1\%$, and $n = 5 \times 12 = 60$. Then

$$A = 50\, s_{\overline{60}|0.01} = 50\frac{(1.01)^{60} - 1}{0.01} = \$4083.50$$

Example 32-5

If a couple can manage to save \$1250 every six months and can earn 9% interest compounded semiannually, how much will they have accumulated toward the purchase of a house after seven years?

We have $P = 1250$, $r = 0.045$, and $n = 14$. So

$$A = 1250\, s_{\overline{14}|0.045} = 1250\frac{(1.045)^{14} - 1}{0.045} = \$23{,}665.00$$

FINDING THE PERIODIC PAYMENTS

Often we know how much the amount A of an annuity is (or how much is to be accumulated after n payments), and we want to determine the size of the payments P. We merely solve the amount formula given above for P. Thus

$$P = \frac{A}{s_{\overline{n}|r}} = \frac{rA}{(1 + r)^n - 1}$$

Example 32-6

How much must be deposited every three months in order for a total of 25 deposits to amount to \$10,000 if the interest paid is 8%, compounded quarterly?

It may be easier to use the formula for A, then to solve for P. Note that $A = 10{,}000$, $r = 0.02$, and $n = 25$.

$$10{,}000 = P\, s_{\overline{25}|0.02} = P\frac{(1.02)^{25} - 1}{0.02}$$

$$P = \frac{10{,}000}{s_{\overline{25}|0.02}} = \frac{10{,}000(0.02)}{(1.02)^{25} - 1} = \$312.20$$

Example 32-7

In order to have \$20,000 toward their son's education, a couple decides to invest a fixed amount annually on each of his first 17 birthdays. If the bank pays 6% compounded annually, what amount should be deposited each year to achieve the stated goal of \$20,000 with the 17th payment?

Here $A = 20{,}000$, $r = 0.06$, $n = 17$, and we seek P.

$$20{,}000 = P\, s_{\overline{17}|0.06} = P\frac{(1.06)^{17} - 1}{0.06}$$

so that

$$P = \frac{20{,}000}{s_{\overline{17}|0.06}} = \frac{20{,}000(0.06)}{(1.06)^{17} - 1} = \$708.90$$

SINKING FUNDS

A *sinking fund* is an account into which equal deposits are made at regular intervals in order to accumulate, with interest, a certain amount at a specified time in the future. It is used by businesses in anticipating such future needs as remodeling, replacing obsolete machinery, or purchasing new equipment, and by individuals who foresee future expenditures. A sinking fund is an annuity of the sort being considered in this section.

Example 32-8

A company will need \$150,000 five years from now to pay off a loan. It decides to set up a sinking fund by making ten equal deposits, one every six months. How big must each deposit be if the company can earn 14% compounded semiannually?

We can use the amount formula again, with $A = 150{,}000$, $r = 0.07$, and $n = 10$. Let P be the size of the periodic deposit. Then

$$150{,}000 = P s_{\overline{10}|0.07} = \frac{(1.07)^{10} - 1}{0.07}$$

$$P = \frac{150{,}000}{s_{\overline{10}|0.07}} = \frac{150{,}000(0.07)}{(1.07)^{10} - 1} = \$10{,}856.64$$

EXERCISES

1. Find the amounts of the following annuities:
   (a) \$250 per month for three years with interest at 18% compounded monthly.
   (b) \$1000 semiannually for five years with interest at 15% compounded semiannually.
2. Suppose \$700 is deposited every three months into an account paying 10% interest compounded quarterly.
   (a) How much will there be in the account after the 40th deposit?
   (b) How much will there be two years after the 40th deposit?
3. What is the periodic payment necessary to pay off a debt of \$25,000 due in five years if there are to be ten such payments, one every six months, and the interest rate is 8% compounded semiannually?
4. A parent plans to make a deposit on each of his daughter's first 18 birthdays. If he wishes to have \$3000 on her 18th birthday,

how much should he deposit each year? Assume the bank pays $7\frac{1}{2}\%$ compounded annually.

5. To pay off a bond issue a city sets up a sinking fund. It needs \$200,000 in ten years and plans to make 20 equal deposits, one every six months. What should be the size of the deposits if the city can earn 12% interest compounded semiannually?

TEST A

1. If \$55 is deposited every two months in a bank paying 9% interest compounded six times a year, how much will be in the account after ten years (60 deposits)?
2. What is the periodic payment necessary to reach a final sum of \$250 if there are to be 20 such payments, one every month, and the interest rate is 9% compounded monthly?

TEST B

1. If \$1200 is deposited every year in an account paying 10% interest compounded annually, how much will be in the account after 25 deposits?
2. To pay off a loan a company sets up a sinking fund. It needs \$75,000 in eight years and plans to make 16 equal deposits, one every six months. If the company receives 10% interest compounded semiannually, what should be the size of the deposits?

SECTION 33. PRESENT VALUE OF AN ANNUITY

In §32 we defined the amount of an annuity. In this section we define the present value of an annuity, then consider some important applications.

The *present value* of an annuity is the sum one must invest (or deposit) today to yield an annuity; that is, to generate a sequence of equal payments (or withdrawals) at regular intervals.

We will see that the present value of an annuity is obtained by adding the present values of all the payments, using today as reference point.

Example 33-1

Suppose you want to make five annual withdrawals, each of $100, making the first withdrawal one year from now. How much should you deposit today if you can earn 9% interest compounded annually?

We draw a time line (Figure 33.1), using today as reference point.

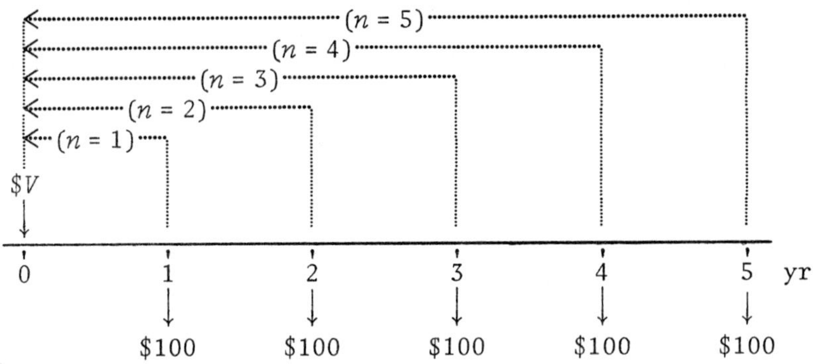

Figure 33.1

Let V be the present value that is the desired answer. Then

$$\begin{aligned} V &= 100(1.09)^{-1} + 100(1.09)^{-2} + 100(1.09)^{-3} + 100(1.09)^{-4} + 100(1.09)^{-5} \\ &= 100[(1.09)^{-1} + (1.09)^{-2} + (1.09)^{-3} + (1.09)^{-4} + (1.09)^{-5}] \qquad (*) \\ &= \$388.97 \end{aligned}$$

With a hand calculator, we found the sum in brackets first, dividing 1 in turn by the successive positive powers of 1.09, then multiplied the sum by 100. (Of course, one may use the negative powers here.)

Example 33-2

Verify that the sum in brackets in line (*) above is equal to

$$\frac{(1.09)^5 - 1}{0.09(1.09)^5}$$

Let

$$Q = (1.09)^{-1} + (1.09)^{-2} + (1.09)^{-3} + (1.09)^{-4} + (1.09)^{-5}$$

We multiply both sides by $(1.09)^5$:

$$(1.09)^5 \cdot Q = (1.09)^4 + (1.09)^3 + (1.09)^2 + (1.09)^1 + 1$$

As was shown in example 32-2, the sum on the right is equal to

$$\frac{(1.09)^5 - 1}{0.09}$$

But since this equals $s_{\overline{5}|0.09}$, we have that

$$Q = \frac{(1.09)^5 - 1}{(0.09)(1.09)^5} = \frac{s_{\overline{5}|0.09}}{(1.09)^5}$$

Example 33-3

Verify that

$$(1+r)^{-1} + (1+r)^{-2} + \ldots + (1+r)^{-n+1} + (1+r)^{-n} = \frac{(1+r)^n - 1}{r(1+r)^n}$$

If we let Q be the sum on the left in the equation and multiply by $(1+r)^n$, we get

$$(1+r)^n \cdot Q = (1+r)^{n-1} + (1+r)^{n-2} + \ldots + (1+r)^1 + 1$$

But, from example 32-3 (p. 311),

$$(1+r)^n \cdot Q = \frac{(1+r)^n - 1}{r}$$

Therefore

$$Q = \frac{(1+r)^n - 1}{r(1+r)^n} = \frac{s_{\overline{n}|r}}{(1+r)^n}$$

NOTATION FOR THIS FUNCTION

The quantity on the right in the last equation is usually denoted by

$$a_{\overline{n}|r}$$

which we read "*a* angle *n* at *r*." It is the present value of an annuity of \$1 per payment for n payments at interest rate r% per interest period. So we can write

$$a_{\overline{n}|r} = \frac{(1+r)^n - 1}{r(1+r)^n} = \frac{1-(1+r)^{-n}}{r}$$

Note that V in example 33-1 is equal to $100\,a_{\overline{5}|0.09}$.

Tables are available for this function for selected values of r and n. The formula at the right above is the simpler of the two for evaluation with a calculator.

FORMULA FOR PRESENT VALUE OF AN ANNUITY

Consider an annuity of n equal withdrawals, each of $\$P$, made at regular intervals. Let the interest be $r\%$ per interest period (which coincides, as usual, with the period between withdrawals). The present value V of the annuity is shown on the time line in Figure 33.2.

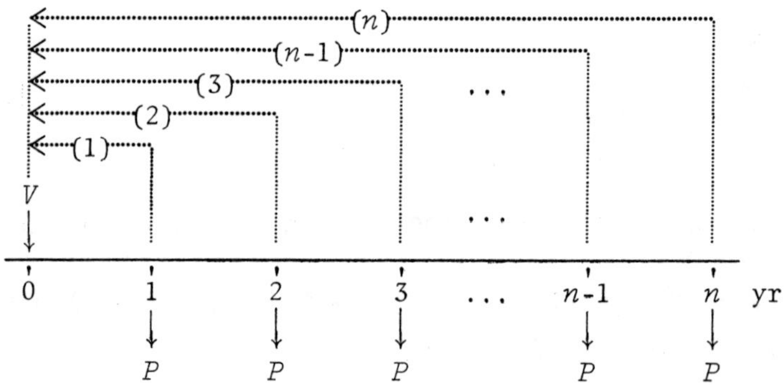

Figure 33.2

We have

$$\begin{aligned} V &= P(1+r)^{-1} + P(1+r)^{-2} + \ldots + P(1+r)^{-1+n} + P(1+r)^{-n} \\ &= P[(1+r)^{-1} + (1+r)^{-2} + \ldots + (1+r)^{-1+n} + (1+r)^{-n}] \end{aligned}$$

So, using the formula of example 33-3, we have

$$V = P\left(\frac{(1+r)^n - 1}{r(1+r)^n}\right) = P\left(\frac{1-(1+r)^{-n}}{r}\right) = Pa_{\overline{n}|r}$$

Example 33-4

Suppose \$50 is to be withdrawn from an account every month for five years. How much must be deposited today if the account can earn 12% compounded monthly?

We have

$$V = 50\,a_{\overline{60}|0.01} = 50\frac{(1.01)^{60} - 1}{(0.01)(1.01)^{60}} = \$2247.76$$

FINDING THE PERIODIC WITHDRAWALS

When we know the present value V of an annuity we can calculate the size of the periodic withdrawals P by solving the last boxed formula for P:

$$P = \frac{V}{a_{\overline{n}|r}} = \frac{Vr(1 + r)^n}{(1 + r)^n - 1}$$

Example 33-5

A man has accumulated \$10,000 in a savings account earning 6% compounded annually. How much can he withdraw at the end of each year so that the balance is zero after ten years?

Using the converted formula, we have

$$P = \frac{10{,}000}{a_{\overline{10}|0.06}} = \frac{10{,}000(0.06)(1.06)^{10}}{(1.06)^{10} - 1} = \$1358.68$$

AMORTIZING A LOAN

When a loan or debt is being repaid by a sequence of equal payments at regular intervals, we say it is being *amortized*. A very common application of amortizing is that in which we know how much the loan or debt is (that is, we know the present value of the annuity) and we want to calculate the periodic payments.

When one buys a car, motorcycle, TV set, furniture, or the like on the installment plan,a loan is being amortized. Usually a cash down-payment is made and the balance is amortized.

Example 33-6

A college student buys a used car for \$1000, pays \$250 down, and borrows the balance from a loan company that charges 20% compounded quarterly. The student agrees to pay off the loan in 12 equal quarterly payments. (a) How much is each payment? (b) Find the total interest to be paid.

Here the present value V is \$750, $r = 0.05$, and $n = 12$. So (a):

$$P = \frac{V}{a_{\overline{n}|r}} = \frac{750}{a_{\overline{12}|0.05}} = \frac{750(0.05)(1.05)^{12}}{(1.05)^{12} - 1} = \$84.62$$

and (b): altogether, the student must pay $12 \times 84.62 = \$1015.44$. So the total interest to be paid is equal to

$$1015.44 - 750 = \$265.44$$

CONSTRUCTING A SCHEDULE

The following table shows how much of each payment in example 33-6 is used to reduce the loan or principal, how much is interest. At the outset, of course, the loan outstanding is \$750.

PAYMENT NUMBER	PAYMENT	INTEREST	PRINCIPAL REPAID (PAYMENT - INTEREST)	LOAN OUTSTANDING AFTER PAYMENT
1	\$84.62	\$37.50	\$47.12	\$702.88
2	84.62	35.14	49.48	653.40
3	84.62	32.67	51.95	601.45
4	84.62	30.07	54.55	546.90
5	84.62	27.35	57.27	489.64
6	84.62	24.48	60.14	429.49
7	84.62	21.47	63.15	366.34
8	84.62	18.32	66.30	300.05
9	84.62	15.00	69.62	230.43
10	84.62	11.52	73.10	157.33
11	84.62	7.87	76.75	80.58
12	84.62	4.03	80.59	0.00

The entries in any given line in the table were found as follows:

The interest equals 5% of the preceding outstanding loan.

The amount of principal repaid equals \$84.62 minus the interest just calculated.

The outstanding loan after a payment is equal to the previous outstanding loan minus the principal repayment just calculated.

Since part of each payment is for the interest due on the outstanding loan at the time of payment, the interest is always greatest at first, but diminishes with successive payments. The part of the payment that can be used to reduce the principal accordingly increases with time. We will return to this point shortly.

OUTSTANDING LOAN; EQUITY

The *outstanding loan* just after a payment is made is defined as the present value of the remaining annuity; that is, the present value of the payments still to be made.

In example 33-6, we can use this definition to calculate the outstanding loan after, say, the fourth payment, by noting that there are eight payments of \$84.62 still to be made. The present value of the remaining annuity thus equals

$$84.62\, a_{\overline{8}|0.05} = 84.62\frac{(1.05)^8 - 1}{(0.05)(1.05)^8} = \$546.92$$

The difference in cents from the amount listed in the table is due to different rounding-off procedures. Still another way of computing the outstanding loan is discussed later as an exercise.

The *equity*, at any time, in property for which a loan is being amortized, is the part of the property already bought at that time. The equity equals the amount of the loan initially made minus the outstanding loan. In example 33-6, thus, the student's equity in the car after the fourth payment is equal to

$$750 - 546.92 = \$203.08$$

Note carefully that equity is defined *only with reference to the loan*; it does not include any down-payment. Thus, in our example, to determine the student's total investment in the car just after the fourth payment we should have to add his down-payment of \$250 to his equity of \$203.08.

MORTGAGES

Strictly, a *mortgage* is a conditional transfer of property pledged by the purchaser of the property as security for the repayment of an amortized loan. We also commonly use the word "mortgage" loosely to mean the loan--as when we speak of "paying off the mortgage." For the purchase of a house, a store, or other large item the loan is usually repaid in monthly installments, and often takes 20 years or more to repay. As in example 33-6, part of each payment is for the interest due on the outstanding loan; the remainder of the payment goes toward the reduction of the loan.

Example 33-7

The Bennetts pay \$25,000 down on a \$75,000 house and arrange for a 25-year mortgage on the balance. The interest is 12% compounded monthly. (a) How much are the monthly mortgage payments? (b) How much interest will the Bennetts pay altogether? (c) What is their equity in the house after ten years? (d) What is their equity after 20 years?

We note that the present value A of the loan is \$50,000, that $r = 1\%$, and that $n = 25 \times 12 = 300$.

For (a) we have*

$$P = \frac{50{,}000}{a_{\overline{300}|0.01}} = \frac{50{,}000(0.01)(1.01)^{300}}{(1.01)^{300} - 1} = \$526.61$$

The monthly payments are therefore \$526.61.

(b) Altogether, after 25 years, the Bennetts will have paid

$$300 \times 526.61 = \$157{,}983$$

So the amount of interest paid during the life of the mortgage is

$$157{,}983 - 50{,}000 = \$107{,}983$$

(c) After ten years, there are still $15 \times 12 = 180$ payments left. The outstanding loan after ten years is the present value of this 180-payment annuity, or

$$526.61\, a_{\overline{180}|0.01} = 526.61 \frac{(1.01)^{180} - 1}{(0.01)(1.01)^{180}} = \$43{,}878.02$$

The equity in the house after ten years is therefore

$$50{,}000 - 43{,}878.02 = \$6{,}121.98$$

(d) After 20 years, $5 \times 12 = 60$ payments remain. The outstanding loan at that time is therefore

$$526.61\, a_{\overline{60}|0.01} = 526.61 \frac{(1.01)^{60} - 1}{(0.01)(1.01)^{60}} = \$23{,}673.83$$

The equity in the house after 20 years is

$$50{,}000 - 23{,}673.83 = \$26{,}326.17$$

The Shock Factor

Young people who take out their first house mortgage loan are often quite surprised to discover how slowly they are paying off the principal. Many are not aware of how a loan of this sort is amortized. The preceding example makes this point patently clear. Note, from the computations above, that

> after ten years, 40% of the term of the mortgage has gone by, but *over 87%* of the loan is still unpaid;
>
> after 20 years, 80% of the term has gone by, but *over 47%* of the loan remains unpaid.

It is especially disturbing to some people paying off a mortgage loan to learn how much--that is, how little!--of the loan is paid off during the first year. In example 33-7 the Bennetts made payments totaling \$6319.32 during the first year of their mortgage, of which *only \$337.70* went toward reducing their loan of \$50,000.

---

*A table of values of $a_{\overline{n}|r}$ for some values of n between 60 and 360 is provided in Appendix 2.

Furthermore, if the interest rate had been 15% (a common rate in the early 1980s), the Bennetts would have had:

> a monthly payment of \$640.42;
> total interest of \$142,126;
> payments totalling \$7,685.04 during the first year, but only \$197.88 of the loan paid off during that year.

Example 33-8

At age 35, Frank Crowell decides he wants to withdraw \$10,000 on each of his 15 birthdays starting at age 61 and ending at age 75. How much must he deposit annually starting at age 36 and ending on his 60th birthday to achieve this retirement goal, if he can earn 8% compounded annually?

To begin with, Crowell must have enough saved up on his 60th birthday to yield a 15-year annuity with equal withdrawals of \$10,000, as shown in Figure 33.3. We let V be the present value of

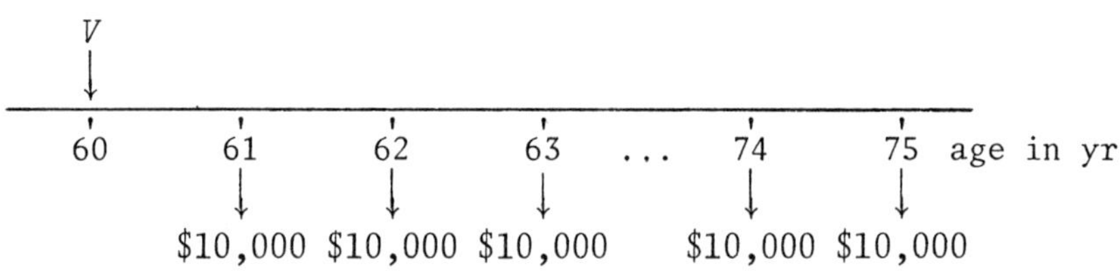

Figure 33.3

the annuity. We have, then, that

$$V = 10,000\,a_{\overline{15}|0.08} = 10,000\frac{(1.08)^{15} - 1}{(0.08)(1.08)^{15}} = \$85,594.80$$

Secondly, Crowell must deposit enough annually for 25 years to accrue to the sum just calculated. Figure 33.4 shows the relevant

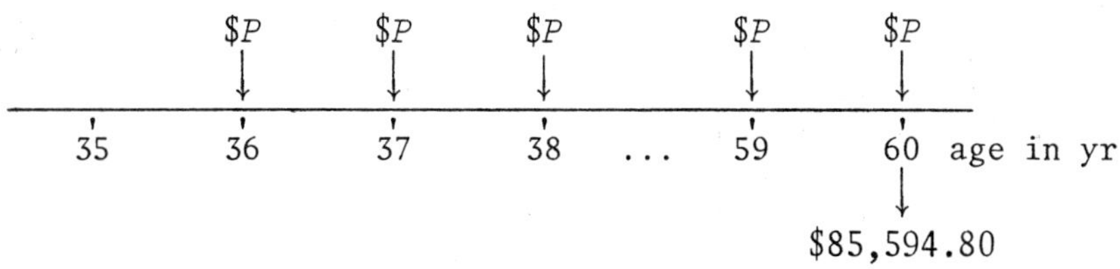

Figure 33.3

time line, where P is the size of the yearly deposit.

We now calculate P:

$$P = \frac{85{,}594.80}{s_{\overline{25}|0.08}} = \frac{85{,}594.80(0.08)}{(1.08)^{25} - 1} = \$1170.83$$

REVIEW OF RELEVANT FORMULAS ON ANNUITIES

An annuity is a sequence of equal periodic payments or withdrawals.

If a problem involves the (final) amount A of an annuity then we use either of the following:

$$A = P s_{\overline{n}|r} = P\left(\frac{(1+r)^n - 1}{r}\right)$$

$$P = \frac{A}{s_{\overline{n}|r}} = \frac{rA}{(1+r)^n - 1}$$

If a problem involves the present value V of an annuity then we use either of these two:

$$V = P a_{\overline{n}|r} = P\left(\frac{(1+r)^n - 1}{r(1+r)^n}\right) = P\left(\frac{1 - (1+r)^{-n}}{r}\right)$$

$$P = \frac{V}{a_{\overline{n}|r}} = V\left(\frac{r(1+r)^n}{(1+r)^n - 1}\right)$$

EXERCISES

1. Suppose you want to withdraw \$250 a month for four years, starting a month from now. How much must you have in your account now, if you can earn 8% compounded monthly?
2. A schoolteacher has savings of \$3500 in an account earning 10% compounded semiannually. She decides to withdraw all of it in 25 equal withdrawals, one each six months. How much can she withdraw at a time?
3. A businessman borrows \$10,000 from a loan company; he plans to pay it off in 36 equal installments, one every month. If the company charges him 18% interest compounded monthly, how much will each payment be? How much interest will he pay altogether on this loan?
4. A couple pays \$25,000 in cash down on a \$100,000 house and gets a 20-year mortgage loan on the balance. If the interest rate is 9% compounded monthly, (a) what will the monthly payments be? (b) how much interest will the couple pay altogether? (c) what

is their equity in the house after ten years? (d) what is their equity after 15 years? (e) how much of the loan is paid off after one year?

5. Do the preceding exercise for an annual interest rate of 15% compounded monthly.
6. A family anticipates that their daughter will need $6000 on each of her birthdays, 18th through 21st, for college expenses. How much must be deposited on each of her first 17 birthdays to meet the anticipated need? Their bank is paying them 7% compounded annually.
7. In example 33-6 we found the outstanding loan after the fourth payment by calculating $\$84.62\, a_{\overline{8}|0.05}$. An alternative procedure is to find

$$750(1.05)^4 - 84.62\, s_{\overline{4}|0.05}$$

Justify this alternative.

TEST A

1. If an investment account is earning 12% compounded quarterly, how much must be in the account now to allow for withdrawals of $125 every three months for the next five years?
2. A nurse borrows $600 from a loan company, which he plans to pay off in ten equal installments, one every month. If the company charges him 15% interest compounded monthly, (a) what will be the size of the payments? (b) how much interest will he pay altogether?

TEST B

1. Suppose that $1500 is to be withdrawn at the end of every year for 20 years. How much must be in the account now if the money is earning 6% compounded annually?
2. Dr Zeller pays $40,000 down on a house costing $120,000 and takes out a 30-year mortgage loan on the balance. If the interest rate is 10% compounded monthly, how much are her monthly payments? How much interest will she have paid when the mortgage is paid off?

SUMMARY OF §§29-33

You should now be able to

(1) use the formula $I = PRt$ for simple interest to find any one variable given the others;
(2) use the formula $A = P(1 + Rt)$ for the amount in simple interest, as in (1);
(3) use the compound-interest formula $A = P(1 + r)^n$ to find any one variable given the others;
(4) find the effective rate of interest for a given investment;
(5) compare investment options;
(6) determine the present or future value of an investment;
(7) use an equation of value when appropriate in applications involving the mathematics of finance;
(8) find the amount and present value of an annuity;
(9) find the periodic payment necessary for an annuity to accumulate to a specified amount at a specified time in the future;
(10) find the periodic payment available from an annuity given its present value;
(11) apply the mathematics of finance to problems involving sinking funds, amortizing a loan, retirement funds, resource management.

REVIEW EXERCISES

1. If you deposit \$300 today at $5\frac{1}{2}\%$ simple interest, how much interest will you earn in two years?
2. How long will it take to double an investment at 6% simple interest?
3. How much must you invest today at 8% simple interest to have \$1000 in your account in four years?
4. Suppose \$5000 is invested at 6%. How much is in the account after one year if the interest is (a) simple? (b) compounded semiannually? (c) compounded quarterly? (d) compounded monthly?
5. Find the effective rate equivalent to 12% compounded (a) semiannually; (b) monthly.

*6. How long will it take to double an investment at 10% compounded quarterly?

*7. What interest rate compounded monthly will double your investment in six years?

8. How much can a businessman borrow today at 14% compounded quarterly if he figures he can pay off \$5000 to cover the loan and all accumulated interest in three years?
9. Find the future value (or amount) of \$1500 at 7% compounded twice a year for three years.
10. Find the present value of a sum worth \$2500 in three years if the interest rate is 9% compounded monthly.
11. How much is in an account today paying 9% interest compounded quarterly if the following deposits were made: \$1000 five years ago, \$500 three years ago, and \$700 one year ago?
12. A family borrows \$5000 for a car and plans to pay it off as follows: \$1000 in one year, \$2000 in two years, and the balance in three years. (a) What will be the final payment if the interest charged is 12%, compounded monthly? (b) How much interest will be paid altogether?
13. A company that promised to make a payment of \$10,000 three years from now wishes instead to settle the obligation by making two equal payments two years from now and four years from now. If they are being charged 12% compounded quarterly, how much must each payment be?
14. Find the amount and present value of an annuity of \$200 a month for $6\frac{1}{4}$ years at 6% compounded monthly.

---

*Solution requires logarithms.

15. How much must be deposited semiannually for five years into an account that earns 10% compounded semiannually to accumulate to \$6000?
16. How much must a county deposit annually into a sinking fund that pays $7\frac{1}{2}\%$ annually to retire a debt of \$40,000 in 10 years?
17. On his 30th birthday a college professor instructs the payroll office to withhold \$25 from each monthly check starting in one month. These amounts are then deposited into a retirement fund that pays 6% compounded monthly. After ten years he increases the amount withheld monthly to \$50. How much is in the retirement fund on his 65th birthday?
18. If a widow has \$18,000 today in an account that pays 10% quarterly, what amount withdrawn quarterly for ten years will deplete the account?
19. What is the initial cash value of a car on which you make a down payment of \$500 and payments of \$100 a month for three years, if you're paying 18% compounded monthly?
20. The Stones put \$20,000 down on an \$80,000 house and get a 25-year mortgage loan on the balance. The bank charges them 12% compounded monthly. (a) How much are the monthly payments? (b) How much interest will the Stones pay altogether? (c) How much equity do they have in the house after ten years?
21. How much must you deposit at the end of every year for the next 30 years to be able to withdraw \$2000 a year for the following 20 years, if money is worth 8% annually?
22. If the college professor in exercise 17 retires at 65, how much can he withdraw every month for ten years before depleting his retirement fund? (Assume the same interest rate prevails.)

CHAPTER TEST

1. If you borrow $700 for one year at 12% simple interest, how much must you repay in a year?
2. How much must you deposit today at $5\frac{1}{2}\%$ simple interest to accumulate $2000 in four years?
3. How much will a deposit of $1000 amount to in five years if the account earns 6% (a) simple interest? (b) compounded semiannually ? (c) compounded monthly?
4. Find the effective rate of interest equivalent to (a) 12% compounded quarterly; (b) 12% compounded monthly.
5. What lump-sum deposit must be made today to accumulate $4500 in six years if you can earn 8% compounded quarterly?
6. How long will it take to double an investment at (a) 9% simple interest? *(b) 9% compounded semiannually? *(c) 9% compounded monthly?
7. A businessman owes $3000 today and $5000 two years from now. If money is worth 10% compounded annually, what single payment one year from now will discharge the debt?
8. Find the present and future values (or amounts) of an annuity of $200 a month for five years at 15% compounded monthly.
9. Starting a month from now, a policeman expects to deposit $40 a month in an account that pays 6% compounded monthly. After five years he plans to transfer his accumulated funds to a 30-month certificate that pays 12%, compounded monthly. If his plans materialize, how much will he have when the certificate matures?
10. How much most a neighborhood grocer who plans to open another store deposit annually for eight years in an account that pays 11% compounded annually to accumulate $40,000?

---

*Solution requires logarithms.

11. Suppose a father decides on his son's birth to be able to provide him with \$8000 a year for college costs on each of his four birthdays, 18th through 21st. How much must the father deposit on each of the boy's first 17 birthdays to insure the needed resources, if he can earn $7\frac{1}{2}\%$ compounded annually?

12. A couple needs a \$40,000 mortgage loan on a house. Bank A offers them a 25-year loan at 12%, bank B offers a 20-year mortgage at 13%, and bank C a 15-year mortgage at 14%, all with monthly compounding. (a) At which bank will the monthly payment be least? (b) Find the total interest to be paid to each bank.

Appendix 1

Basics of set theory

Sets arise frequently in this text because of their usefulness in clarifying mathematical concepts. In this appendix we summarize the basics of set theory.

DEFINITIONS; NOTATION; TERMINOLOGY

A *set* is any collection of objects (people, things, numbers, letters, ideas, etc.). We usually use capital letters to denote sets. The objects in the set are called its *elements*, often listed (or described) within a pair of braces:

A = {1,2,3,4,5} is the set of the first five positive integers.
B = {A,E,I,O,U} is the set of vowel-letters in the English alphabet.
C = {all the positive odd integers} is the infinite set 1,3,5,...
D = {HH,HT,TH,TT} is the set of possible outcomes when a coin is tossed twice, with H for "heads" and T for "tails." TH thus means "tails on first toss, heads on second."

The notation

$$\{x \mid x \text{ is an integer, } 1 \leqslant x \leqslant 100\}$$

is read "the set of all x such that x is an integer between 1 and 100 inclusive." This is known as *set-builder notation.*

Note that the set just named is *equal* to the set

$$\{1,2,3,\ldots,100\}$$

because the two have exactly the same elements. Thus

$$\{p,q,r,s\} = \{q,r,s,p\} \quad \text{but} \quad \{p,r,s\} \neq \{p,q,s\}$$

It is necessary to recognize a special set that contains no elements at all. This is called the *empty set*, and is denoted by $\emptyset$. It arises frequently in both theory and applications.

The set of all elements of any possible relevance in a given problem is called the *universal set*, usually denoted by U. When considering the students who are taking accounting or mathematics or both at a certain college, we could take the universal set to be all the students at the college, or just all those enrolled in the Business School of the college. If our concern is the racial extraction of the population of Honolulu, we could choose as the universal set all the residents of the city, or all those of the state of Hawaii. There is almost always some range of choice in our definition of the universal set for a problem; the crucial consideration is to be sure it is *large enough.*

INTERSECTION; UNION; COMPLEMENT

Sets can be used to form new sets.

The *intersection* of sets A and B, denoted by $A \cap B$, is the set of elements common to A AND B. The *union* of sets A and B, denoted by $A \cup B$, is the set of elements in A OR in B OR in both. If, for example,

$$A = \{b,c,e,g,u\} \quad \text{and} \quad B = \{b,d,g,u,w\}$$

then

$$A \cap B = \{b,g,u\} \quad \text{and} \quad A \cup B = \{b,c,d,e,g,u,w\}$$

Or if

$$A = \{x \mid x \text{ is an integer, } 2 \leqslant x \leqslant 7\}$$

and

$$B = \{x \mid x \text{ is a positive integer, } x > 5\}$$

then

$$A \cap B = \{6,7\} \quad \text{and} \quad A \cup B = \{x \mid x \text{ is an integer, } x \geqslant 2\}$$

We often read $A \cap B$ as "A intersect B" and $A \cup B$ as "A union B."

Note that if x is an element of $A \cup B$ then x belongs *at least* to one of A and B. Similarly, $A \cup B \cup C$ is the set of elements each of which belongs to *at least* one of the sets A, B, and C. The set $A \cap B \cap C$ contains only those elements which are common to A, B, and C--which belong to *all three* of those sets.

Another way to form a new set from a given set A is by taking its *complement*. The set of elements in U (the universal set for the given problem) which are *not* in A is the complement of A, denoted by A' and read "A prime." If, for example, U is the set of students in a particular high school and A is the set of girls at the school, then its complement A' is the set of boys enrolled in the school. If B is the set of vowel-letters in the English alphabet, then B' is the set of consonant-letters. If U is the set of all integers and P is the set of all positive integers, then P' consists of all the negative integers and the number zero.

SUBSETS

Set A is called a *subset* of B if every element of A is also an element of B. We write $A \subseteq B$. Thus

$$\{3,5\} \subset \{3,5,9\}$$

If $B = \{3,5,9\}$ then all the following are subsets of B:

{3} {5} {9} {3,5} {3,9} {5,9} {3,5,9}

There is still one more subset of B: the empty set $\emptyset$. It qualifies as a subset because every element of $\emptyset$ is in fact also an element of B. Indeed, the empty set is a subset of *every* set.

It also follows from the definition above that every set is a subset of itself.

DISJOINT SETS

Sets A and B are said to be *disjoint* if they have no element in common; that is, if $A \cap B = \emptyset$.

In the English alphabet, the set of vowel-letters and the set of consonant-letters are disjoint. Also, the set of even integers and the set of odd integers are disjoint.

VENN DIAGRAMS

All of the definitions given above can be illustrated using figures called *Venn diagrams* (named after the British logician John Venn, 1834-1923).

In such a diagram the universal set is usually represented by the interior of a rectangle, and the sets by the interiors of other geometric figures, often circles. Regions of the diagram then corre-

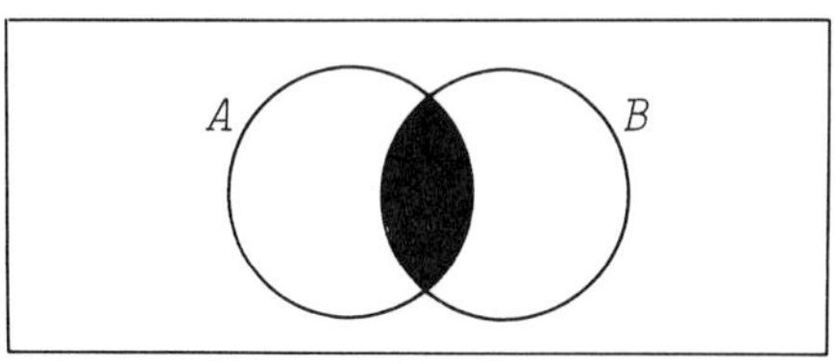

$A \cap B$
The intersection of A and B is all the elements in both A and B

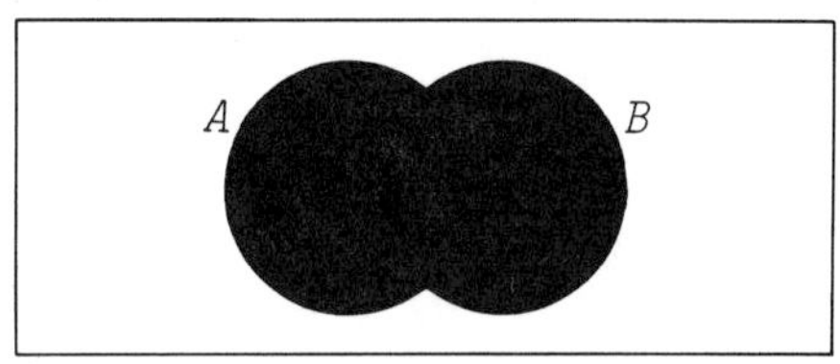

$A \cup B$
The union of A and B is the set of elements in A or B or both

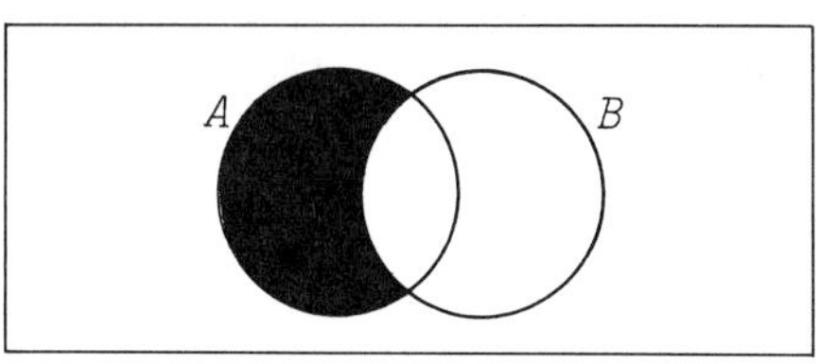

$A \cap B'$
The intersection of A with the complement of B is the set of elements in A but not in B

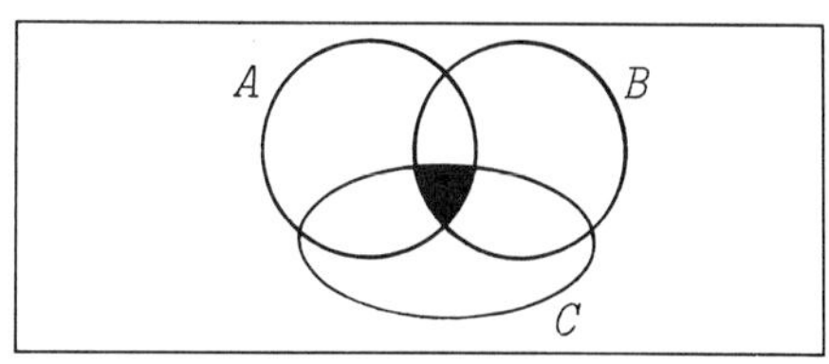

$A \cap B \cap C$
The intersection of sets A, B, and C is the set of elements in all three sets

Figure A1.1

spond to unions, intersections or complements of the sets. Figure A1.1 shows Venn diagrams for four sets obtained from sets A, B, and C by union, intersection, and complementation. Note that the letter "A" denotes the interior of the entire circle at the left, "B" of the entire circle at the right. The shaded region in each diagram represents the set specified beneath the diagram.

The diagrams given in Figure A1.2 illustrate other definitions given above.

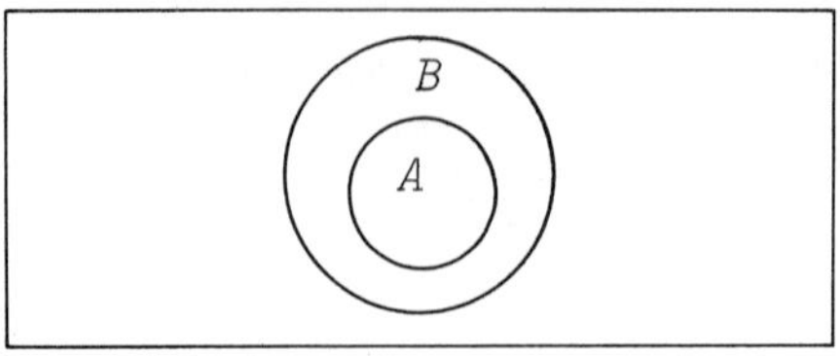

$A \subset B$
A is a subset of B.

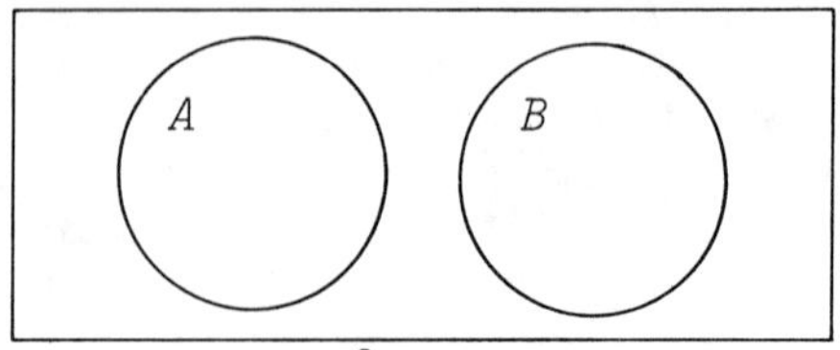

$A \cap B = \emptyset$
Sets A and B are disjoint.

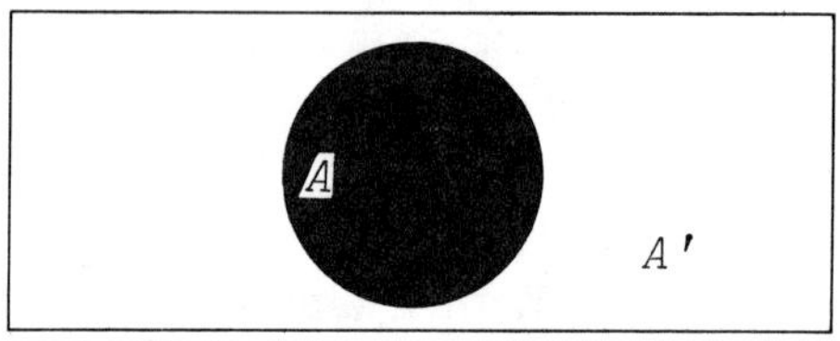

Set A' is the complement of A.

Figure A1.2

PARTITION OF A SET

Instead of using shading, as above, to identify a set, it is often more convenient to number the regions into which a universal set may be divided. In the following diagrams, the sets represented by the numbered regions are both disjoint and *exhaustive*: that is, every element of U is in one *and only one* of the numbered regions. A decomposition of U into disjoint, exhaustive sets is called a *partition* of U.

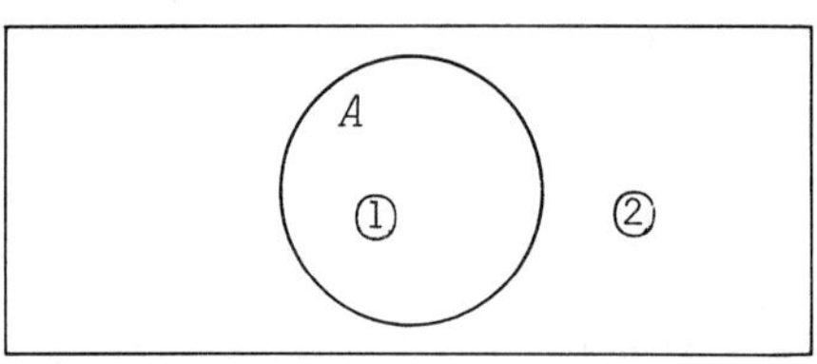

Set ① = A Set ② = A'
Note that
$A \cap A' = \emptyset$ $A \cup A' = U$

Figure A1.3

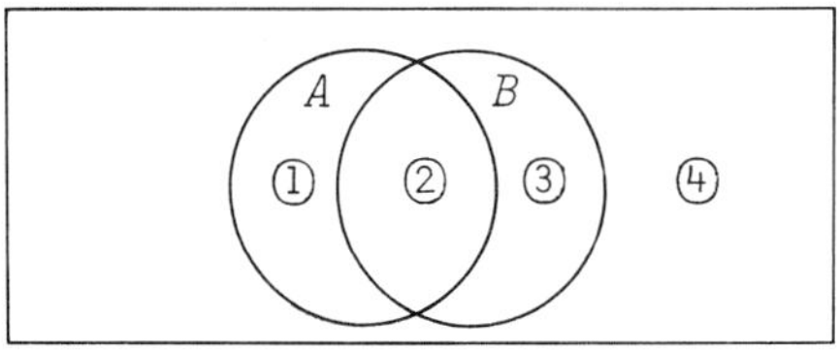

Set ① = $A \cap B'$
Set ② = $A \cap B$
Set ③ = $A' \cap B$
Set ④ = $A' \cap B'$

← Figure A1.4

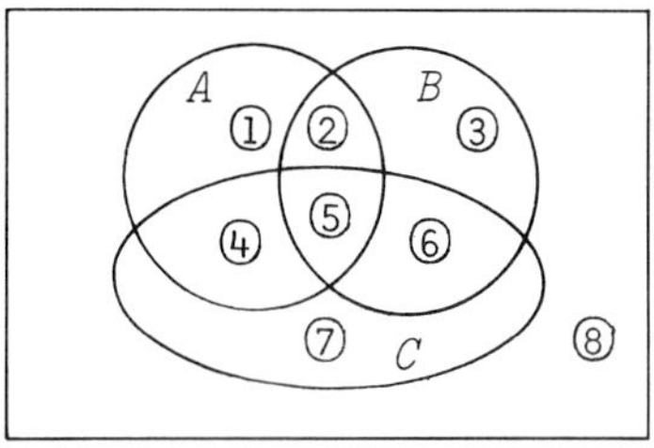

Set
① = $A \cap B' \cap C'$
② = $A \cap B \cap C'$
③ = $A' \cap B \cap C'$
④ = $A \cap B' \cap C$
⑤ = $A \cap B \cap C$
⑥ = $A' \cap B \cap C$
⑦ = $A' \cap B' \cap C$
⑧ = $A' \cap B' \cap C'$

← Figure A1.5

To ascertain quickly which intersection of sets is represented by a numbered region, we decide whether an element in the region is in each of the given sets or in its complement. For example, take region ⑥. An element in ⑥ is not in A (it is therefore in A'), it is in B, and it is in C. Thus the element is in $A' \cap B \cap C$. An element in ① is in A, but is not in B nor in C. Region ① is therefore the set $A \cap B' \cap C'$.

VERIFYING THEOREMS

Example A1-1

Use a Venn diagram to verify the following *distributive law*:

$$A \cup (B \cap C) = (A \cup B) \cap (A \cup C)$$

We use Figure A1.5 above, looking first at the set listed on the left in our proposed law. A consists of the regions with numbers 1, 2, 4, and 5. $B \cap C$ consists of regions 5 and 6. So

$$A \cup (B \cap C) = (1,2,4,5) \cup (5,6) = (1,2,4,5,6)$$

Now we look at the set listed on the right in the proposed law. $A \cup B$ = 1,2,3,4,5,6; $A \cup C$ = 1,2,4,5,6,7; so

$$(A \cup B) \cap (A \cup C) = (1,2,3,4,5,6) \cap (1,2,4,5,6,7) = (1,2,4,5,6)$$

We have thus shown that

$$A \cup (B \cap C) = (A \cup B) \cap (A \cup C)$$

This distributive law is true for all sets A, B, C.

The student should verify the following distributive law:

$$A \cap (B \cup C) = (A \cap B) \cup (A \cap C)$$

Example A1-2

Use a Venn diagram to verify the following theorem (one of *De Morgan's laws*, named after Augustus De Morgan, 1806-1871, British mathematician and logician):

$$(A \cup B)' = A' \cap B'$$

Using Figure A1.4 above, we have

$$A \cup B = 1,2,3$$
$$(A \cup B)' = (1,2,3)' = 4$$
$$A' \cap B' = (3,4) \cap (1,4) = 4$$

Since both sides of the equation name region 4, the law is verified.

The student should verify the other of De Morgan's laws, namely:

$$(A \cap B)' = A' \cup B'$$

The technique illustrated above, of numbering the regions of a Venn diagram, is useful in problems involving counting.

THE NUMBER OF ELEMENTS IN A UNION OF SETS

The following theorem is used repeatedly in Chapter 1 and is the basis of the addition law for the probability of a union of events in Chapter 2.

We denote the number of elements in set X by $n(X)$. If A and B are finite sets, then

$$n(A \cup B) = n(A) + n(B) - n(A \cap B)$$

Refer to Figure A1.4 above.

Suppose $A = \{c,e,f,g\}$ and $B = \{a,e,g,h,m\}$. Then

$$A \cup B = \{a,c,e,f,g,h,m\}$$

Since the two elements e and g are in both A and B, the sum $n(A) + n(B)$ counts those elements twice. Indeed, whatever sets A and B may be, the sum $n(A) + n(B)$ counts every element in $A \cap B$ twice. To compensate, we subtract $n(A \cap B)$, getting the law of addition presented above. In our example, we have

$$n(A \cup B) = n\{a,c,e,f,g,h,m\} = 7$$
$$n(A) = 4 \qquad n(B) = 5$$
$$n(A \cap B) = n\{e,g\} = 2$$

So

$$n(A) + n(B) - n(A \cap B) = 4 + 5 - 2 = 7$$

Appendix 2

Tables

TABLE 1

$(1+r)^n$

The next eight pages (pp. 338-345) present the values of $(1+r)^n$ for n from 1 to 100 and for r from 0.005 to 0.08 by steps of 0.005 (that is, from 0.5% to 8% by steps of 0.5%).

The values were generated on a Vax-11/750 under Vax/Vms 3.4, using Vax-11 PASCAL 2.2 with double precision (14 decimal digits used in computations). The print-out shows five decimal positions. The programmer was Randel Morgan.

The computations were also made, independently, on an IBM-PC using IBM-PC BASIC 1.10, with double precision. The values agreed exactly.

$n \downarrow$	$r \rightarrow$ 0.005	0.01	0.015	0.02	0.025	0.03	0.035	0.04
1	1.00500	1.01000	1.01500	1.02000	1.02500	1.03000	1.03500	1.04000
2	1.01003	1.02010	1.03023	1.04040	1.05062	1.06090	1.07123	1.08160
3	1.01508	1.03030	1.04568	1.06121	1.07689	1.09273	1.10872	1.12486
4	1.02015	1.04060	1.06136	1.08243	1.10381	1.12551	1.14752	1.16986
5	1.02525	1.05101	1.07728	1.10408	1.13141	1.15927	1.18769	1.21665
6	1.03038	1.06152	1.09344	1.12616	1.15969	1.19405	1.22926	1.26532
7	1.03553	1.07214	1.10984	1.14869	1.18869	1.22987	1.27228	1.31593
8	1.04071	1.08286	1.12649	1.17166	1.21840	1.26677	1.31681	1.36857
9	1.04591	1.09369	1.14339	1.19509	1.24886	1.30477	1.36290	1.42331
10	1.05114	1.10462	1.16054	1.21899	1.28008	1.34392	1.41060	1.48024
11	1.05640	1.11567	1.17795	1.24337	1.31209	1.38423	1.45997	1.53945
12	1.06168	1.12683	1.19562	1.26824	1.34489	1.42576	1.51107	1.60103
13	1.06699	1.13809	1.21355	1.29361	1.37851	1.46853	1.56396	1.66507
14	1.07232	1.14947	1.23176	1.31948	1.41297	1.51259	1.61869	1.73168
15	1.07768	1.16097	1.25023	1.34587	1.44830	1.55797	1.67535	1.80094
16	1.08307	1.17258	1.26899	1.37279	1.48451	1.60471	1.73399	1.87298
17	1.08849	1.18430	1.28802	1.40024	1.52162	1.65285	1.79468	1.94790
18	1.09393	1.19615	1.30734	1.42825	1.55966	1.70243	1.85749	2.02582
19	1.09940	1.20811	1.32695	1.45681	1.59865	1.75351	1.92250	2.10685
20	1.10490	1.22019	1.34686	1.48595	1.63862	1.80611	1.98979	2.19112
21	1.11042	1.23239	1.36706	1.51567	1.67958	1.86029	2.05943	2.27877
22	1.11597	1.24472	1.38756	1.54598	1.72157	1.91610	2.13151	2.36992
23	1.12155	1.25716	1.40838	1.57690	1.76461	1.97359	2.20611	2.46472
24	1.12716	1.26973	1.42950	1.60844	1.80873	2.03279	2.28333	2.56330
25	1.13280	1.28243	1.45095	1.64061	1.85394	2.09378	2.36324	2.66584

n ↓	r → 0.005	0.01	0.015	0.02	0.025	0.03	0.035	0.04
26	1.13846	1.29526	1.47271	1.67342	1.90029	2.15659	2.44596	2.77247
27	1.14415	1.30821	1.49480	1.70689	1.94780	2.22129	2.53157	2.88337
28	1.14987	1.32129	1.51722	1.74102	1.99650	2.28793	2.62017	2.99870
29	1.15562	1.33450	1.53998	1.77584	2.04641	2.35657	2.71188	3.11865
30	1.16140	1.34785	1.56308	1.81136	2.09757	2.42726	2.80679	3.24340
31	1.16721	1.36133	1.58653	1.84759	2.15001	2.50008	2.90503	3.37313
32	1.17304	1.37494	1.61032	1.88454	2.20376	2.57508	3.00671	3.50806
33	1.17891	1.38869	1.63448	1.92223	2.25885	2.65234	3.11194	3.64838
34	1.18480	1.40258	1.65900	1.96068	2.31532	2.73191	3.22086	3.79432
35	1.19073	1.41660	1.68388	1.99989	2.37321	2.81386	3.33359	3.94609
36	1.19668	1.43077	1.70914	2.03989	2.43254	2.89828	3.45027	4.10393
37	1.20266	1.44508	1.73478	2.08069	2.49335	2.98523	3.57103	4.26809
38	1.20868	1.45953	1.76080	2.12230	2.55568	3.07478	3.69601	4.43881
39	1.21472	1.47412	1.78721	2.16474	2.61957	3.16703	3.82537	4.61637
40	1.22079	1.48886	1.81402	2.20804	2.68506	3.26204	3.95926	4.80102
41	1.22690	1.50375	1.84123	2.25220	2.75219	3.35990	4.09783	4.99306
42	1.23303	1.51879	1.86885	2.29724	2.82100	3.46070	4.24126	5.19278
43	1.23920	1.53398	1.89688	2.34319	2.89152	3.56452	4.38970	5.40050
44	1.24539	1.54932	1.92533	2.39005	2.96381	3.67145	4.54334	5.61652
45	1.25162	1.56481	1.95421	2.43785	3.03790	3.78160	4.70236	5.84118
46	1.25788	1.58046	1.98353	2.48661	3.11385	3.89504	4.86694	6.07482
47	1.26417	1.59626	2.01328	2.53634	3.19170	4.01190	5.03728	6.31782
48	1.27049	1.61223	2.04348	2.58707	3.27149	4.13225	5.21359	6.57053
49	1.27684	1.62835	2.07413	2.63881	3.35328	4.25622	5.39606	6.83335
50	1.28323	1.64463	2.10524	2.69159	3.43711	4.38391	5.58493	7.10668

n ↓	r → 0.005	0.01	0.015	0.02	0.025	0.03	0.035	0.04
51	1.28964	1.66108	2.13682	2.74542	3.52304	4.51542	5.78040	7.39095
52	1.29609	1.67769	2.16887	2.80033	3.61111	4.65089	5.98271	7.68659
53	1.30257	1.69447	2.20141	2.85633	3.70139	4.79041	6.19211	7.99405
54	1.30908	1.71141	2.23443	2.91346	3.79392	4.93412	6.40883	8.31381
55	1.31563	1.72852	2.26794	2.97173	3.88877	5.08215	6.63314	8.64637
56	1.32221	1.74581	2.30196	3.03117	3.98599	5.23461	6.86530	8.99222
57	1.32882	1.76327	2.33649	3.09179	4.08564	5.39165	7.10559	9.35191
58	1.33546	1.78090	2.37154	3.15362	4.18778	5.55340	7.35428	9.72599
59	1.34214	1.79871	2.40711	3.21670	4.29248	5.72000	7.61168	10.11503
60	1.34885	1.81670	2.44322	3.28103	4.39979	5.89160	7.87809	10.51963
61	1.35559	1.83486	2.47987	3.34665	4.50978	6.06835	8.15382	10.94041
62	1.36237	1.85321	2.51707	3.41358	4.62253	6.25040	8.43921	11.37803
63	1.36918	1.87174	2.55482	3.48186	4.73809	6.43791	8.73458	11.83315
64	1.37603	1.89046	2.59314	3.55149	4.85654	6.63105	9.04029	12.30648
65	1.38291	1.90937	2.63204	3.62252	4.97796	6.82998	9.35670	12.79874
66	1.38982	1.92846	2.67152	3.69497	5.10241	7.03488	9.68419	13.31068
67	1.39677	1.94774	2.71160	3.76887	5.22997	7.24593	10.02313	13.84311
68	1.40376	1.96722	2.75227	3.84425	5.36072	7.46331	10.37394	14.39684
69	1.41078	1.98689	2.79355	3.92114	5.49473	7.68721	10.73703	14.97271
70	1.41783	2.00676	2.83546	3.99956	5.63210	7.91782	11.11283	15.57162
71	1.42492	2.02683	2.87799	4.07955	5.77291	8.15536	11.50177	16.19448
72	1.43204	2.04710	2.92116	4.16114	5.91723	8.40002	11.90434	16.84226
73	1.43920	2.06757	2.96498	4.24436	6.06516	8.65202	12.32099	17.51595
74	1.44640	2.08825	3.00945	4.32925	6.21679	8.91158	12.75222	18.21659
75	1.45363	2.10913	3.05459	4.41584	6.37221	9.17893	13.19855	18.94525

n ↓ r →	0.005	0.01	0.015	0.02	0.025	0.03	0.035	0.4
76	1.46090	2.13022	3.10041	4.50415	6.53151	9.45429	13.66050	19.70306
77	1.46821	2.15152	3.14692	4.59424	6.69480	9.73792	14.13862	20.49119
78	1.47555	2.17304	3.19412	4.68612	6.86217	10.03006	14.63347	21.31083
79	1.48292	2.19477	3.24203	4.77984	7.03372	10.33096	15.14564	22.16327
80	1.49034	2.21672	3.29066	4.87544	7.20957	10.64089	15.67574	23.04980
81	1.49779	2.23888	3.34002	4.97295	7.38981	10.96012	16.22439	23.97179
82	1.50528	2.26127	3.39012	5.07241	7.57455	11.28892	16.79224	24.93066
83	1.51281	2.28388	3.44097	5.17386	7.76392	11.62759	17.37997	25.92789
84	1.52037	2.30672	3.49259	5.27733	7.95801	11.97642	17.98827	26.96500
85	1.52797	2.32979	3.54498	5.38288	8.15696	12.33571	18.61786	28.04360
86	1.53561	2.35309	3.59815	5.49054	8.36089	12.70578	19.26948	29.16535
87	1.54329	2.37662	3.65213	5.60035	8.56991	13.08695	19.94392	30.33196
88	1.55101	2.40038	3.70691	5.71235	8.78416	13.47956	20.64195	31.54524
89	1.55876	2.42439	3.76251	5.82660	9.00376	13.88395	21.36442	32.80705
90	1.56655	2.44863	3.81895	5.94313	9.22886	14.30047	22.11218	34.11933
91	1.57439	2.47312	3.87623	6.06200	9.45958	14.72948	22.88610	35.48411
92	1.58226	2.49785	3.93438	6.18324	9.69607	15.17137	23.68712	36.90347
93	1.59017	2.52283	3.99339	6.30690	9.93847	15.62651	24.51616	38.37961
94	1.59812	2.54806	4.05329	6.43304	10.18693	16.09530	25.37423	39.91479
95	1.60611	2.57354	4.11409	6.56170	10.44160	16.57816	26.26233	41.51139
96	1.61414	2.59927	4.17580	6.69293	10.70264	17.07551	27.18151	43.17184
97	1.62221	2.62527	4.23844	6.82679	10.97021	17.58777	28.13286	44.89872
98	1.63032	2.65152	4.30202	6.96333	11.24447	18.11540	29.11751	46.69466
99	1.63848	2.67803	4.36655	7.10259	11.52558	18.65887	30.13663	48.56245
100	1.64667	2.70481	4.43205	7.24465	11.81372	19.21863	31.19141	50.50495

n ↓ r →	0.045	0.05	0.055	0.06	0.065	0.07	0.075	0.08
1	1.04500	1.05000	1.05500	1.06000	1.06500	1.07000	1.07500	1.08000
2	1.09203	1.10250	1.11302	1.12360	1.13423	1.14490	1.15563	1.16640
3	1.14117	1.15762	1.17424	1.19102	1.20795	1.22504	1.24230	1.25971
4	1.19252	1.21551	1.23882	1.26248	1.28647	1.31080	1.33547	1.36049
5	1.24618	1.27628	1.30696	1.33823	1.37009	1.40255	1.43563	1.46933
6	1.30226	1.34010	1.37884	1.41852	1.45914	1.50073	1.54330	1.58687
7	1.36086	1.40710	1.45468	1.50363	1.55399	1.60578	1.65905	1.71382
8	1.42210	1.47746	1.53469	1.59385	1.65500	1.71819	1.78348	1.85093
9	1.48610	1.55133	1.61909	1.68948	1.76257	1.83846	1.91724	1.99900
10	1.55297	1.62889	1.70814	1.79085	1.87714	1.96715	2.06103	2.15892
11	1.62285	1.71034	1.80209	1.89830	1.99915	2.10485	2.21561	2.33164
12	1.69588	1.79586	1.90121	2.01220	2.12910	2.25219	2.38178	2.51817
13	1.77220	1.88565	2.00577	2.13293	2.26749	2.40985	2.56041	2.71962
14	1.85194	1.97993	2.11609	2.26090	2.41487	2.57853	2.75244	2.93719
15	1.93528	2.07893	2.23248	2.39656	2.57184	2.75903	2.95888	3.17217
16	2.02237	2.18287	2.35526	2.54035	2.73901	2.95216	3.18079	3.42594
17	2.11338	2.29202	2.48480	2.69277	2.91705	3.15882	3.41935	3.70002
18	2.20848	2.40662	2.62147	2.85434	3.10665	3.37993	3.67580	3.99602
19	2.30786	2.52695	2.76565	3.02560	3.30859	3.61653	3.95149	4.31570
20	2.41171	2.65330	2.91776	3.20714	3.52365	3.86968	4.24785	4.66096
21	2.52024	2.78596	3.07823	3.39956	3.75268	4.14056	4.56644	5.03383
22	2.63365	2.92526	3.24754	3.60354	3.99661	4.43040	4.90892	5.43654
23	2.75217	3.07152	3.42615	3.81975	4.25639	4.74053	5.27709	5.87146
24	2.87601	3.22510	3.61459	4.04893	4.53305	5.07237	5.67287	6.34118
25	3.00543	3.38635	3.81339	4.29187	4.82770	5.42743	6.09834	6.84848

$n \downarrow$	$r \rightarrow$ 0.045	0.05	0.055	0.06	0.065	0.07	0.075	0.08
26	3.14068	3.55567	4.02313	4.54938	5.14150	5.80735	6.55572	7.39635
27	3.28201	3.73346	4.24440	4.82235	5.47570	6.21387	7.04739	7.98806
28	3.42970	3.92013	4.47784	5.11169	5.83162	6.64884	7.57595	8.62711
29	3.58404	4.11614	4.72412	5.41839	6.21067	7.11426	8.14414	9.31727
30	3.74532	4.32194	4.98395	5.74349	6.61437	7.61226	8.75496	10.06266
31	3.91386	4.53804	5.25807	6.08810	7.04430	8.14511	9.41158	10.86767
32	4.08998	4.76494	5.54726	6.45339	7.50218	8.71527	10.11745	11.73708
33	4.27403	5.00319	5.85236	6.84059	7.98982	9.32534	10.87625	12.67605
34	4.46636	5.25335	6.17424	7.25103	8.50916	9.97811	11.69197	13.69013
35	4.66735	5.51602	6.51383	7.68609	9.06225	10.67658	12.56887	14.78534
36	4.87738	5.79182	6.87209	8.14725	9.65130	11.42394	13.51154	15.96817
37	5.09686	6.08141	7.25005	8.63609	10.27864	12.22362	14.52490	17.24563
38	5.32622	6.38548	7.64880	9.15425	10.94675	13.07927	15.61427	18.62528
39	5.56590	6.70475	8.06949	9.70351	11.65829	13.99482	16.78534	20.11530
40	5.81636	7.03999	8.51331	10.28572	12.41607	14.97446	18.04424	21.72452
41	6.07810	7.39199	8.98154	10.90286	13.22312	16.02267	19.39756	23.46248
42	6.35162	7.76159	9.47553	11.55703	14.08262	17.14426	20.85237	25.33948
43	6.63744	8.14967	9.99668	12.25045	14.99799	18.34435	22.41630	27.36664
44	6.93612	8.55715	10.54650	12.98548	15.97286	19.62846	24.09752	29.55597
45	7.24825	8.98501	11.12655	13.76461	17.01110	21.00245	25.90484	31.92045
46	7.57442	9.43426	11.73851	14.59049	18.11682	22.47262	27.84770	34.47409
47	7.91527	9.90597	12.38413	15.46592	19.29441	24.04571	29.93628	37.23201
48	8.27146	10.40127	13.06526	16.39387	20.54855	25.72891	32.18150	40.21057
49	8.64367	10.92133	13.78385	17.37750	21.88421	27.52993	34.59511	43.42742
50	9.03264	11.46740	14.54196	18.42015	23.30668	29.45703	37.18975	46.90161

n ↓	$r \rightarrow$ 0.045	0.05	0.055	0.06	0.065	0.07	0.075	0.08
51	9.43910	12.04077	15.34177	19.52536	24.82161	31.51902	39.97898	50.65374
52	9.86386	12.64281	16.18557	20.69689	26.43502	33.72535	42.97740	54.70604
53	10.30774	13.27495	17.07577	21.93870	28.15329	36.08612	46.20071	59.08252
54	10.77159	13.93870	18.01494	23.25502	29.98326	38.61215	49.66576	63.80913
55	11.25631	14.63563	19.00576	24.65032	31.93217	41.31500	53.39069	68.91386
56	11.76284	15.36741	20.05108	26.12934	34.00776	44.20705	57.39499	74.42696
57	12.29217	16.13578	21.15389	27.69710	36.21827	47.30155	61.69962	80.38112
58	12.84532	16.94257	22.31735	29.35893	38.57245	50.61265	66.32709	86.81161
59	13.42336	17.78970	23.54481	31.12046	41.07966	54.15554	71.30162	93.75654
60	14.02741	18.67919	24.83977	32.98769	43.74984	57.94643	76.64924	101.25706
61	14.65864	19.61315	26.20596	34.96695	46.59358	62.00268	82.39793	109.35763
62	15.31828	20.59380	27.64729	37.06497	49.62216	66.34286	88.57778	118.10624
63	16.00760	21.62349	29.16789	39.28887	52.84760	70.98686	95.22111	127.55474
64	16.72794	22.70467	30.77212	41.64620	56.28270	75.95595	102.36270	137.75912
65	17.48070	23.83990	32.46459	44.14497	59.94107	81.27286	110.03990	148.77985
66	18.26733	25.03190	34.25014	46.79367	63.83724	86.96196	118.29289	160.68223
67	19.08936	26.28349	36.13390	49.60129	67.98666	93.04930	127.16486	173.53681
68	19.94839	27.59766	38.12126	52.57737	72.40580	99.56275	136.70222	187.41976
69	20.84606	28.97755	40.21793	55.73201	77.11217	106.53214	146.95489	202.41334
70	21.78414	30.42643	42.42992	59.07593	82.12446	113.98939	157.97650	218.60641
71	22.76442	31.94775	44.76356	62.62049	87.46255	121.96865	169.82474	236.09492
72	23.78882	33.54513	47.22556	66.37772	93.14762	130.50646	182.56160	254.98251
73	24.85932	35.22239	49.82296	70.36038	99.20221	139.64191	196.25372	275.38111
74	25.97799	36.98351	52.56323	74.58200	105.65036	149.41684	210.97275	297.41160
75	27.14700	38.83269	55.45420	79.05692	112.51763	159.87602	226.79570	321.20453

$n \downarrow$	$r \rightarrow$ 0.045	0.05	0.0055	0.06	0.065	0.07	0.075	0.08
76	28.36861	40.77432	58.50418	83.80034	119.83128	171.06734	243.80538	346.90089
77	29.64520	42.81304	61.72191	88.82836	127.62031	183.04205	262.09078	374.65296
78	30.97923	44.95369	65.11662	94.15806	135.91563	195.85500	281.74759	404.62520
79	32.37330	47.20137	68.69803	99.80754	144.75015	209.56485	302.87866	436.99522
80	33.83010	49.56144	72.47643	105.79599	154.15891	224.23439	325.59456	471.95483
81	35.35245	52.03951	76.46263	112.14375	164.17924	239.93079	350.01415	509.71122
82	36.94331	54.64149	80.66807	118.87238	174.85089	256.72595	376.26521	550.48812
83	38.60576	57.37356	85.10482	126.00472	186.21619	274.69677	404.48510	594.52717
84	40.34302	60.24224	89.78558	133.56500	198.32025	293.92554	434.82149	642.08934
85	42.15846	63.25435	94.72379	141.57890	211.21106	314.50033	467.43310	693.45649
86	44.05559	66.41707	99.93360	150.07364	224.93978	336.51535	502.49058	748.93301
87	46.03809	69.73792	105.42995	159.07806	239.56087	360.07143	540.17737	808.84765
88	48.10980	73.22482	111.22859	168.62274	255.13232	385.27643	580.69068	873.55546
89	50.27474	76.88606	117.34617	178.74010	271.71592	412.24578	624.24248	943.43990
90	52.53711	80.73037	123.80021	189.46451	289.37746	441.10298	671.06066	1018.91509
91	54.90128	84.76688	130.60922	200.83238	308.18699	471.98019	721.39021	1100.42830
92	57.37183	89.00523	137.79272	212.88232	328.21915	505.01880	775.49448	1188.46256
93	59.95356	93.45549	145.37132	225.65526	349.55339	540.37012	833.65657	1283.53956
94	62.65148	98.12826	153.36675	239.19458	372.27436	578.19603	896.18081	1386.22273
95	65.47079	103.03468	161.80192	253.54625	396.47220	618.66975	963.39437	1497.12055
96	68.41698	108.18641	170.70102	268.75903	422.24289	661.97663	1035.64895	1616.89019
97	71.49574	113.59573	180.08958	284.88457	449.68868	708.31499	1113.32262	1746.24141
98	74.71305	119.27552	189.99451	301.97765	478.91844	757.89704	1196.82181	1885.94072
99	78.07514	125.23929	200.44420	320.09631	510.04814	810.94984	1286.58345	2036.81598
100	81.58852	131.50126	211.46864	339.30208	543.20127	867.71633	1383.07721	2199.76126

TABLE 2

$a_{\overline{n}|r}$ FOR SELECTED VALUES OF n AND r

n ↓	$r \to \frac{2}{3}\%$	$\frac{3}{4}\%$	$\frac{5}{6}\%$	1%	$1\frac{1}{4}\%$	$1\frac{1}{2}\%$
60	49.31843	48.17337	47.06537	44.95504	42.03459	39.38027
120	82.42148	78.94169	75.67116	69.70052	61.98285	55.49845
180	104.64059	98.59341	93.05744	83.32166	71.44964	62.09556
240	119.55429	111.14495	103.62462	90.81942	75.94228	54.79573
300	129.56452	119.16162	110.04723	94.94655	78.07434	65.90090
360	136.28349	124.28187	113.95082	97.21833	79.06814	66.35324

Answers

Answers are given below to Section Exercises (*odd-numbered exercises only*), Section Tests A (all problems), Chapter Review Exercises (*odd-numbered exercises only*), and Chapter Tests (all problems).

CHAPTER 1

SECTION 1

SECTION EXERCISES (pp. 6-8)

1. 105 3. 450 5. 50 7. 4
9. 70 11(a). 13 11(b). 12

TEST A (pp. 8-9)

1.

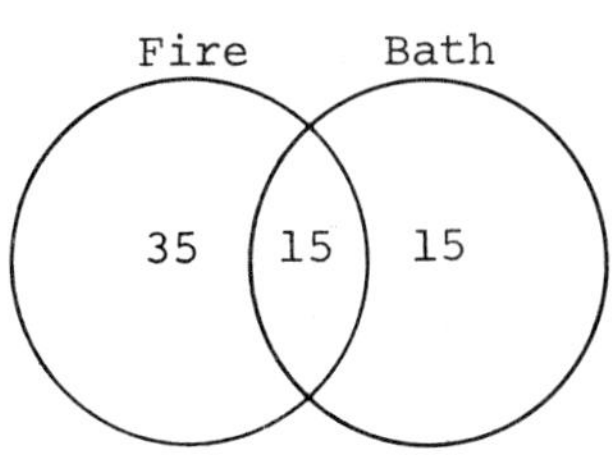

$35 + 15 + 15 = 65$

2.

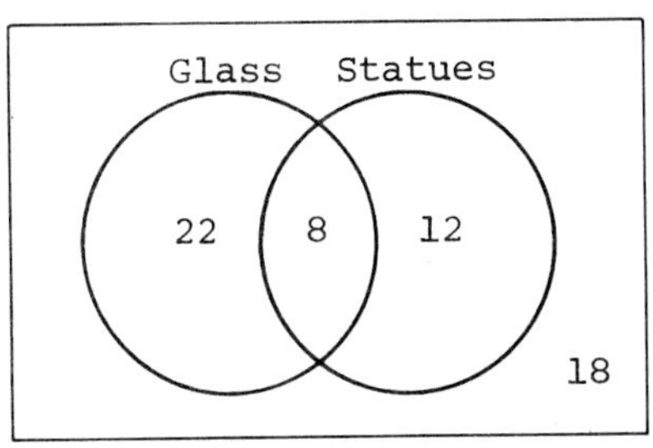

$60 - (22 + 8 + 12) = 18$

3.

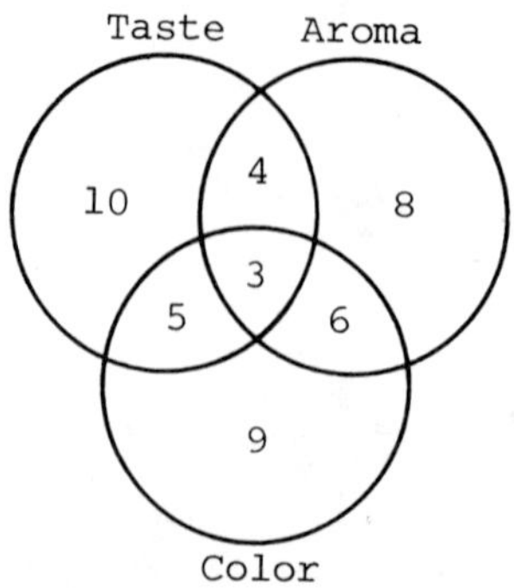

The number commended only for taste is 10; only for aroma, 8; only for color, 9.

SECTION 2

SECTION EXERCISES (pp. 13-15)

1.

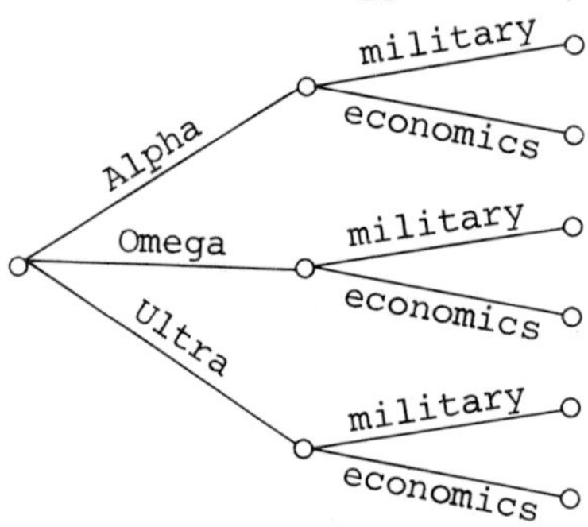

3.

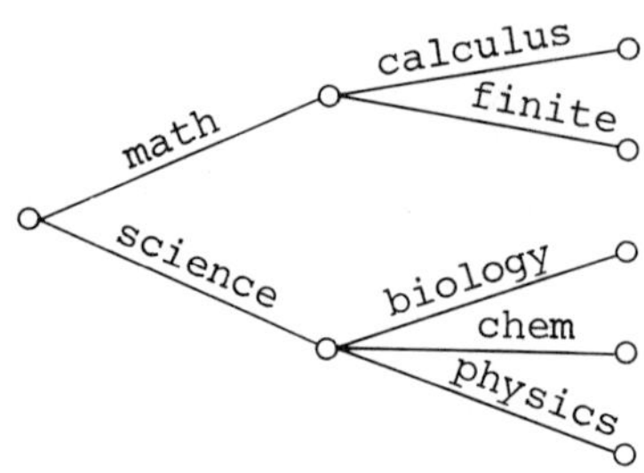

5.

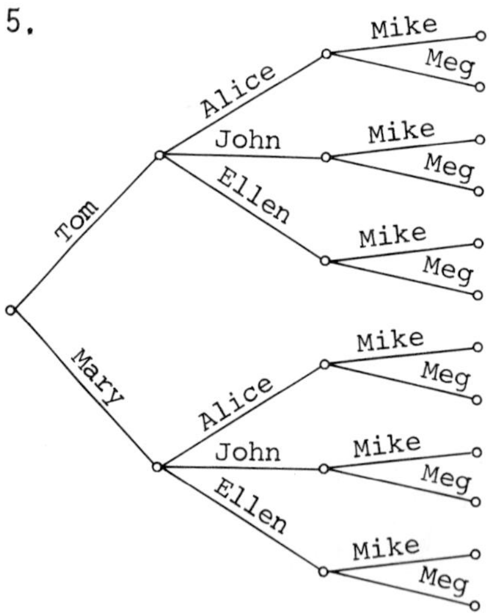

7.

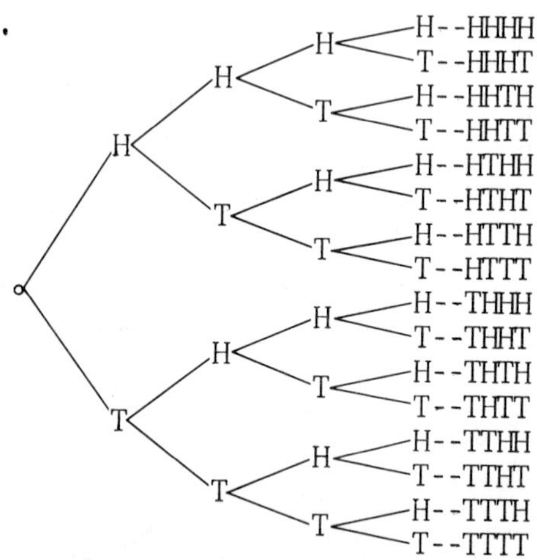

9.

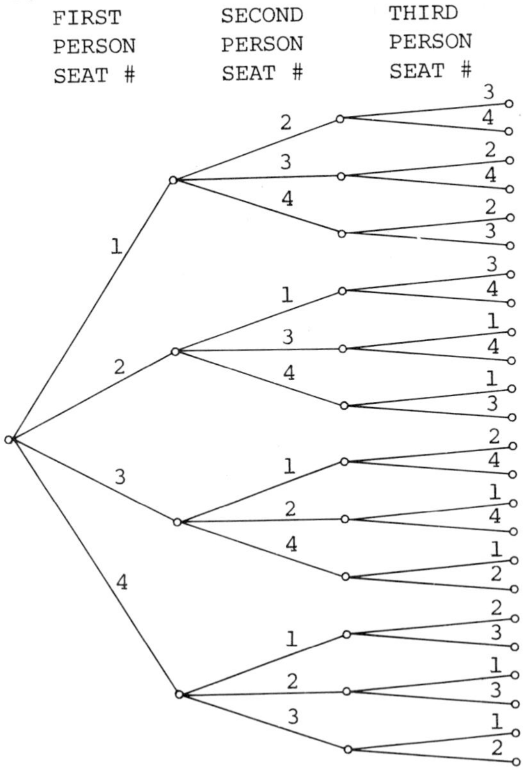

There are $4 \times 3 \times 2 = 24$ different seating arrangements.

11.

(a)

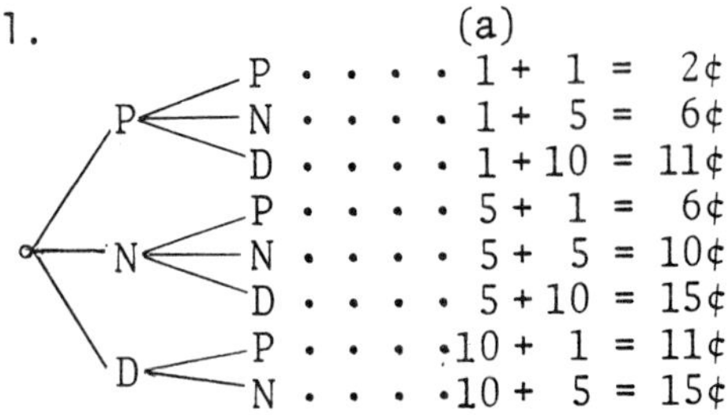

(b)

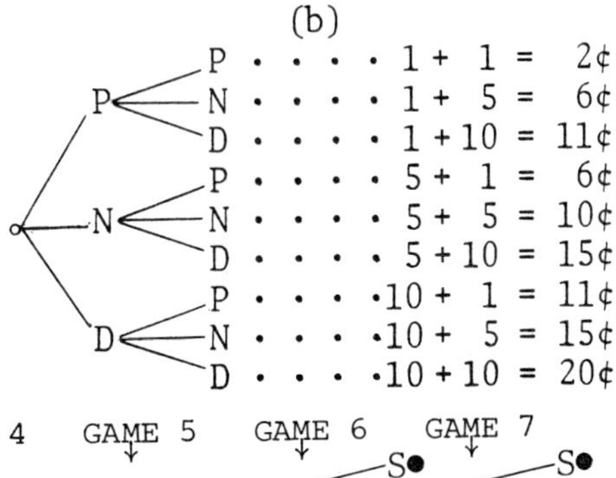

13.

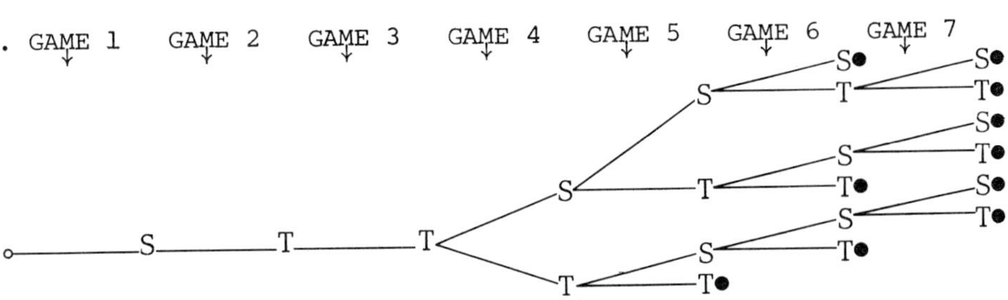

TEST A (p. 15)

1.

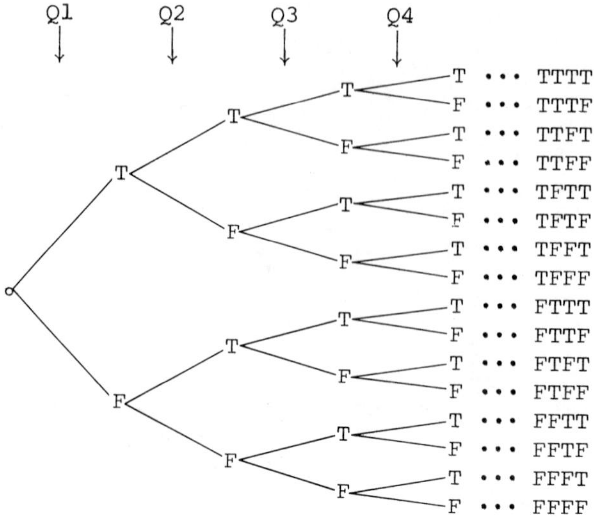
Q1
Q2
Q3
Q4
TTTT
TTTF
TTFT
TTFF
TFTT
TFTF
TFFT
TFFF
FTTT
FTTF
FTFT
FTFF
FFTT
FFTF
FFFT
FFFF

2.

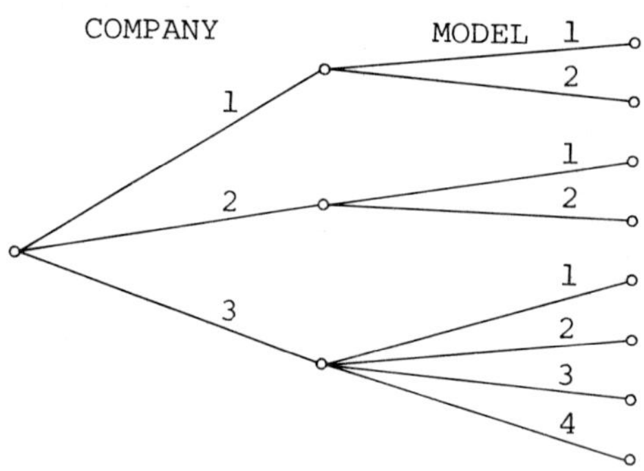
COMPANY
MODEL

3. The numbers on the tree are for the books:

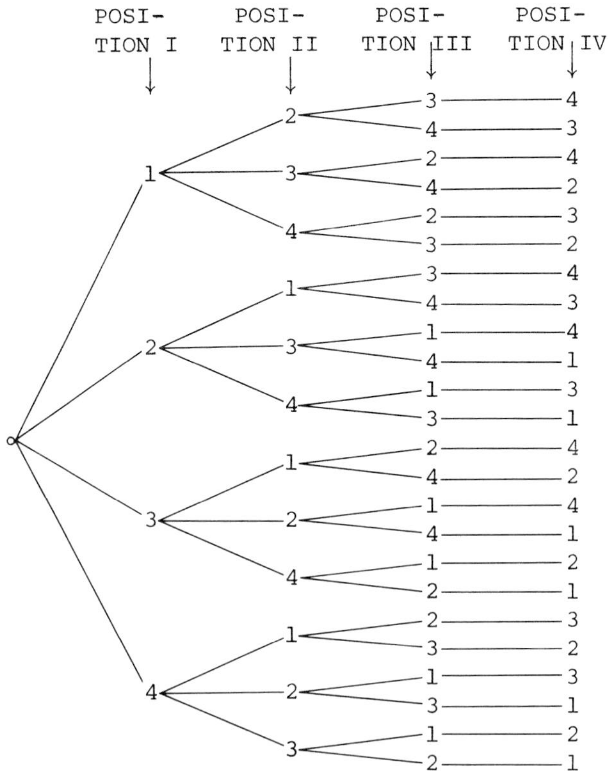

SECTION 3

Section Exercises (p. 19)

1. $10 \times 8 \times 12 = 960$
3. $2 \times 2 \times 2 \times 2 \times 2 \times 2 \times 2 \times 2 \times 2 \times 2 = 2^{10} = 1024$
5. $2 \times 2 \times 2 \times 2 \times 4 \times 4 \times 4 = 1024$
7. (a) $26 \times 25 \times 24 \times 10 \times 9 \times 8 = 11,232,000$
   (b) $26 \times 26 \times 26 \times 10 \times 10 \times 10 = 17,576,000$
9. (a) $10 \times 9 \times 8 \times 7 \times 6 \times 5 = 151,200$
   (b) $6 \times 5 \times 4 \times 3 \times 2 \times 1 = 720$

Test A (pp. 19-20)

1. $2 \times 4 \times 3 \times 3 = 72$
2. $4 \times 4 \times 4 \times 4 \times 4 \times 2 \times 2 \times 2 \times 2 \times 2 = 32,768$
3. $5 \times 4 \times 3 \times 2 \times 1 = 120$
4. $20 \times 19 = 380$

SECTION 4

Section Exercises (pp. 26-27)

1. (a) 720 (b) 5040 (c) 40,320
   (d) $20 \cdot 19 \cdot 18 \cdot 17 = 116{,}280$
   (e) $\dfrac{10 \cdot 9 \cdot 8 \cdot 7 \cdot 6 \cdot \cancel{5 \cdot 4 \cdot 3 \cdot 2 \cdot 1}}{5 \cdot 4 \cdot 3 \cdot 2 \cdot 1 \cdot \cancel{5 \cdot 4 \cdot 3 \cdot 2 \cdot 1}} = 252$
3. (a) n (b) $n(n-1)$ (c) 1 (d) $n!$
5. $P(4,4) = 4! = 24$
7. $P(5,5) = 5! = 120$
9. $P(9,4) = \dfrac{9!}{5!} = 9 \cdot 8 \cdot 7 \cdot 6 = 3024$
11. $P(5,2) \times P(6,3) = \dfrac{5!}{3!} \times \dfrac{6!}{3!} = 5 \cdot 4 \cdot 6 \cdot 5 \cdot 4 = 2400$
13. $P(7,3) + P(6,3) = (7 \cdot 6 \cdot 5) + (6 \cdot 5 \cdot 4) = 210 + 120 = 330$

Test A (p, 27)

1. $P(4,4) = 4! = 24$
2. $P(10,4) = \dfrac{10!}{6!} = 10 \cdot 9 \cdot 8 \cdot 7 = 5040$
3. $P(7,2) \times P(5,2) = \dfrac{7!}{5!} \times \dfrac{5!}{3!} = 7 \cdot 6 \cdot 5 \cdot 4 = 850$

SECTION 5

Section Exercises (pp. 33-35)

1. (a) 28 (b) 6 (c) 120 (d) 1
3. $C(n,n-r) = \dfrac{n!}{(n-r)!\,[n-(n-r)]!} = \dfrac{n!}{(n-r)!\,(n-n+r)!}$
   $= \dfrac{n!}{(n-r)!\,r!} = \dfrac{n!}{r!\,(n-r)!} = C(n,r)$
5. (a) $C(50,2) = \dfrac{50 \cdot 49}{2 \cdot 1} = 1225$ (b) $P(50,2) = 50 \cdot 49 = 2450$
7. (a) $C(1000,3) = 166{,}670{,}000$ (b) $P(1000,3) = 997{,}002{,}000$
9. $C(30,3) \times C(10,4) = 852{,}600$
11. $C(5,3) + C(10,5) = 10 + 252 = 262$
13. $C(25,20) \times C(5,5) = 53{,}130 \times 1 = 53{,}130$
15. $C(10,2) \times C(8,3) \times C(5,5) = \dfrac{10 \cdot 9}{2 \cdot 1} \times \dfrac{8 \cdot 7 \cdot 6}{3 \cdot 2 \cdot 1} \times \dfrac{5!}{5!} = \dfrac{10!}{2!3!5!} = 2520$
17. $6! - 1 = 719$
19. $C(12,4) \times C(8,2) \times C(14,3) = 5{,}045{,}040$
21. (a) $C(6,2) = 15$
    (b) $15 + 4 + 4 + 1 = 24$ *or* $1 \cdot C(4,2) + 2 \cdot (4 + 4 + 1) = 24$

Test A (p. 35)

1. $C(20,3) = \dfrac{20 \cdot 19 \cdot 18}{3 \cdot 2 \cdot 1} = 1140$
2. $C(10,4) \times C(10,3) = 25{,}200$
3. $C(12,3) \times C(9,3) \times C(6,3) = \dfrac{12!}{3!3!3!3!} = 369{,}600$

REVIEW EXERCISES (pp. 37-38)

1. 80
3. (a)

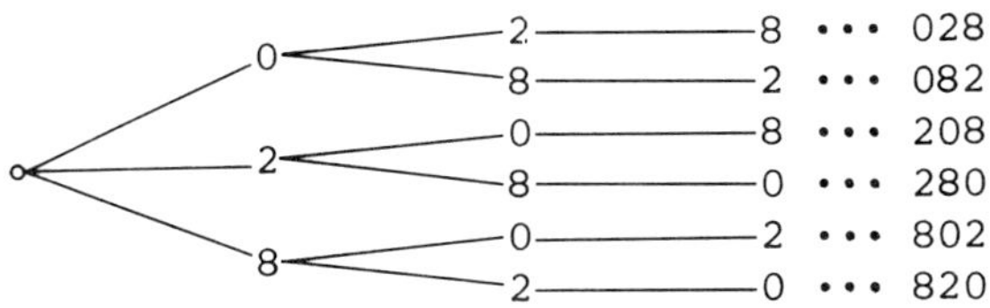

(a) 6 (b) 27

5. (a) $C(12,4) = 495$ (b) $C(11,3) = 165$ 7. $10! - 1$
9. (a) 10^5 (b) 100
11. (a) $C(4,2)\cdot C(4,2)\cdot C(4,3) = 144$ (b) $[C(4,2)]^3\cdot C(40,7)$
13. (a) $2^7\cdot 4^3$ (b) 1 (c) 3^3
15. (a) 64
 (b) $C(6,0) + C(6,1) + C(6,2) + C(6,3) + C(6,4) + C(6,5) + C(6,6)$
 (c) 2^6. Visualize a tree. We can either order mushrooms or not (2 alternatives); for each of these we can either order pepperoni or not (so far we have $2 \times 2 = 2^2$ alternatives), and so on. For 6 ingredients, we have $2 \times 2 \times 2 \times 2 \times 2 \times 2 = 2^6$ alternatives; $2^6 = 64$.

CHAPTER TEST (pp. 39-40)

1.

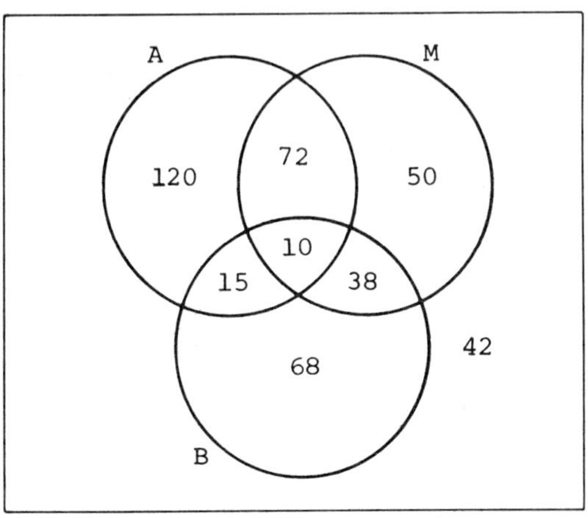

A = accounting; M = mathematics; B = biology

(a) 42 (b) 120

2.

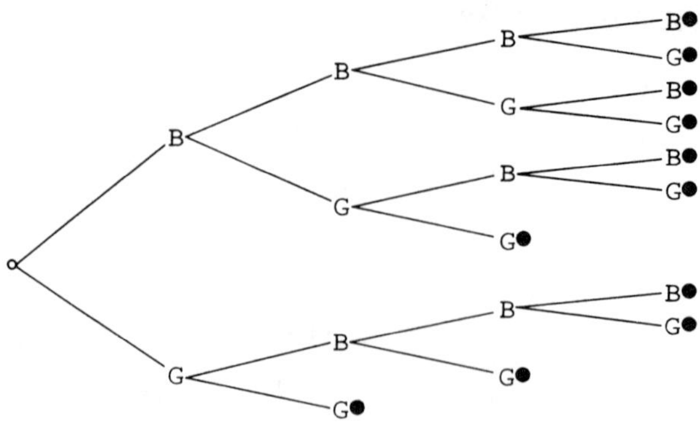

3. See Figure 2.3 (p. 12). (a) 4 (b) $C(3,2)+C(3,3) = 4$
4. $2\cdot 8\cdot 3\cdot 4 = 192$
5. $P(11,3) = 11\cdot 10\cdot 9 = 990$
6. $C(6,2)\cdot C(4,1)\cdot C(7,2) = 1260$
7. $8\cdot 2\cdot 8 = 128$
8. (a) $C(5,3) = 10$ (b) $P(5,3) = 60$
9. (a) $C(16,5) = 4368$ (b) $C(10,5)+C(6,5) = 258$
10. $C(10,3)+C(8,4) = 190$

CHAPTER 2

SECTION 6

SECTION EXERCISES (pp. 49-51)

1. (a) {0,1,2,3} (b) {HHH, HHT, HTH, HTT, THH, THT, TTH, TTT}
3. (a) {0,1,2,3,4}
   (b) {HHHH, HHHT, HHTH, HHTT, HTHH, HTHT, HTTH, HTTT, THHH, THHT, THTH, THTT, TTHH, TTHT, TTTH, TTTT}
5. (a) {2,3,4,5,6,7,8,9,10,11,12}
   (b) The sample space consists of 36 simple events, each of the form (i,j), where i and j are each any number from 1 to 6 inclusive. For a full display see S_2 on page 54.
7. The distributions in rows 1, 3, and 4 are valid.
9. (a) $p(e_1) = \frac{12}{25}$, $p(e_2) = \frac{6}{25}$, $p(e_3) = \frac{4}{25}$, $p(e_4) = \frac{3}{25}$
   (b) $p(e_1) = \frac{1}{10}$, $p(e_2) = \frac{1}{5}$, $p(e_3) = \frac{3}{10}$, $p(e_4) = \frac{2}{5}$
11. $\frac{1}{21}$, $\frac{2}{21}$, $\frac{1}{7}$, $\frac{4}{21}$, $\frac{5}{21}$, $\frac{2}{7}$.
13. $\frac{3}{10}$ 15. $\frac{3}{8}$ 17. $\frac{1}{4}$ 19. $\frac{15}{16}$ 21. 0.1

23. 51%, 49%, 0.3%, 45.3%

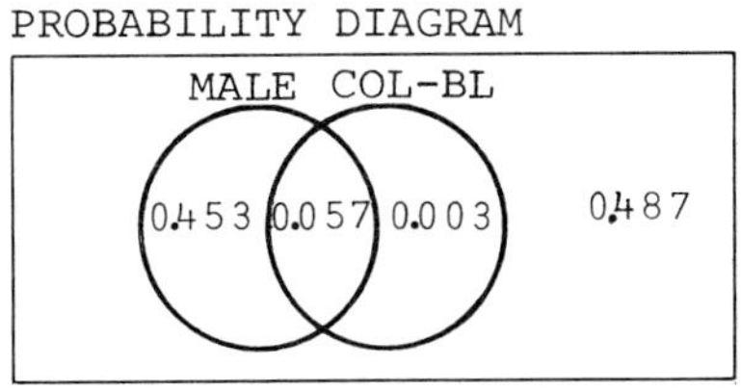

Test A (p. 51)

1. (a) {0,1,2,3}
   (b) {PPP, PPN, PNP, PNN, NPP, NPN, NNP, NNN} (P-passed, N=failed)
2. Letting $p(e_2) = x$ we have $p(e_1) = p(e_3) = 2x$. Then $2x + x + 2x = 1$ and $x = \frac{1}{5}$; so $p(e_1) = \frac{2}{5}$, $p(e_2) = \frac{1}{5}$, $p(e_3) = \frac{2}{5}$.
3. $0.5 = 0.3 + 0.4 - p(X \cap Y)$; so $p(X \cap Y) = 0.2$ and $p[(X \cap Y)'] = 1 - 0.2 = 0.8$.
4. (a) $0.02 + 0.88 + 0.07 = 0.97$
   (b) 0.02
   (c) $1 - 0.97 = 0.03$

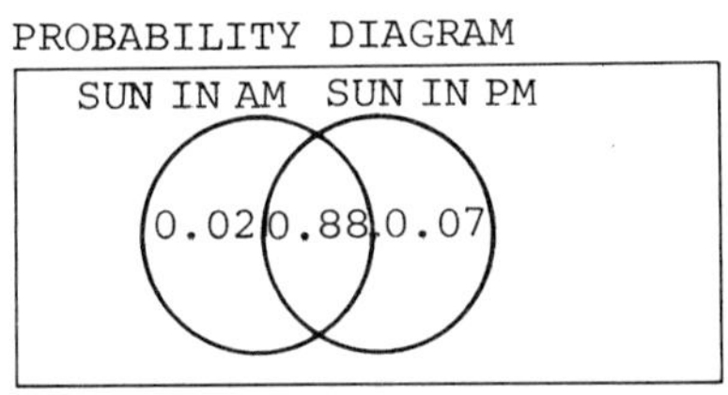

SECTION 7

Section Exercises (pp. 58-60)

1. (a) {HH, HT, TH, TT}; number of heads: {0,1,2}.
   (b) {HHH, HHT, HTH, HTT, THH, THT, TTH, TTT}: number of heads: {0,1,2,3}.
   (c) {HHHH, HHHT, HHTH, HHTT, HTHH, HTHT, HTTH, HTTT, THHH, THHT, THTH, THTT, TTHH, TTHT, TTTH, TTTT}: {0,1,2,3,4}.
   (d) {1,2,3,4,5,6}; {≤ 4, > 4}.
   (e) See the display on page 54; {2,3,4,5,6,7,8,9,10,11,12}.
   (f) Like (c) above with "T" and "F" instead of "H" and "T" respectively; number of correct answers: {0,1,2,3,4}.
   (g) {AA, AB, BA, BB, CA, CB}; number of A's: {≥1, 0}.
3. (a) $\frac{1}{6}$ (b) $\frac{5}{12}$ (c) $\frac{35}{36}$ (d) $\frac{7}{9}$ (e) 0
5. (a) $\frac{1}{37}$ (b) $\frac{1}{37}$ (c) $\frac{36}{37}$ (d) $\frac{18}{37}$ (e) $\frac{18}{37}$ (f) $\frac{12}{37}$
7. (a) $\frac{C(10,4)}{C(12,4)} = \frac{14}{33}$ (b) $\frac{C(7,4)}{C(12,4)} = \frac{7}{99}$

9. (a) $\frac{C(12,3)}{C(20,3)} = \frac{11}{57}$ (b) $\frac{C(8,3)}{C(20,3)} = \frac{14}{285}$

11. (a) $\frac{14}{33}$ (b) $\frac{92}{99}$

13. (a) $\frac{1}{15}$ (b) $\frac{7}{15}$ (c) $\frac{7}{15}$

15. (a) $\frac{5}{42}$ (b) 1 (c) $\frac{1}{21}$

TEST A (p. 60)

1. (a) {(1,2),(1,3),(1,4),(2,1),(2,3),(2,4), (3,1),(3,2),(3,4),(4,1),(4,2),(4,3)}
   (b) 0
   (c) The sum is 5 for 4 simple events: {(1,4),(2,3),(3,2),(4,1)}. The probability is $\frac{4}{12} = \frac{1}{3}$.

2. (a) $\frac{C(3,2)}{C(8,2)} = \frac{3}{28}$ (b) $\frac{C(5,2)}{C(8,2)} = \frac{5}{14}$

3. $\frac{C(3,1) \times C(7,1)}{C(10,2)} + \frac{C(3,2) \times C(7,0)}{C(10.2)} = \frac{8}{15}$

SECTION 8

SECTION EXERCISES (pp. 67-69)

1. (a) $\frac{2}{11}$ (b) $\frac{1}{2}$ 3. (a) $\frac{1}{4}$ (b) $\frac{3}{7}$

5. $\frac{1}{3} \div \frac{2}{5} = \frac{5}{6}$

7. (a) $\frac{667}{1127} = 0.59$ (b) $\frac{49}{1127} = 0.04$ (c) $\frac{29}{1127} = 0.03$
   (d) $\frac{440}{1078} = 0.41$ (e) $\frac{638}{667} = 0.96$

9. 0.5 11. (a) $\frac{7}{18}$ (b) $\frac{7}{8}$ 13. $\frac{1}{5}$

TEST A (p. 69)

1. (a) $\frac{13}{26} = \frac{1}{2}$ (b) $\frac{2}{12} = \frac{1}{6}$

2. $p(\text{smokes}|\text{cancer}) = \frac{p(\text{smokes} \cap \text{cancer})}{p(\text{cancer})} = \frac{0.015}{0.02} = 0.75$

3. $p(\text{major} \cap \text{subsidiary}) = p(\text{subsidiary}|\text{major}) \cdot p(\text{major}) = \frac{1}{2} \cdot \frac{1}{3} = \frac{1}{6}$

SECTION 9

SECTION EXERCISES (pp. 74-75)

1. a, c, d, and e 3. no

5. (a) $\frac{1}{108}$ (b) $\frac{1}{23328}$ 7. (a) $\frac{1}{7}$ (b) 0.1 (c) 0.4

9. yes

TEST A (p. 75)

1. $p(E) = \frac{1}{6}$; $p(F) = \frac{1}{6}$; $p(G) = \frac{1}{36}$; $p(E \cap F) = \frac{1}{36}$; $p(E \cap G) = \frac{1}{36}$
   (a) $p(E)p(F) = p(E \cap F)$; so E and F are independent.
   (b) $p(E)p(G) \neq p(E \cap G)$; so E and G are not independent.
2. $(0.6)(0.65) = 0.39 \neq 0.55$; so the two events are not independent.
3. (a) $p(E \cap F) = p(E)p(F)$; so $p(E \cup F) = 0.2 + 0.1 - (0.2)(0.1)$, which equals 0.28.
   (b) $p(E \cap F) = 0$, so $p(E \cup F) = 0.2 + 0.1 - 0 = 0.3$.

SECTION 10

SECTION EXERCISES (pp. 83-85)

1. (a) 0.007 (b) $\frac{9}{14}$ (c) $\frac{5}{14}$
3. (a) 0.2825 (b) $\frac{40}{113}$ (c) $\frac{60}{113}$ (d) $\frac{13}{113}$
5. (a) 0.017 (b) $\frac{5}{17}$ (c) $\frac{6}{17}$ (d) $\frac{6}{17}$
7. (a) $\frac{1}{50}$ (b) $\frac{1}{4}$ (c) $\frac{1}{3}$ (d) $\frac{5}{12}$
9. (a) 0.53 (b) $\frac{32}{53}$ (c) $\frac{15}{53}$ (d) $\frac{5}{53}$ (e) $\frac{1}{53}$ (f) $\frac{47}{53}$
11.

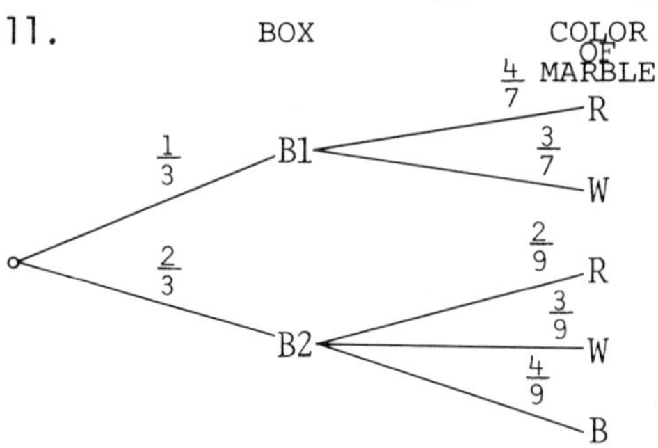

(a) $\frac{64}{189}$ (b) $\frac{14}{23}$ (c) 1

13. In the tree, I = pass test one, II = pass test two, and III = pass test three.
   (a) $p(\text{II} \cap \text{I}') = (0.90)(0.20) = 0.18$
   (b) Since $p(\text{II}' \cap \text{I}') = (0.10)(0.20) = 0.02$
   $p(\text{III} \cap (\text{II}' \cap \text{I}')) = (0.50)(0.02) = 0.01$.

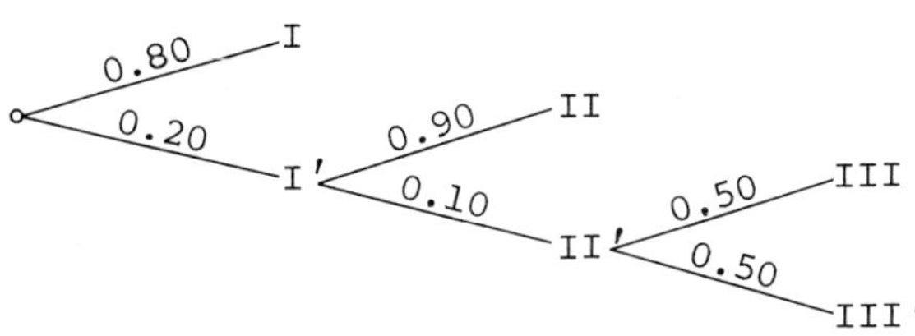

TEST A (p. 85)

1. In the tree, R = return and R′ = no return.

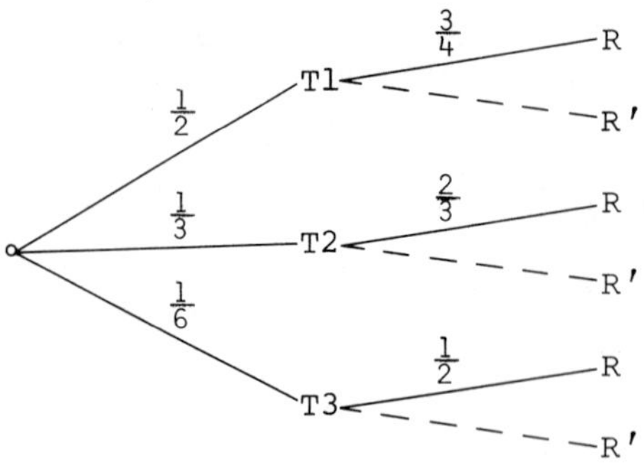

$\frac{1}{2} \times \frac{3}{4} = \frac{3}{8}$

$\frac{1}{3} \times \frac{2}{3} = \frac{2}{9}$

$\frac{1}{6} \times \frac{1}{2} = \frac{1}{12}$

(a) $p(\mathrm{R}) = \frac{3}{8} + \frac{2}{9} + \frac{1}{12} = \frac{49}{72}$ (b) $p(\mathrm{T1}|\mathrm{R}) = \frac{3}{8} \div \frac{49}{72} = \frac{27}{49}$

(c) $p(\mathrm{T2}|\mathrm{R}) = \frac{2}{9} \div \frac{49}{72} = \frac{16}{49}$ (d) $p(\mathrm{T3}|\mathrm{R}) = \frac{1}{12} \div \frac{49}{72} = \frac{6}{49}$

SECTION 11

SECTION EXERCISES (pp. 90-91)

1. (a) $\frac{1}{3}$ (b) $\frac{3}{8}$
3. (a) $\frac{2}{5}$ (b) $\frac{3}{5}$
5. (a) $\frac{11}{18}$ (b) $\frac{7}{18}$
7. (a) $\frac{1}{3}$ (b) $\frac{8}{15}$
9. $\frac{19}{48}$
11. (a) $\frac{1}{2}$ (b) 0 (c) $\frac{1}{4}$ (d) 0 (e) $\frac{1}{2^n}$

TEST A (pp. 91-92)

1.

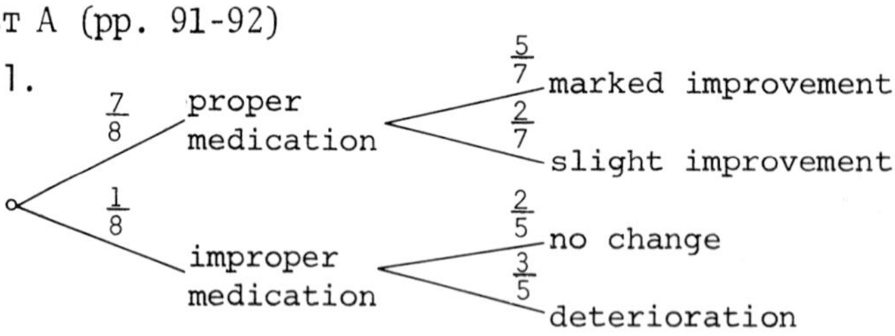

(a) $\frac{7}{8} \times \frac{5}{7} = \frac{5}{8}$

(b) $\frac{7}{8} \times \frac{2}{7} = \frac{1}{4}$

(c) $\frac{1}{8} \times \frac{2}{5} = \frac{1}{20}$

(d) $\frac{1}{8} \times \frac{3}{5} = \frac{3}{40}$

2.

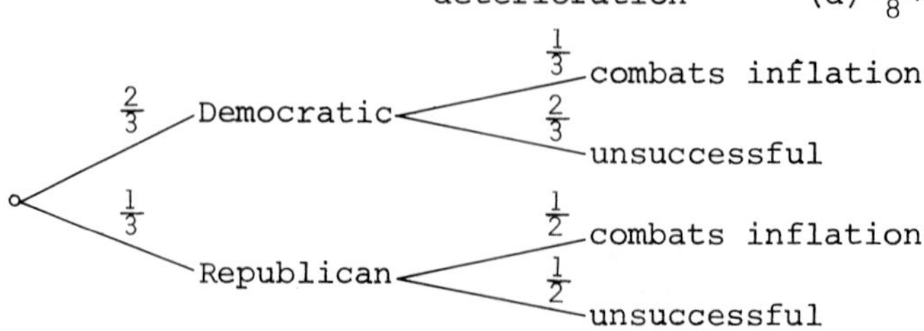

2. (cont'd) $(\frac{2}{3}\times\frac{1}{3}) + (\frac{1}{3}\times\frac{1}{2}) = \frac{2}{9}+\frac{1}{6} = \frac{7}{18}$

SECTION 12

Section Exercises (pp. 97-98)

1. (a) $C(6,4)(\frac{3}{4})^4(\frac{1}{4})^2 = \frac{1215}{4096}$ (b) $C(5,5)(\frac{3}{4})^5 = \frac{243}{1024}$
   (c) $1 - (\frac{1}{4})^7 = \frac{16383}{16384}$
3. 0.8192 5. 0.68256 7. $C(6,1)(\frac{1}{6})^1(\frac{5}{6})^5 \approx 0.4$
9. $[C(8,0) + C(8,1) + C(8,2) + C(8,3)](\frac{1}{2})^8 \approx 0.363$
11. $C(10,9)(\frac{1}{2})^{10} + C(10,10)(\frac{1}{2})^{10} \approx 0.011$
13. The probability of getting at least one tail equals 1 minus the probability of getting all heads. On n tosses, the latter equals $(\frac{1}{2})^n$.

$1 - (\frac{1}{2})^1 = \frac{1}{2};\ 1 - (\frac{1}{2})^2 = \frac{3}{4};\ 1 - (\frac{1}{2})^3 = \frac{7}{8};\ \ldots\ ;\ 1 - (\frac{1}{2})^5 = \frac{31}{32} > 0.95$

15. $\frac{112}{125}$

Test A (pp. 98-99)

1. $C(5,3)(\frac{3}{5})^3(\frac{2}{5})^2 = 0.3456;\ 0.3456 + 5(\frac{3}{5})^4(\frac{2}{5}) + (\frac{3}{5})^5 = 0.68256;$
   $1 - (\frac{3}{5})^5 = 0.92224$
2. $1 - (0.97)^5 \approx 0.141$
3. $C(5,3)(\frac{1}{4})^3(\frac{3}{4})^2 + C(5,4)(\frac{1}{4})^4(\frac{3}{4}) + C(5,5)(\frac{1}{4})^5 = \frac{53}{512}$

SECTION 13

Section Exercises (pp. 105-106)

1. \$12,600 3. (a) Let be the answer, in dollars. Then

$$\tfrac{1}{20}(500 - c) + \tfrac{1}{4}(60 - c) + \tfrac{7}{10}(10 - c) = 0 \quad\text{and}\quad c = 47$$

   (b) \$1300
5. 1.6 7. $1\frac{7}{13}$ 9. 2.5
11. A nondefective calculator can be located only on the first, second, or third draw. The probabilities for these are respectively

$$(\tfrac{10}{12}),\ (\tfrac{2}{12})(\tfrac{10}{11}),\quad\text{and}\quad (\tfrac{2}{12})(\tfrac{1}{11})(\tfrac{10}{10})$$

Therefore you can expect to have to test

$$1(\tfrac{10}{12}) + 2(\tfrac{2}{12})(\tfrac{10}{11}) + 3(\tfrac{2}{12})(\tfrac{1}{11})(\tfrac{10}{10}) = \tfrac{156}{132}$$

or about 1.2 calculators.

TEST A (p. 106)

1. $\$1000(\frac{1}{10000}) + \$100(\frac{10}{10000}) + \$1(\frac{100}{10000}) = \$0.21$

2. (a) $2\left(\frac{C(3,2)}{C(7,2)}\right) + 1\left(\frac{C(3,1)C(4,1)}{C(7,2)}\right) + 0\left(\frac{C(4,2)}{C(7,2)}\right) = 2\left(\frac{1}{7}\right) + 1\left(\frac{4}{7}\right) = \frac{6}{7}$

   (b) Let k be your loss in dollars when at least 1 member is a Populist. Then the probabilities computed in part (a) yield
$$5\left(\frac{C(4,2)}{C(7,2)}\right) - k\left(\frac{1}{7} + \frac{4}{7}\right) = 0 \quad \text{and} \quad k = 2$$

3. $30(0.1) = 3$

REVIEW EXERCISES (pp. 108-110)

1. (a) {0,1,2,3} (b) {HHH, HHT, HTH, HTT, THH, THT, TTH, TTT}
3. (a) $\frac{1}{6}$ (b) $\frac{1}{2}$ (c) $\frac{5}{6}$
5. (a) 0.9 (b) 0.1 (c) 0.15 (d) 0.25
7. (a) $\frac{C(15,3)}{C(20,3)} = \frac{91}{228}$ (b) $\frac{91}{228} + \frac{C(5,1)C(15,2)}{C(20,3)} = \frac{49}{57}$
9. (a) $\frac{1}{2}$ (b) 1 (c) $\frac{1}{6}$ (d) 0 (e) $\frac{3}{13}$
11. $\frac{1}{6}$ 13. No. $(0.55)(0.5) = 0.275 \neq 0.35$.
15. (a) 0.585 (b) $\frac{10}{65} = \frac{2}{13}$ (c) $\frac{8}{39}$

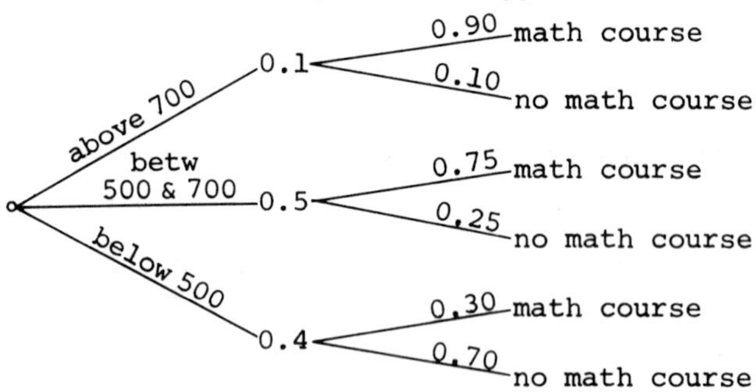

17. $C(10,0)(\frac{1}{2})^{10} + C(10,1)(\frac{1}{2})^{10} = \frac{11}{1024}$
19. $C(4,3)(0.4)^1(0.6)^3 = 0.3456$
21. $2\left(\frac{C(3,2)}{C(12,2)}\right) + 1\left(\frac{C(3,1)C(9,1)}{C(12,2)}\right) + 0\left(\frac{C(9,2)}{C(12,2)}\right) = \frac{1}{2}$

CHAPTER TEST (pp. 111-112)

1. See page 54. (a) $\frac{5}{36}$ (b) $\frac{5}{36}$ (c) $\frac{11}{12}$

2. 0.8

3. (a) 0.73 (b) 0.95

4. (a) $\frac{1}{4}$ (b) $\frac{11}{26}$ (c) $\frac{9}{13}$ (d) $\frac{3}{13}$

5. (a) $\frac{1}{120}$ (b) $1 - \frac{C(7,3)}{C(10,3)} = \frac{17}{24}$ (c) $\frac{C(3,3)}{C(10,3)} \div \frac{17}{24} = \frac{1}{85}$

6. $\frac{0.06}{0.07} = \frac{6}{7}$

7. $(0.40)(0.82) = 0.328$

8. See Figure 2.3 (p. 12) and note that $p(E) = \frac{3}{8}$; $p(F) = \frac{1}{8}$; $p(G) = \frac{1}{2}$; $p(H) = \frac{1}{2}$. So

(a) $p(E \cap F) = 0 \neq p(E) \cdot p(F)$ (b) $p(G \cap H) = \frac{1}{4} = p(G) \cdot p(H)$

(c) $p(G \cap H) \neq 0$ (d) $p(H|G) = \frac{p(H \cap G)}{p(G)} = \frac{1}{4} \div \frac{1}{2} = \frac{1}{2}$

9. (NG = no good)

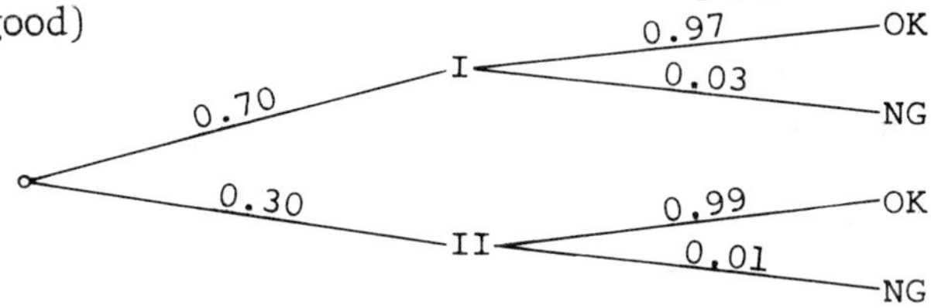

(a) $p(\text{NG}) = (0.70)(0.03) + (0.30)(0.01) = 0.024$

(b) $p(\text{II}|\text{NG}) = \frac{(0.30)(0.01)}{0.024} = 0.125$

10.

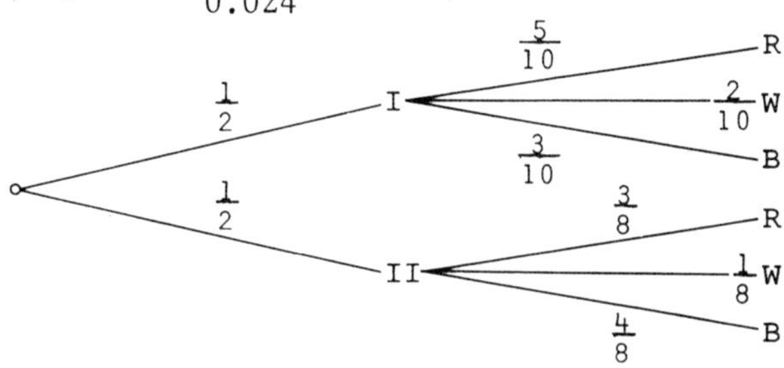

(a) $p(\text{B}) = \frac{1}{2} \cdot \frac{3}{10} + \frac{1}{2} \cdot \frac{4}{8} = \frac{2}{5}$ (b) $p(\text{B} \cap \text{II}) = \frac{1}{2} \cdot \frac{4}{8} = \frac{1}{4}$

(c) $p(\text{II}|\text{B}) = \frac{1}{4} \div \frac{2}{5} = \frac{5}{8}$ (d) $p(\text{W}|\text{I}) = \frac{2}{10} = \frac{1}{5}$

11. (a) $C(5,5)(\frac{1}{2})^5 = \frac{1}{32}$ (b) $1 - 1 \cdot (\frac{1}{2})^5 - 5 \cdot (\frac{1}{2})^5 = \frac{13}{16}$

12. If he is willing to pay $\$x$ to play, then $5(\frac{1}{11}) = x$ and $x = \frac{5}{11}$ or about 45 cents.

13. $C(20,17) \cdot (\frac{1}{4})^{17} (\frac{3}{4})^3 + C(20,18) \cdot (\frac{1}{4})^{18} (\frac{3}{4})^2 + C(20,19) \cdot (\frac{1}{4})^{19} (\frac{3}{4}) + C(20,20) \cdot (\frac{1}{4})^{20}$

14. $5 \cdot (\frac{1}{4}) \cdot 20 = 25\%$

15. $1 - C(5,5)(0.60)^5 = 0.92$

CHAPTER 3

SECTION 14

Goal Exercises (p. 118)

1. The descriptions define functions from S into T in (a), (b), and (c), but from T into S only in (b).
3. 520, 1000, 2500. As x increases the cost increases without bound. $C(0) = 500$; this may be interpreted as the fixed cost: a constant amount (for rent, maintenance, insurance, machinery, and so on) that does not depend on the number of items produced.
5. $f(-1) = 2$; $f(0) = 0$; $f(1) = 2$; $f(5) = 50$; $f(a) = 2a^2$.
7. $g(2) = 49$; $g(-2) = 9$; $g(0) = 25$; $g(-5) = 0$.
9. (a) F (b) F (c) T (d) F (e) T

Test A (p. 119)

1. (a) R (b) $x \neq -2$ (c) $x \geqslant -3$ (d) $x \neq 0,4$
2.

	$f(x)$	$f(1)$	$f(-1)$	$f(0)$
(a)	4	4	4	4
(b)	$x^2 - 1$	0	0	-1
(c)	$\frac{1}{x-2}$	-1	$-\frac{1}{3}$	$-\frac{1}{2}$
(d)	$\sqrt{2-x}$	1	$\sqrt{3}$	$\sqrt{2}$

SECTION 15

Goal Exercises (pp. 127-128)

1. (a) 3 (b) -2 (c) no slope (d) $\frac{3}{2}$
   (e) no slope (f) 0 (g) 2 (h) 3
3. (a) $y = 3x - 3$ (b) $y = -2x + 1$ (c) $x = 2$
   (d) $y = \frac{3}{2}x + 3$ (e) $x = -1$ (f) $y = -2$
   (g) $y = 2x$ (h) $y = 3x - 2$
5. (See top of facing page.)
7. (a), (d), and (f) are increasing; (c) is decreasing; (b) and (e) are constant functions.
9. (a) $v = 4000 - 400t$ (b) $v = 6500 - 260t$
   (c) $v = 15{,}000 - 300t$ (d) $v = 120 - 20t$
11. $n(p) = -2.5p + 200$; 125
13. $S(p) = 2000p - 30{,}000$; 14,000
15. (a) 0°, 37°, 100° (b) $F = \frac{9}{5}C + 32$ (c) 104°, 122°, 212°

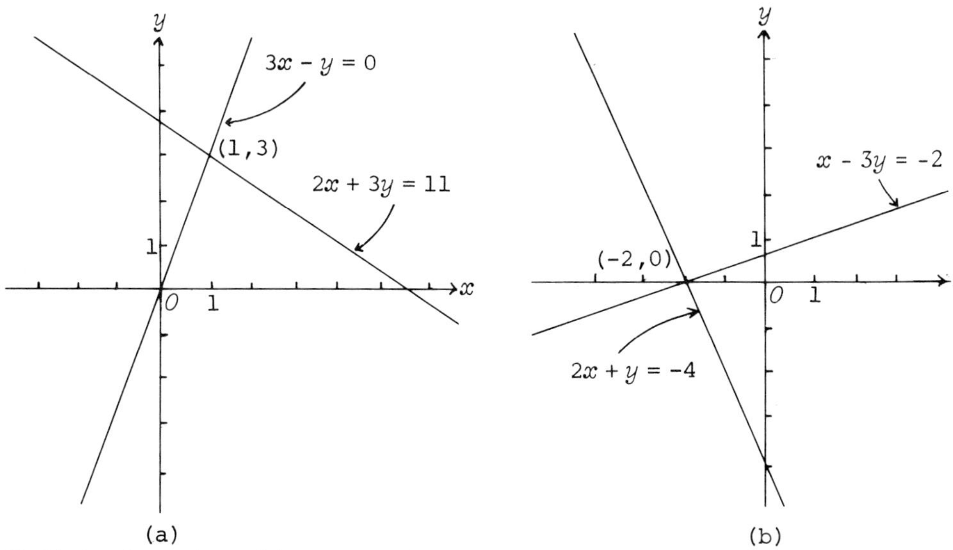

(a) (1,3) is the point of intersection

(b) (-2,0) is the point of intersection

FIGURE FOR SECTION 15 EXERCISE 5

TEST A (pp. 128-129)

1. (a) 2 (b) -2 (c) $-\frac{1}{2}$
2. (a) $y = 2$ (b) $y = -2x + 5$ (c) $y = 2x + 5$
3.

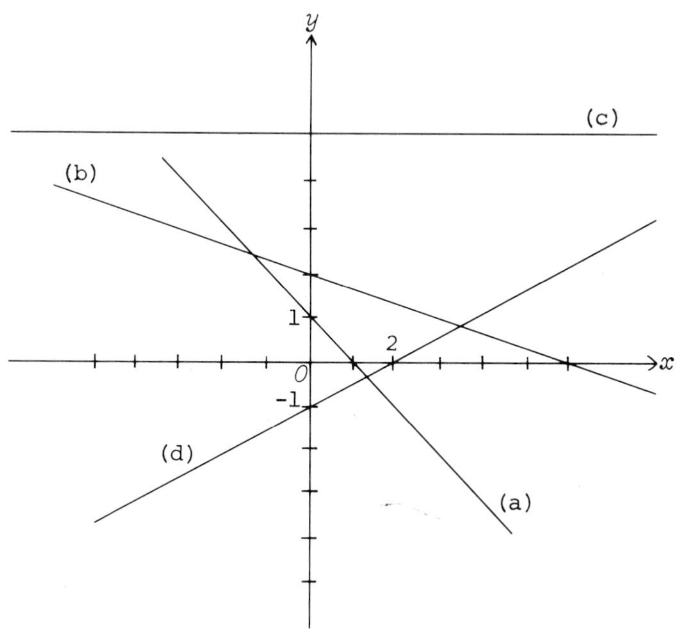

SECTION 16

Section Exercises (pp. 135-136)

1. (a), (d), (e), (f) 3. $y = 2x + 80$, $y = 9x$ 5. $y = 0.01p$, $y = 1000 - 0.03p$

7. If x is the number of stripped-down models, y that of standard models, and z of luxury models, then

$$\begin{aligned} x + y + z &= 20 \\ 6500x + 8500y + 11{,}000z &= 157{,}500 \end{aligned}$$

9. If w is the number of grams of B-1, x of B-2, y of B-6, and z of B-12, then

$$\begin{aligned} w + x + y + z &= 10{,}000 \\ w &= x \\ y &= 10w \\ y &= 400z \end{aligned}$$

Test A (p. 136)

1. (b) and (c)
2. (a) are identical; (b) intersect; (c) are parallel
3. $y = 0.05p$, $y = 200 - 0.15p$
4. If x is the number of apple trees, y of pear trees, and z of cherry trees, then

$$\begin{aligned} x + y + z &= 40 \\ 10x + 12y + 15z &= 470 \end{aligned}$$

SECTION 17

Section Exercises (pp. 142-143)

1. $x = 2,\ y = 5$ 3. $x = 5,\ y = -1$ 5. $x = 1,\ y = \frac{1}{3}$

7. $x = -2,\ y = 1,\ z = 3$ 9. $x = \frac{17}{7},\ y = \frac{11}{7},\ z = -\frac{29}{7}$

11. 120 units 13. 20 cars

15. 75 mg demerol, 25 mg phenergen

17. \$12,000 in bank, \$8000 in stocks, \$4000 in bonds

Test A (p. 143)

1. (a) $x = 3,\ y = 2$ (b) The system has an infinite number of solutions since the equations are identical. (c) $x = 0,\ y = 1,\ z = 2$

2. 400 tires 3. 125 and 150

SECTION 18

Section Exercises (pp. 154-155)

1.

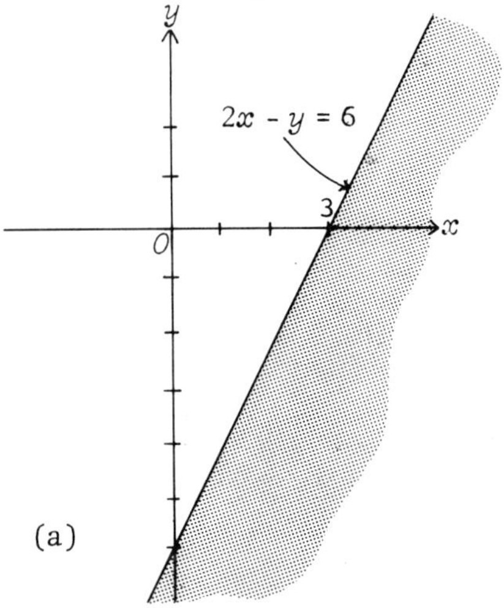

(a)

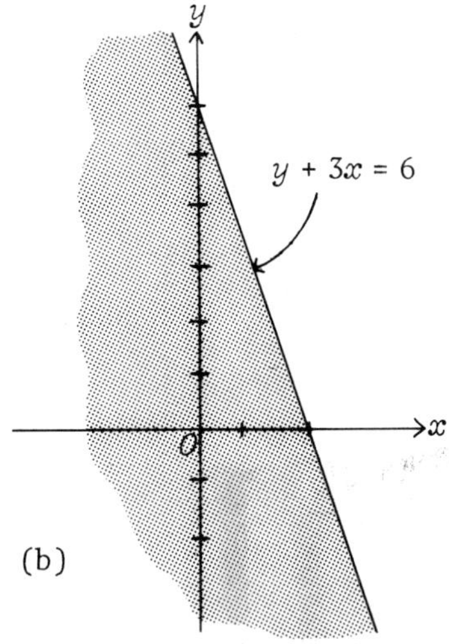

(b)

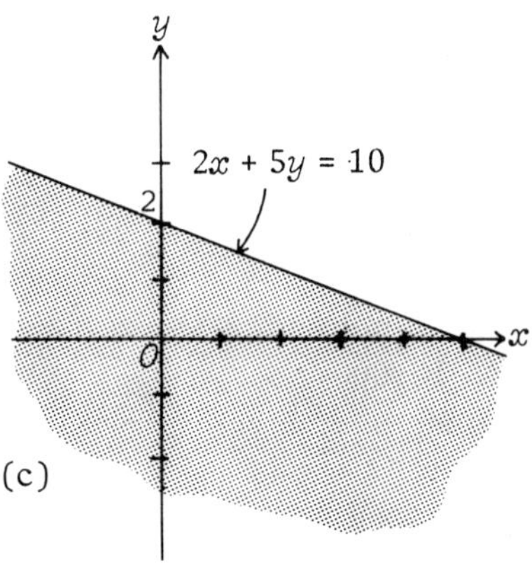

(c)

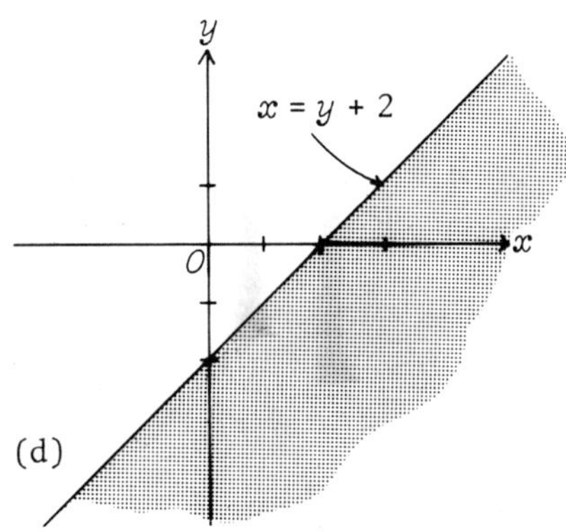

(d)

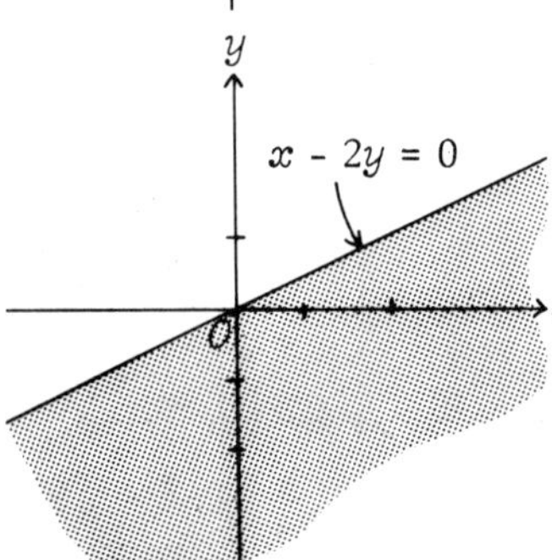

(e)

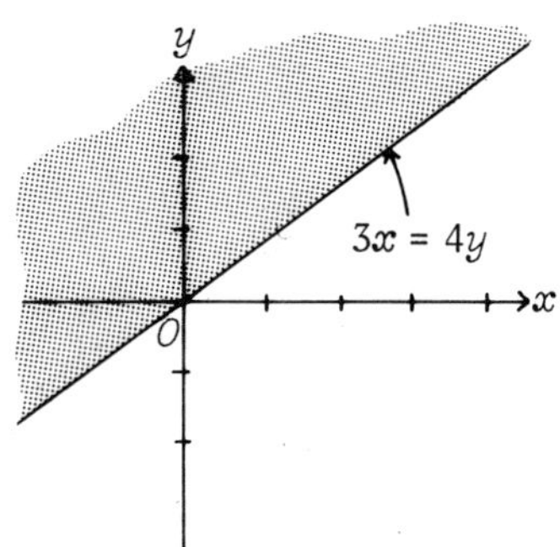

(f)

1. (cont'd)

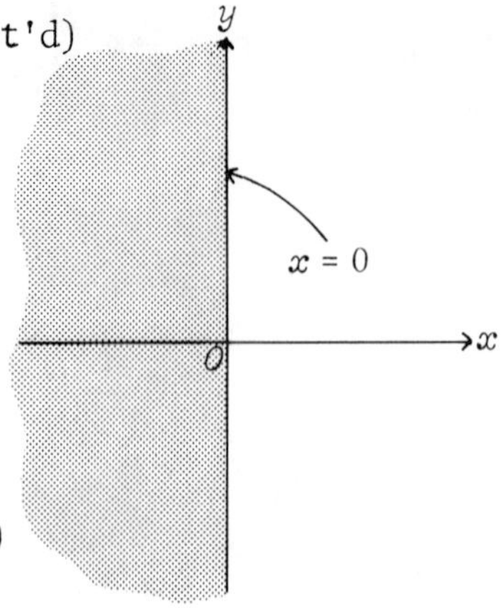

(g)

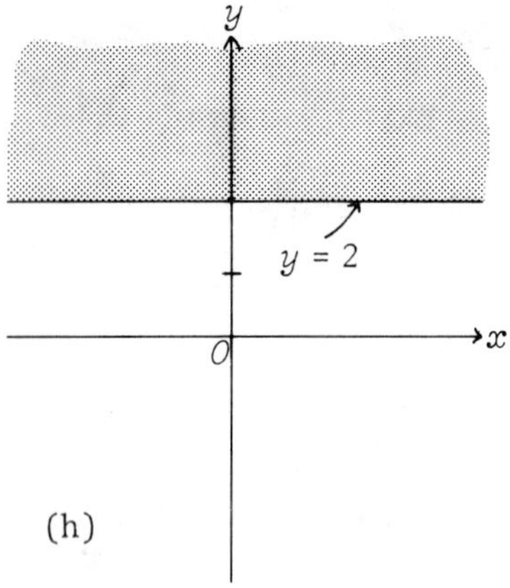

(h)

3. x = number of first model to be produced, y = number of second:

$$\begin{aligned} x,y &\geqslant 0 \\ 5x+7y &\leqslant 2000 \\ 3x+3y &\leqslant 1000 \end{aligned}$$

5. x = number of first pill, y = number of second:

$$\begin{aligned} x,y &\geqslant 0 \\ 1000x+1500y &\geqslant 5000 \\ 100x+\ \ 75y &\geqslant \ \ 400 \end{aligned}$$

7. x = amount (in dollars) invested in stocks, y = amount in bounds, and z = amount in certificates:

$$\begin{aligned} x,y,z &\geqslant 0 \\ x+y+z &\geqslant \ \ 5{,}000 \\ x+y+z &\leqslant 10{,}000 \\ x &\geqslant y \\ z &\geqslant 2y \end{aligned}$$

TEST A (pp. 155-156)

1.

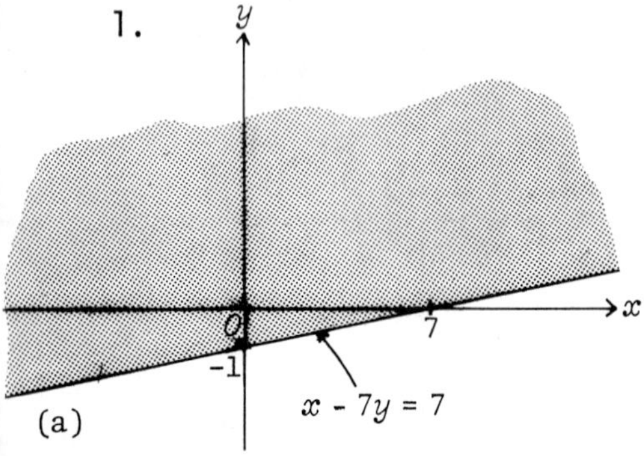

(a)

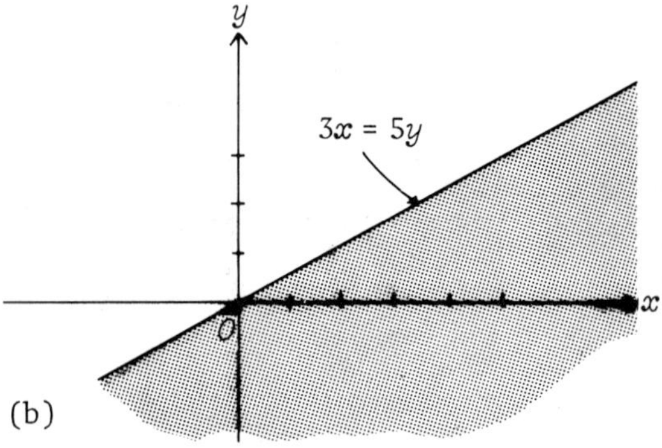

(b)

2.

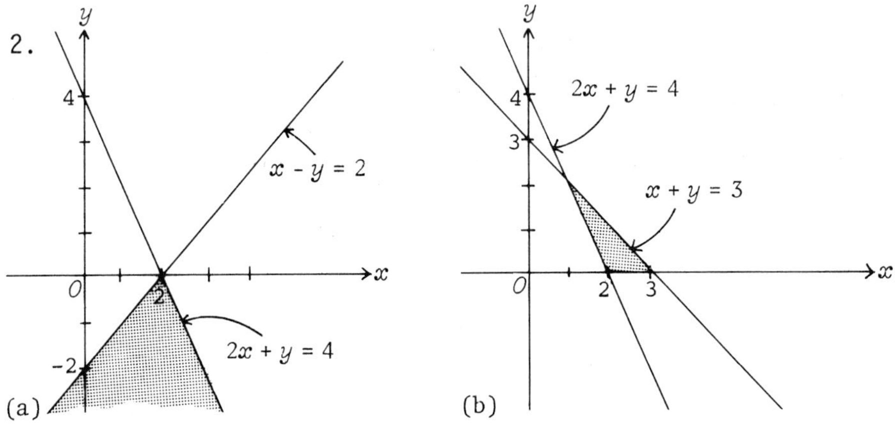

3. x = the number of black-and-white sets, y = number of color sets:

$$x, y \geqslant 0$$
$$x + 2y \leqslant 1000$$
$$2x + 3y \leqslant 1500$$

REVIEW EXERCISES (pp. 158-159)

1. only (a)

3.

	$f(0)$	$f(2)$	$f(-1)$
(a)	-1	5	-4
(b)	2	1	4
(c)	2	6	3
(d)	1	-7	-1

5. (a) $y = 3x - 2$ (b) $y = -\frac{1}{2}x + 5$ (c) $3x - 2y = -2$

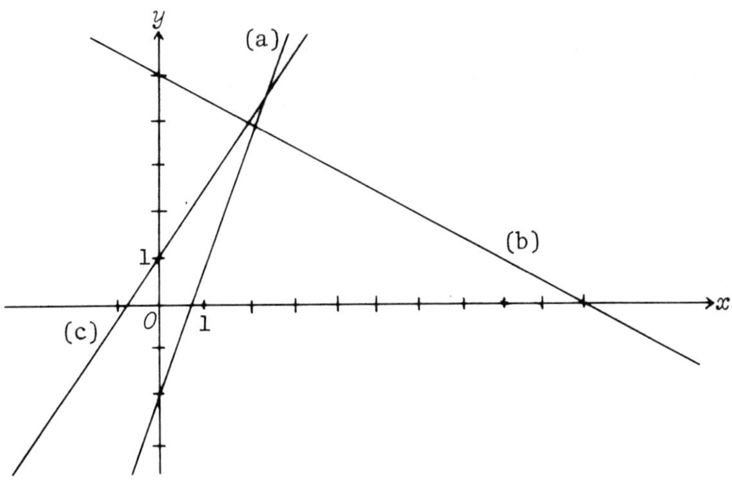

7. If p is the nightly rent and n the number that can be rented at p dollars each, then $n(p) = -6p + 250$.
9. (a) intersect (b) parallel (c) identical
11. (a) $x = 4$, $y = 3$ (b) $x = 2$, $y = -1$, $z = 1$ (c) $x = -2$, $y = 2$, $z = 5$
13. If x, y, and z are the number respectively of shares at \$50, \$100, and \$600, then the system of equations is

$$\begin{aligned} x + y + z &= 60 \\ 50x + 100y + 600z &= 10{,}000 \\ y &= x + z \end{aligned}$$

The solution is 20 shares at \$50 each, 30 at \$100 each, and 10 at \$600 each.

15.

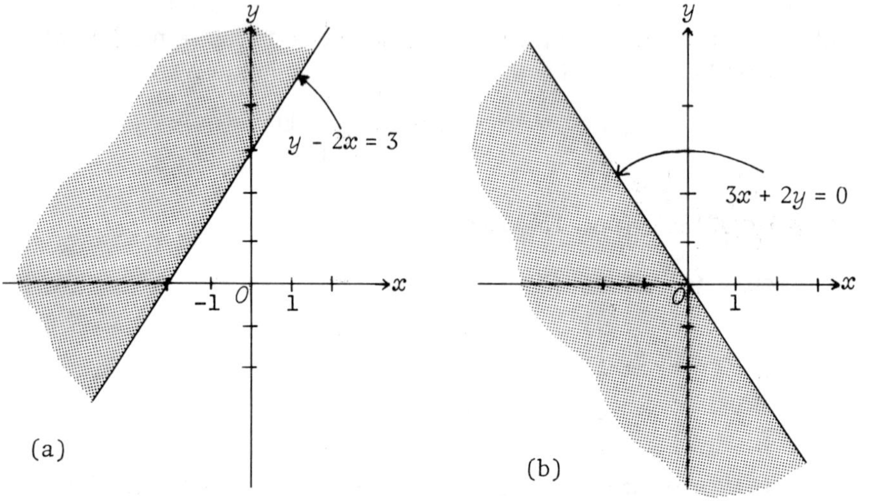

17. If x, y, and z are the amounts invested respectively in stocks, notes, and certificates of deposit, then

$$\begin{aligned} x, y, z &\geqslant 0 \\ x + y + z &\geqslant 10{,}000 \\ x + y + z &\leqslant 20{,}000 \\ x &\geqslant y + z \\ y &\leqslant \tfrac{1}{2} z \end{aligned}$$

CHAPTER TEST (p. 160)

1. $x \neq 0, 3$ 2. $f(0) = 0$, $f(-2) = 0$ 3. $y = -2x - 1$
4. $x + 2y = -3$ (See top of facing page.)
5. (a) $y = -500x + 6000$ (b) 3500
6. (c); $x = -2$, $y = 5$ 7. \$10

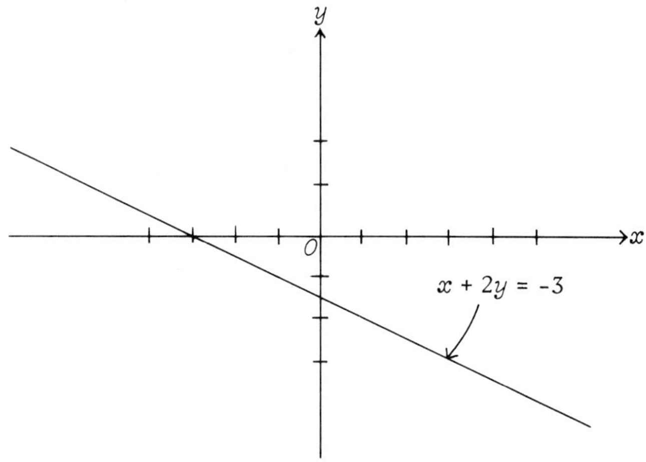

FIGURE FOR CHAPTER 3 TEST PROBLEM 4

8. 1250 paperbacks and 750 hard-covers 9. $x = 3$, $y = -1$, $z = 1$

10.

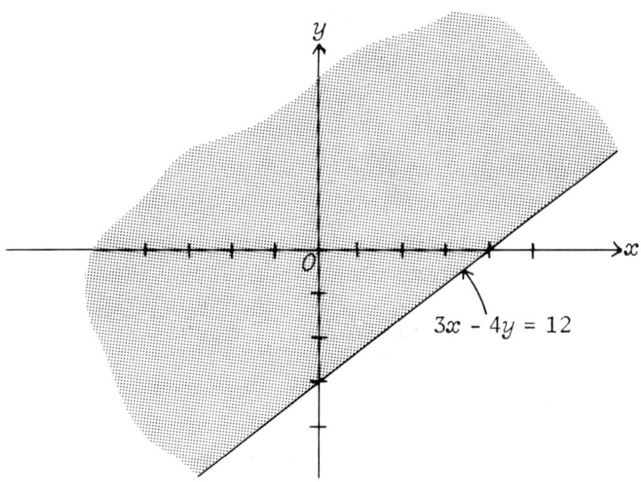

11. (See overleaf.)

12. Let x be number of A, y number of B. Then

$$\begin{aligned} x,y &\geqslant 0 \\ 4x + 4y &\leqslant 24 \\ 10x + 2y &\leqslant 36 \end{aligned}$$

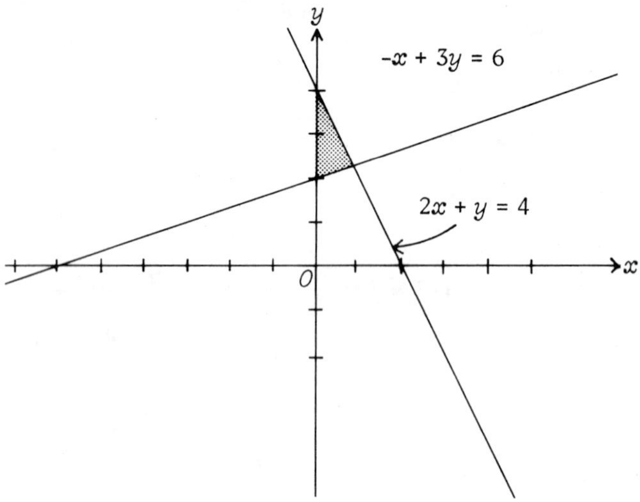

FIGURE FOR CHAPTER 3 TEST PROBLEM 11

CHAPTER 4

SECTION 19

SECTION EXERCISES (pp. 168-169)

1. A is 2×3, B is 4×3, C is 1×4, D is 5×1.
3. (a) transition, permutation, symmetric
   (b) diagonal, symmetric (c) transition
   (d) -- (e) symmetric
   (f) transition, permutation, diagonal, symmetric
   (g) -- (h) --
5. $\begin{pmatrix} \frac{3}{4} & \frac{1}{5} & \frac{1}{20} \end{pmatrix}$
7. $\begin{pmatrix} 8 & 7\frac{1}{2} & 8 & 8 & 10 \\ 6 & 4 & 5 & 4\frac{1}{2} & 8 \\ 4 & 8 & 8 & 0 & 0 \end{pmatrix}$
9. $$\begin{array}{ll} & \text{SWITCH TO} \\ & \begin{array}{ccc} \text{A} & \text{B} & \text{C} \end{array} \\ \text{NOW USING A} \\ \text{B} \\ \text{C} \end{array} \begin{pmatrix} \frac{3}{4} & \frac{1}{6} & \frac{1}{12} \\ \frac{3}{10} & \frac{1}{2} & \frac{1}{5} \\ \frac{1}{2} & \frac{1}{2} & 0 \end{pmatrix}$$

   This is a transition matrix
11. $x = -1$, $y = 2$

TEST A (pp. 169-170)

1. (a) $(\frac{7}{12}\ \ 0\ \ \frac{1}{3}\ \ \frac{1}{12}\ \ 0)$ or any other vector with nonnegative elements whose sum is 1.

(b) $\begin{pmatrix} 1 & 0 & 0 \\ 0 & 1 & 0 \\ 0 & 0 & 1 \end{pmatrix}$ or $\begin{pmatrix} \frac{1}{2} & 0 & \frac{1}{4} & \frac{1}{4} \\ 0 & 0 & 0 & 1 \\ \frac{1}{2} & \frac{1}{6} & \frac{1}{6} & \frac{1}{6} \\ 1 & 0 & 0 & 0 \end{pmatrix}$ or any other square matrix each of whose rows is a probability vector.

2. A is 1×4, B is 3×2, C is 2×2.

3.

	PRICE IN DOLLARS OF	
	HARD-COVER	SOFT-COVER
PUBLISHER I	18	11
II	12	8
III	19	12
IV	14	8
V	15	9

SECTION 20

SECTION EXERCISES (p. 175)

1. (0 5)

3. $\begin{pmatrix} 0 \\ 5 \end{pmatrix}$

5. $\begin{pmatrix} 9 & 17 \\ 4 & -6 \end{pmatrix}$

7. $\begin{pmatrix} 11 & -6 \\ 20 & 49 \end{pmatrix}$

9. $\begin{pmatrix} 52 & 14 \\ 43 & 1 \\ 9 & 6 \end{pmatrix}$

TEST A (p. 175)

1. $\begin{pmatrix} 3 & 0 \\ 6 & 9 \end{pmatrix}$

2. $\begin{pmatrix} 11 \\ 52 \end{pmatrix}$

3. not defined

SECTION 21

SECTION EXERCISES (pp. 184-184)

1. (a) 1 (b) 8 (c) $\begin{pmatrix} 4 & 12 \\ 5 & 15 \end{pmatrix}$ (d) $\begin{pmatrix} 1 & 0 & 0 \\ 0 & 0 & 0 \end{pmatrix}$

(e) (3 18 17) (f) not defined (g) not defined

(h) $\begin{pmatrix} 11 \\ 10 \\ 28 \end{pmatrix}$ (i) (1 -2 -1) (j) $\begin{pmatrix} 4 & 5 \\ 5 & 10 \end{pmatrix}$

1. (cont'd)

(k) $\begin{pmatrix} 4 & 5 \\ 14 & 20 \end{pmatrix}$ (l) not defined (m) $\begin{pmatrix} 7 & 0 & 4 \\ 0 & 2 & 1 \\ 7 & -3 & -2 \end{pmatrix}$

3. (a) $\begin{pmatrix} 3 & 2 \\ 1 & -1 \end{pmatrix}\left[\begin{pmatrix} 0 & 2 \\ 4 & 1 \end{pmatrix}\begin{pmatrix} -1 & 0 \\ 1 & 2 \end{pmatrix}\right] = \begin{pmatrix} 3 & 2 \\ 1 & -1 \end{pmatrix}\begin{pmatrix} 2 & 4 \\ -3 & 2 \end{pmatrix} = \begin{pmatrix} 0 & 16 \\ 5 & 2 \end{pmatrix}$

$\left[\begin{pmatrix} 3 & 2 \\ 1 & -1 \end{pmatrix}\begin{pmatrix} 0 & 2 \\ 4 & 1 \end{pmatrix}\right]\begin{pmatrix} -1 & 0 \\ 1 & 2 \end{pmatrix} = \begin{pmatrix} 8 & 8 \\ -4 & 1 \end{pmatrix}\begin{pmatrix} -1 & 0 \\ 1 & 2 \end{pmatrix} = \begin{pmatrix} 0 & 16 \\ 5 & 2 \end{pmatrix}$

(b) $\begin{pmatrix} 2 & -1 \\ 0 & 1 \end{pmatrix}\left[\begin{pmatrix} 3 & 0 & 1 \\ 2 & 4 & -1 \end{pmatrix}\begin{pmatrix} 1 & 2 & 5 \\ 0 & 3 & 1 \\ 1 & 4 & -1 \end{pmatrix}\right] = \begin{pmatrix} 2 & -1 \\ 0 & 1 \end{pmatrix}\begin{pmatrix} 4 & 10 & 14 \\ 1 & 12 & 15 \end{pmatrix} = \begin{pmatrix} 7 & 8 & 13 \\ 1 & 12 & 15 \end{pmatrix}$

(b) $\left[\begin{pmatrix} 2 & -1 \\ 0 & 1 \end{pmatrix}\begin{pmatrix} 3 & 0 & 1 \\ 2 & 4 & 1 \end{pmatrix}\right]\begin{pmatrix} 1 & 2 & 5 \\ 0 & 3 & 1 \\ 1 & 4 & -1 \end{pmatrix} = \begin{pmatrix} 4 & -4 & 3 \\ 2 & 4 & -1 \end{pmatrix}\begin{pmatrix} 1 & 2 & 5 \\ 0 & 3 & 1 \\ 1 & 4 & -1 \end{pmatrix}$

$= \begin{pmatrix} 7 & 8 & 13 \\ 1 & 12 & 15 \end{pmatrix}$

5. (a) $\begin{pmatrix} 1 & 0 \\ 0 & 4 \end{pmatrix}, \begin{pmatrix} 1 & 0 \\ 0 & 8 \end{pmatrix}$ (b) $\begin{pmatrix} 1 & 0 \\ 0 & 1 \end{pmatrix}, \begin{pmatrix} 0 & 1 \\ 1 & 0 \end{pmatrix}$ (c) $\begin{pmatrix} 4 & 0 \\ 0 & 9 \end{pmatrix}, \begin{pmatrix} 8 & 0 \\ 0 & 27 \end{pmatrix}$

(d) $\begin{pmatrix} 1 & 0 \\ 15 & 16 \end{pmatrix}, \begin{pmatrix} 1 & 0 \\ 63 & 64 \end{pmatrix}$ (e) $\begin{pmatrix} 11 & 14 \\ 7 & 18 \end{pmatrix}, \begin{pmatrix} 47 & 78 \\ 39 & 86 \end{pmatrix}$

(f) $\begin{pmatrix} 1 & 0 & 0 \\ 0 & 1 & 0 \\ 0 & 0 & 1 \end{pmatrix}, \begin{pmatrix} 1 & 0 & 0 \\ 0 & 0 & 1 \\ 0 & 1 & 0 \end{pmatrix}$ (g) $\begin{pmatrix} 9 & 0 & 0 \\ 0 & 4 & 0 \\ 0 & 0 & 16 \end{pmatrix}, \begin{pmatrix} 27 & 0 & 0 \\ 0 & 8 & 0 \\ 0 & 0 & 64 \end{pmatrix}$

(h) $\begin{pmatrix} 1 & 0 & 6 \\ 12 & 1 & 12 \\ 24 & 8 & 25 \end{pmatrix}, \begin{pmatrix} 25 & 2 & 30 \\ 60 & 23 & 63 \\ 124 & 40 & 149 \end{pmatrix}$

7. Store I took in \$4175, II took in \$2950.

TEST A (pp. 184-185)

1. (a) 2 (b) not defined (c) not defined

(d) $\begin{pmatrix} a + c \\ 2a + 3c \end{pmatrix}$ (e) $\begin{pmatrix} 4 & 0 \\ 1 & 1 \end{pmatrix}$

2. (a) $\begin{pmatrix} 720 \\ 566 \end{pmatrix}$ (b) \$566

SECTION 22

Section Exercises (pp. 197-198)

1. (a) $x = 4,\ y = 1,\ z = 2$ (b) $x = -1,\ y = 3,\ z = 1$
   (c) $x = 1,\ y = -2,\ z = 3$ (d) $x = 0,\ y = 1,\ z = -1$
   (e) $x = 2,\ y = 0,\ z = 1$ (f) $x = 4,\ y = 5,\ z = -4$
   (g) $x = 2,\ y = 1$ (h) $x = -2,\ y = 3$
   (i) $w = -1,\ x = 0,\ y = 1,\ z = 2$ (j) $w = 4,\ x = 3,\ y = 2,\ z = 1$

Test A (p. 198)

1. $$\left(\begin{array}{ccc|c} 1 & -1 & 3 & 7 \\ 2 & 0 & 1 & 11 \\ 0 & 1 & -2 & -1 \end{array}\right) \to \left(\begin{array}{ccc|c} 1 & -1 & 3 & 7 \\ 0 & 2 & -5 & -3 \\ 0 & 1 & -2 & -1 \end{array}\right) \to \left(\begin{array}{ccc|c} 1 & -1 & 3 & 7 \\ 0 & 1 & -2 & -1 \\ 0 & 2 & -5 & -3 \end{array}\right) \to$$
$$\left(\begin{array}{ccc|c} 1 & 0 & 1 & 6 \\ 0 & 1 & -2 & -1 \\ 0 & 0 & -1 & -1 \end{array}\right) \to \left(\begin{array}{ccc|c} 1 & 0 & 1 & 6 \\ 0 & 1 & -2 & -1 \\ 0 & 0 & 1 & 1 \end{array}\right) \to \left(\begin{array}{ccc|c} 1 & 0 & 0 & 5 \\ 0 & 1 & 0 & 1 \\ 0 & 0 & 1 & 1 \end{array}\right).$$
$x = 5,\ y = 1,\ z = 1$ is the unique solution.

2. $$\left(\begin{array}{ccc|c} 0 & 2 & 1 & 3 \\ 1 & 1 & 1 & 2 \\ 1 & -1 & 0 & 1 \end{array}\right) \to \left(\begin{array}{ccc|c} 1 & 1 & 1 & 2 \\ 0 & 2 & 1 & 3 \\ 1 & -1 & 0 & 1 \end{array}\right) \to \left(\begin{array}{ccc|c} 1 & 1 & 1 & 2 \\ 0 & 2 & 1 & 3 \\ 0 & -2 & -1 & -1 \end{array}\right) \to$$
$$\left(\begin{array}{ccc|c} 1 & 1 & 1 & 2 \\ 0 & 1 & \frac{1}{2} & \frac{3}{2} \\ 0 & -2 & -1 & -1 \end{array}\right) \to \left(\begin{array}{ccc|c} 1 & 0 & \frac{1}{2} & \frac{1}{2} \\ 0 & 1 & \frac{1}{2} & \frac{3}{2} \\ 0 & 0 & 0 & 2 \end{array}\right).$$ There is no solution.

3. $$\left(\begin{array}{ccc|c} 0 & 1 & -1 & 4 \\ 2 & 1 & 0 & 3 \\ 1 & 0 & 1 & -1 \end{array}\right) \to \left(\begin{array}{ccc|c} 1 & 0 & 1 & -1 \\ 2 & 1 & 0 & 3 \\ 0 & 1 & -1 & 4 \end{array}\right) \to \left(\begin{array}{ccc|c} 1 & 0 & 1 & -1 \\ 0 & 1 & -2 & 5 \\ 0 & 1 & -1 & 4 \end{array}\right) \to$$
$$\left(\begin{array}{ccc|c} 1 & 0 & 1 & -1 \\ 0 & 1 & -2 & 5 \\ 0 & 0 & 1 & -1 \end{array}\right) \to \left(\begin{array}{ccc|c} 1 & 0 & 0 & 0 \\ 0 & 1 & 0 & 3 \\ 0 & 0 & 1 & -1 \end{array}\right).$$ $x = 0,\ y = 3,\ z = -1$ is the unique solution.

4. $$\left(\begin{array}{ccc|c} 2 & 1 & -1 & 1 \\ 1 & -1 & 1 & 2 \\ 3 & 1 & -1 & 2 \end{array}\right) \to \left(\begin{array}{ccc|c} 1 & -1 & 1 & 2 \\ 2 & 1 & -1 & 1 \\ 3 & 1 & -1 & 2 \end{array}\right) \to \left(\begin{array}{ccc|c} 1 & -1 & 1 & 2 \\ 0 & 3 & -3 & -3 \\ 0 & 4 & -4 & -4 \end{array}\right) \to$$
$$\left(\begin{array}{ccc|c} 1 & -1 & 1 & 2 \\ 0 & 1 & -1 & -1 \\ 0 & 1 & -1 & -1 \end{array}\right) \to \left(\begin{array}{ccc|c} 1 & 0 & 0 & 1 \\ 0 & 1 & -1 & -1 \\ 0 & 0 & 0 & 0 \end{array}\right).$$ There are many solutions.

SECTION 23

Section Exercises (pp. 205-206)

1. (a) $\begin{pmatrix} 5 & -3 \\ -3 & 2 \end{pmatrix}$ (b) $\begin{pmatrix} -\frac{1}{5} & \frac{4}{5} \\ \frac{2}{5} & -\frac{3}{5} \end{pmatrix}$ (c) no inverse

   (d) $\begin{pmatrix} -\frac{3}{4} & \frac{1}{4} \\ 1 & 0 \end{pmatrix}$ (e) $\begin{pmatrix} -1 & 0 \\ \frac{1}{2} & \frac{1}{2} \end{pmatrix}$ (f) $\begin{pmatrix} 5 & 6 & -2 \\ -2 & -2 & 1 \\ -2 & -3 & 1 \end{pmatrix}$

   (g) $\begin{pmatrix} 1 & -1 & -5 \\ 1 & -2 & -5 \\ -\frac{1}{2} & 1 & 3 \end{pmatrix}$ (h) no inverse (i) $\begin{pmatrix} \frac{1}{2} & -\frac{1}{2} & -\frac{1}{2} & \frac{1}{2} \\ -\frac{3}{8} & \frac{13}{8} & \frac{9}{8} & -\frac{5}{8} \\ \frac{5}{8} & -\frac{3}{8} & -\frac{7}{8} & \frac{3}{8} \\ \frac{1}{4} & \frac{1}{4} & \frac{1}{4} & \frac{1}{4} \end{pmatrix}$

3. (a) $\begin{pmatrix} x \\ y \end{pmatrix}$ equals, respectively, $\begin{pmatrix} 1 \\ 4 \end{pmatrix}$, $\begin{pmatrix} -1 \\ 0 \end{pmatrix}$, and $\begin{pmatrix} 5 \\ -8 \end{pmatrix}$.

   (b) $\begin{pmatrix} x \\ y \end{pmatrix}$ equals, respectively, $\begin{pmatrix} 1 \\ 2 \end{pmatrix}$, $\begin{pmatrix} 10 \\ -1 \end{pmatrix}$, and $\begin{pmatrix} 0 \\ 1 \end{pmatrix}$.

   (c) $\begin{pmatrix} x \\ y \end{pmatrix}$ equals, respectively, $\begin{pmatrix} 6 \\ 3 \end{pmatrix}$, $\begin{pmatrix} 1 \\ 1 \end{pmatrix}$, and $\begin{pmatrix} 15 \\ 5 \end{pmatrix}$.

   (d) $\begin{pmatrix} x \\ y \\ z \end{pmatrix}$ equals, respectively, $\begin{pmatrix} 1 \\ 2 \\ 3 \end{pmatrix}$, $\begin{pmatrix} 0 \\ 1 \\ -2 \end{pmatrix}$, and $\begin{pmatrix} 8 \\ 5 \\ 1 \end{pmatrix}$.

   (e) $\begin{pmatrix} x \\ y \\ z \end{pmatrix}$ equals, respectively, $\begin{pmatrix} 0 \\ 1 \\ 2 \end{pmatrix}$, $\begin{pmatrix} -1 \\ -1 \\ -1 \end{pmatrix}$, and $\begin{pmatrix} 1 \\ 2 \\ 3 \end{pmatrix}$.

Test A (p. 206)

1. In each of the following the required inverse is the part to the right of the vertical bar in the last augmented matrix:

   (a) $\left(\begin{array}{cc|cc} 1 & 3 & 1 & 0 \\ 2 & -1 & 0 & 1 \end{array}\right) \rightarrow \left(\begin{array}{cc|cc} 1 & 3 & 1 & 0 \\ 0 & -7 & -2 & 1 \end{array}\right) \rightarrow \left(\begin{array}{cc|cc} 1 & 3 & 1 & 0 \\ 0 & 0 & \frac{2}{7} & -\frac{1}{7} \end{array}\right) \rightarrow \left(\begin{array}{cc|cc} 1 & 0 & \frac{1}{7} & \frac{3}{7} \\ 0 & 1 & \frac{2}{7} & -\frac{1}{7} \end{array}\right)$

   (b) $\left(\begin{array}{cc|cc} 0 & -1 & 1 & 0 \\ 2 & 0 & 0 & 1 \end{array}\right) \rightarrow \left(\begin{array}{cc|cc} 2 & 0 & 0 & 1 \\ 0 & -1 & 1 & 0 \end{array}\right) \rightarrow \left(\begin{array}{cc|cc} 1 & 0 & 0 & \frac{1}{2} \\ 0 & -1 & 1 & 0 \end{array}\right) \rightarrow \left(\begin{array}{cc|cc} 1 & 0 & 0 & \frac{1}{2} \\ 0 & 1 & -1 & 0 \end{array}\right)$

   (c) $\left(\begin{array}{ccc|ccc} 2 & 0 & 1 & 1 & 0 & 0 \\ 1 & 0 & -1 & 0 & 1 & 0 \\ 0 & 2 & 3 & 0 & 0 & 1 \end{array}\right) \rightarrow \left(\begin{array}{ccc|ccc} 1 & 0 & -1 & 0 & 1 & 0 \\ 0 & 2 & 3 & 0 & 0 & 1 \\ 2 & 0 & 1 & 1 & 0 & 0 \end{array}\right) \rightarrow$

1. (c) (cont'd) $\left(\begin{array}{ccc|ccc} 1 & 0 & -1 & 0 & 1 & 0 \\ 0 & 1 & \frac{3}{2} & 0 & 0 & \frac{1}{2} \\ 0 & 0 & 3 & 1 & -2 & 0 \end{array}\right) \rightarrow \left(\begin{array}{ccc|ccc} 1 & 0 & 0 & \frac{1}{3} & \frac{1}{3} & 0 \\ 0 & 1 & 0 & -\frac{1}{2} & 1 & \frac{1}{2} \\ 0 & 0 & 1 & \frac{1}{3} & -\frac{2}{3} & 0 \end{array}\right)$

2. (a) $\begin{pmatrix} 3 & 7 \\ 1 & 2 \end{pmatrix}^{-1} = \frac{1}{6-7}\begin{pmatrix} 2 & -7 \\ -1 & 3 \end{pmatrix} = \begin{pmatrix} -2 & 7 \\ 1 & -3 \end{pmatrix}$; $\begin{pmatrix} -2 & 7 \\ 1 & -3 \end{pmatrix}\begin{pmatrix} -1 \\ 0 \end{pmatrix} = \begin{pmatrix} 2 \\ -1 \end{pmatrix}$;

$\begin{pmatrix} -2 & 7 \\ 1 & -3 \end{pmatrix}\begin{pmatrix} 33 \\ 10 \end{pmatrix} = \begin{pmatrix} 4 \\ 3 \end{pmatrix}$; $\begin{pmatrix} -2 & 7 \\ 1 & -3 \end{pmatrix}\begin{pmatrix} -10 \\ -3 \end{pmatrix} = \begin{pmatrix} -1 \\ -1 \end{pmatrix}$

(b) $\left(\begin{array}{ccc|ccc} 1 & -1 & 0 & 1 & 0 & 0 \\ 0 & 1 & 2 & 0 & 1 & 0 \\ 3 & 0 & -1 & 0 & 0 & 1 \end{array}\right) \rightarrow \left(\begin{array}{ccc|ccc} 1 & -1 & 0 & 1 & 0 & 0 \\ 0 & 1 & 2 & 0 & 1 & 0 \\ 0 & 3 & -1 & -3 & 0 & 1 \end{array}\right) \rightarrow$

$\left(\begin{array}{ccc|ccc} 1 & 0 & 2 & 1 & 1 & 0 \\ 0 & 1 & 2 & 0 & 1 & 0 \\ 0 & 0 & -7 & -3 & -3 & 1 \end{array}\right) \rightarrow \left(\begin{array}{ccc|ccc} 1 & 0 & 0 & \frac{1}{7} & \frac{1}{7} & \frac{2}{7} \\ 0 & 1 & 0 & -\frac{6}{7} & \frac{1}{7} & \frac{2}{7} \\ 0 & 0 & 1 & \frac{3}{7} & \frac{3}{7} & -\frac{1}{7} \end{array}\right)$;

$\begin{pmatrix} \frac{1}{7} & \frac{1}{7} & \frac{2}{7} \\ -\frac{6}{7} & \frac{1}{7} & \frac{2}{7} \\ \frac{3}{7} & \frac{3}{7} & -\frac{1}{7} \end{pmatrix}\begin{pmatrix} 0 \\ 4 \\ 5 \end{pmatrix} = \begin{pmatrix} 2 \\ 2 \\ 1 \end{pmatrix}$; $\begin{pmatrix} \frac{1}{7} & \frac{1}{7} & \frac{2}{7} \\ -\frac{6}{7} & \frac{1}{7} & \frac{2}{7} \\ \frac{3}{7} & \frac{3}{7} & -\frac{1}{7} \end{pmatrix}\begin{pmatrix} -5 \\ 25 \\ -10 \end{pmatrix} = \begin{pmatrix} 0 \\ 5 \\ 10 \end{pmatrix}$

SECTION 24

SECTION EXERCISES (pp. 213-214)

Some answers may differ in final digits because of accumulated rounding errors. We used a hand calculator and kept decimals to three places in most computations.

5. 15.66; 20.34

7. (a) $R = \begin{array}{c} \\ A \\ B \\ C \end{array}\begin{array}{c} \begin{array}{ccc} A & B & C \end{array} \\ \begin{pmatrix} .20 & .20 & .20 \\ .30 & 0 & .50 \\ 0 & .20 & .20 \end{pmatrix} \end{array}$; $D = \begin{pmatrix} 5 \\ 6 \\ 4.5 \end{pmatrix}$

(b) $X = RX + D$

(c) A \$12,100; B \$14,220; C \$9,180

(d) A \$9,400; B \$9,080; C \$8520

TEST A (pp. 214-215)

1. Since $I - R = \begin{pmatrix} .8 & -.5 \\ -.4 & .7 \end{pmatrix}$ and this matrix is 2×2, we use the formula on page 203 to find its inverse:

1. (cont'd)

$$(I - R)^{-1} = \frac{1}{.56 - .20}\begin{pmatrix} .7 & .5 \\ .4 & .8 \end{pmatrix} = \frac{1}{.36}\begin{pmatrix} .7 & .3 \\ .4 & .8 \end{pmatrix} = \begin{pmatrix} 1.944 & 1.389 \\ 1.111 & 2.222 \end{pmatrix}$$

2. $$\begin{pmatrix} x \\ y \end{pmatrix} = \begin{pmatrix} 1.944 & 1.389 \\ 1.111 & 2.222 \end{pmatrix}\begin{pmatrix} 12 \\ 20 \end{pmatrix} = \begin{pmatrix} 51.108 \\ 57.772 \end{pmatrix}$$

3. The input-output matrix R and the external demand vector D are

$$R = \begin{matrix} & \text{B} & \text{A} \\ \text{A} & 0 & .20 \\ \text{B} & 15 & 0 \end{matrix} \qquad D = \begin{pmatrix} 9{,}700 \\ 12{,}125 \end{pmatrix}$$

Then

$$I - R = \begin{pmatrix} 1 & -.2 \\ -.15 & 1 \end{pmatrix} \quad \text{and} \quad (I - R)^{-1} = \begin{pmatrix} 1.031 & .206 \\ .155 & 1.031 \end{pmatrix}$$

So

$$\begin{pmatrix} x \\ y \end{pmatrix} = \begin{pmatrix} 1.031 & .206 \\ .155 & 1.031 \end{pmatrix}\begin{pmatrix} 15 \\ 8 \end{pmatrix} = \begin{pmatrix} 17.113 \\ 10.573 \end{pmatrix}$$

SECTION 25

SECTION EXERCISES (pp. 225-229)

1. (a), (c), (e), and (f) are regular.
3. (a) Rounded to three decimals:

GAME FROM NOW	PROBABILITY OF WIN	LOSS
1 (next)	(0.600	0.400)
2	(0.560	0.440)
3	(0.556	0.444)
4	"	"
5	"	"
10	"	"
15	"	"
20	"	"

Expected percentage win for the season: 55.6%

(b) Rounded to three decimals:

$$T^2 = \begin{pmatrix} 0.56 & 0.44 \\ 0.55 & 0.45 \end{pmatrix} \qquad T^3 = \begin{pmatrix} 0.556 & 0.444 \\ 0.555 & 0.445 \end{pmatrix}$$

$$T^4 = \begin{pmatrix} 0.556 & 0.444 \\ 0.556 & 0.444 \end{pmatrix} = T^5 = T^{10} = T^{15} = T^{20}$$

(c) $\left(\frac{5}{9}\ \frac{4}{9}\right)$, which, rounded to three decimal places, yields $(0.556\ \ 0.444)$.

5. (a)

THIS MONTH: CUSTOMERS WHO:	NEXT MONTH: CUSTOMERS WHO: CHARGE	DON'T CHARGE
CHARGE	0.60	0.40
DON'T CHARGE	0.20	0.80

5. (cont'd)

   (b) $\frac{1}{3}$

7. 57.1%, 42.9%   9. 0.5 (no matter what the given percentages).
11. 8.70%, 86.96%

TEST A (p. 229)

1. (a) is regular because $\begin{pmatrix} 0.9 & 0.1 \\ 1 & 0 \end{pmatrix}^2 = \begin{pmatrix} 0.91 & 0.09 \\ 0.9 & 0.1 \end{pmatrix}$, which has no zeros. (c) is regular because it has no zeros. (b) is irregular.
2. The matrix is:

		NEXT DAY: RAIN	NO RAIN
ONE DAY:	RAIN	0.8	0.2
	NO RAIN	0.5	0.5

   From $x + y = 1$ and $(x\ y)\begin{pmatrix} 0.8 & 0.2 \\ 0.5 & 0.5 \end{pmatrix} = (x\ y)$ we get

$$\begin{aligned} x + \quad y &= 1 \\ 0.8x + 0.5y &= x \\ 0.2x + 0.5y &= y \end{aligned}$$

   which has a solution $x = \frac{5}{7}$, $y = \frac{2}{7}$. So it is expected that $\frac{5}{7}$ of the days will be rainy.
3. The matrix is:

		TO: UPPER	MIDDLE	LOWER
FROM:	UPPER	0.85	0.15	0
	MIDDLE	0.1	0.8	0.1
	LOWER	0.05	0.20	0.75

   From $x + y + z = 1$ and $(x\ y\ z)\begin{pmatrix} 0.85 & 0.15 & 0 \\ 0.1 & 0.8 & 0.1 \\ 0.05 & 0.20 & 0.75 \end{pmatrix} = (x\ y\ z)$ we get

$$\begin{aligned} x + \quad y + \quad z &= 1 \\ 0.85x + 0.10y + 0.05z &= x \\ 0.15x + 0.80y + 0.20z &= y \\ 0.10y + 0.75z &= z \end{aligned}$$

   which has a solution $x = \frac{4}{11}$, $y = \frac{5}{11}$, $z = \frac{2}{11}$. So, eventually, 36.36% of the population will be upper class, 45.46% will be middle class, and 18.18% lower class.

REVIEW EXERCISES (pp. 232-234)

1. A is 3×2, B is 1×5, C is 4×4, D is 5×1.
3. C is a transition, permutation, and symmetric matrix.
5. (a) $\begin{pmatrix} -1 & -5 & 1 \\ 5 & -2 & 2 \end{pmatrix}$ (b) $\begin{pmatrix} 2 & 4 & 7 \\ -7 & 16 & 2 \end{pmatrix}$ (c) not defined

5. (cont'd)
(d) not defined (e) $\begin{pmatrix} 1 & 0 \\ 0 & 1 \end{pmatrix}$ (f) $\begin{pmatrix} x + \frac{1}{2}y - 2z \\ -2x + z \\ -x + y - 3z \end{pmatrix}$

(g) not defined (h) $(-x + 5y + 3z)$

(i) $\begin{pmatrix} -2y + 3z \\ x + 4y + 2z \end{pmatrix}$ (j) $\begin{pmatrix} 1 & 4 & 2 \\ 0 & -2 & 3 \end{pmatrix}$

7. (a) $x = 1,\ y = 0,\ z = -2$ (b) many solutions
(c) $x = 5,\ y = \frac{1}{2},\ z = -6$

9. (a) $x = 3,\ y = -2;\ x = \frac{19}{5},\ y = \frac{2}{5};\ x = -\frac{28}{5},\ y = \frac{11}{5}$.
(b) $x = 1,\ y = 0,\ z = 0;\ x = 3,\ y = -3,\ z = 0$.

11. \$38,480; \$28,120 13. (a) and (c) are regular.

15. (a) $\frac{4}{5}$ (b) $\frac{1}{5}$

CHAPTER TEST (pp. 235-236)

1. A is 2×4; B is 1×5; C is 4×2.
2. (a) A probability vector is one whose elements are all non-negative and add up to one.
   (b) A transition matrix is a square matrix in which each row is a probability vector.
3. (a) not defined (b) $\begin{pmatrix} -4 & 0 & -2 \\ 17 & 3 & 10 \\ -3 & -1 & -2 \end{pmatrix}$ (c) not defined

   (d) $\begin{pmatrix} 1 & 0 \\ 0 & 1 \end{pmatrix}$ (e) not defined (f) $\begin{pmatrix} -1 & 0 \\ 2 & -3 \\ 0 & 1 \end{pmatrix}$

   (g) not defined
4. (a)

	PORT.	APT.	STAND.	DEL.
PITTSBURGH	500	450	460	300
NEW ORLEANS	460	340	450	290
SAN DIEGO	520	480	400	340

   (b)

	SALES PRICE (\$)
PORTABLE	65
APT. SIZE	175
STANDARD	300
DELUXE	500

5.

	REVENUE (\$)
PITTSBURGH	399,250
NEW ORLEANS	369,400
SAN DIEGO	407,800

San Diego had the largest revenue.

6. (a) no solution (b) $x = 2,\ y = -1,\ z = 3$
7. (a) $\begin{pmatrix} \frac{2}{13} & \frac{3}{26} \\ -\frac{1}{13} & \frac{5}{26} \end{pmatrix}$ (b) no inverse

8. The inverse of the coefficient matrix is $\begin{pmatrix} \frac{1}{2} & \frac{1}{2} & -\frac{1}{2} \\ -\frac{1}{8} & -\frac{3}{8} & \frac{5}{8} \\ \frac{3}{4} & \frac{1}{4} & -\frac{3}{4} \end{pmatrix}$

(a) $x = 2$, $y = -1$, $z = 3$ (b) $x = -\frac{1}{2}$, $y = -\frac{9}{8}$, $z = -\frac{9}{4}$

9. (a) I, $10,780; II, $3840 (b) I, $14,630; II, $5570
10. (a) and (b)
11. Democrats $\frac{58}{89}$ or 65.2%; Republicans $\frac{19}{89}$ or 21.3%; Independents $\frac{12}{89}$ or 13.5%

CHAPTER 5

SECTION 26

Section Exercises (pp. 252-253)

	MAXIMUM	MINIMUM
1.	19 at (5,3)	0 at (0,0)
3.	10 at (5,0)	3 at (0,1)
5.	17 at (7,1)	0 at (0,0)
7.	36 at (4,4)	26 at (2,4)
9.	--	6 at (1,4)
11.	--	33 at (3,7)

13. $4375 with 5 oils and 25 finger-paintings.
15. 240 animals: 180 of the first species, 60 of the second.
17. 40 booth-size tables yield maximum profit of $320.

Test A (p. 253)

1.

(x,y)	$f(x,y) = x + 2y$
(0,0)	0 MIN
(0,3)	6 MAX
$(\frac{5}{2},\frac{3}{2})$	$5\frac{1}{2}$
(4,0)	4

2.

(x,y)	$f(x,y) = 4x + 5y$
(0,6)	30
(1,3)	19 MIN
(5,0)	20

3. Maximize $f(x,y) = 410x + 560y$ subject to

$$x,y \geq 0$$
$$5x + 7y \leq 350$$
$$3x + 4y \leq 204$$

(x,y)	$P(x,y) = 410x + 560y$
(0,50)	28,000
(28,30)	28,280 MAX
(68,0)	27,880
(0,0)	0

SECTION 27

Section Exercises (pp. 270-271)

1. (a) Maximize $f(z,y) = -2z + 3y$ subject to $z \geqslant 0$, $y \geqslant 0$, $-3z - y \leqslant 4$, $-5z + 2y \leqslant 8$.
   (b) Already in standard form.
   (c) Maximize $f(x,y) = x + y$ subject to $x \geqslant 0$, $y \geqslant 0$, $-3x + y \leqslant 4$, $x + y \leqslant 4$.

3. (a)

x	y	s	t	f	
4	1	1	0	0	2
3	-9	0	1	0	5
-2	1	0	0	1	0

   (b)

x	y	s	t	f		
2	0	-1	1	0	0	11
0	4	2	0	1	0	7
1	-1	2	0	0	1	0

   (c)

x	y	s	t	u	v	f	
1	3	1	0	0	0	0	5
2	-1	0	1	0	0	0	6
4	1	0	0	1	0	0	7
-1	2	0	0	0	1	0	8
-1	-7	0	0	0	0	1	0

5. 10 at (0,5) 7. 52 at (3,8) 9. 6 at (0,3,0)
11. $12\frac{1}{2}$ at $(\frac{1}{2},0,3)$ 13. 17 at (6,11) 15. $\frac{17}{7}$ at $(0,\frac{4}{7},\frac{9}{7},0)$
17. $4375 with 5 oils and 25 finger-paintings.

Test A (p. 271)

1.

x	y	z	s	t	u	f	
4	1	0	1	0	0	0	5
3	0	-1	0	1	0	0	8
0	2	1	0	0	1	0	9
-3	5	-1	0	0	0	1	0

2.

x	y	s	t	f	
5	2	1	0	0	20
5	1	0	1	0	15
-3	-2	0	0	1	0

$20 \div 5 = 4$
$15 \div 5 = 3$ ✓

x	y	s	t	f	
0	**1**	0	-1	0	5
1	$\frac{1}{5}$	0	$\frac{1}{5}$	0	3
0	$-\frac{7}{5}$	0	$\frac{3}{5}$	1	9

$5 \div 1 = 5$ ✓
$3 \div \frac{1}{5} = 15$

x	y	s	t	f	
0	1	1	-1	1	5
1	0	$-\frac{1}{5}$	**$\frac{2}{5}$**	0	2
0	0	$\frac{7}{5}$	$\frac{2}{5}$	1	16

2. (cont'd)

x	y	s	t	f		
$\frac{5}{2}$	1	$\frac{1}{2}$	0	0	10	($\leftarrow y$)
$\frac{5}{2}$	0	$-\frac{1}{2}$	1	0	5	
2	0	1	0	1	20	($\leftarrow f$)

So f has maximum value 20 at (0,10).

3.

x	y	z	w	s	t	u	f	
2	1	3	2	1	0	0	0	7
3	4	1	5	0	1	0	0	2
1	2	4	3	0	0	1	0	9
-1	-1	-1	-1	0	0	0	1	0

$7 \div 3 = 2\frac{1}{4}$
$2 \div 1 = 2$ ✓
$9 \div 4 = 2\frac{1}{4}$

x	y	z	w	s	t	u	f		
-7	-11	0	-13	1	-3	0	0	1	
3	4	1	5	0	1	0	0	2	($\leftarrow z$)
-11	-14	0	-17	0	-4	1	0	1	
2	3	0	4	0	1	0	1	2	($\leftarrow f$)

So f has maximum value 2 at (0,0,2,0).

4. Maximize $f = 50x + 80y$, where x and y are the number respectively of dresses and of gowns, subject to $2x + 2y \leqslant 20$, $x + 3y \leqslant 18$, $x + 2y \leqslant 15$.

x	y	s	t	u	f	
2	2	1	0	0	0	20
1	3	0	1	0	0	18
1	2	0	0	1	0	15
-50	-80	0	0	0	1	0

$20 \div 2 = 10$
$18 \div 3 = 6$ ✓
$15 \div 2 = 7\frac{1}{2}$

x	y	s	t	u	f	
$\frac{4}{3}$	0	1	$-\frac{2}{3}$	0	0	8
$\frac{1}{3}$	1	0	$\frac{1}{3}$	0	0	6
$\frac{1}{3}$	0	0	$-\frac{2}{3}$	1	0	3
$\frac{70}{3}$	0	0	$\frac{80}{3}$	0	1	480

$8 \div \frac{4}{3} = 6$ ✓
$6 \div \frac{1}{3} = 18$
$3 \div \frac{1}{3} = 9$

x	y	s	t	u	f		
1	0	$\frac{3}{4}$	$-\frac{1}{2}$	0	0	6	($\leftarrow x$)
0	1	$-\frac{1}{4}$	$\frac{1}{2}$	0	0	4	($\leftarrow y$)
0	0	$-\frac{1}{4}$	$-\frac{1}{2}$	1	0	1	
0	0	$\frac{35}{2}$	$\frac{45}{3}$	0	1	620	($\leftarrow f$)

So maximum revenue, equal to $620, is obtained when 6 dresses and 4 gowns are made (and sold).

SECTION 28

SECTION EXERCISES (pp. 278-279)

1. Maximize $f(x,y) = 2x + 5y$ subject to $4x + 7y \leqslant 5$, $x - 3y \leqslant 3$.
3. Maximize $f(x,y,z) = 2x + 4y + 5z$ subject to $x + 9y + 3z \leqslant 3$, $x - y + 7z \leqslant 4$.
5. Maximize $f(x,y,z) = 5x + 7y + 6z$ subject to $4x + 3y + z \leqslant 4$, $-x - z \leqslant 2$, $y + 2z \leqslant 5$, $x - y + 3z \leqslant 3$.
7. Minimum 12 at (3,6).
9. Minimum 22 at (2,8).
11. Minimum 12 at (0,6,0).
13. Minimum $12\frac{1}{2}$ at $(0,\frac{1}{2},\frac{5}{2})$.
15. Minimum 20 at $(0,\frac{1}{3},2)$
17. $560,000

TEST A (p. 279)

1. Matrix:

1	1	2
1	2	3
3	4	0

Transpose:

1	1	3
1	2	4
2	3	0

Initial tableau:

x	y	s	t	f	
1	1	1	0	0	3
1	2	0	1	0	4
-2	-3	0	0	1	0

Final tableau:

x	y	s	t	f	
1	0	2	-1	0	2
0	1	-1	1	0	1
0	0	1	1	1	7
			↑ p	↑ q	

$g(p,q)$ is minimized at (1,1), where its value is 7.

2. Matrix:

2	1	0	1
1	-1	2	2
2	1	5	0

Transpose:

2	1	2
1	-1	1
0	2	5
1	2	0

Initial tableau:

x	y	s	t	u	f	
2	1	1	0	0	0	2
1	-1	0	1	0	0	1
0	2	0	0	1	0	5
-1	-2	0	0	0	1	0

Final tableau:

x	y	s	t	u	f	
2	1	1	0	0	0	2
3	0	1	1	0	0	3
-4	0	-2	0	1	0	1
3	0	2	0	0	1	4
		↑ p	↑ q	↑ r		

So $g(p,q,r)$ is minimized at (2,0,0), where its value is 4.

3. Minimize $800p + 600q$, where p and q are the number of days of operating mines I and II respectively. The constraints are $10p \geqslant 300$, $20p + 30q \geqslant 1800$, and $40p + 30q \geqslant 3000$; or, more simply, they are $p \geqslant 30$, $2p + 3q \geqslant 180$, and $4p + 3q \geqslant 300$.

Matrix:

1	0	30
2	3	180
4	3	300
800	600	0

Transpose:

1	2	4	800
0	3	3	600
30	180	300	0

3. (cont'd)

Initial tableau:

x	y	r	s	t	f	
1	2	4	1	0	0	800
0	3	3	0	1	0	600
-30	-180	-300	0	0	1	0

800 ÷ 4 = 200
600 ÷ 3 = 200

x	y	r	s	t	f	
1	-2	0	1	$-\frac{4}{3}$	0	0
0	1	1	0	$\frac{1}{3}$	0	200
-30	120	0	0	100	1	60,000

x	y	r	s	t	f	
1	-2	0	1	$-\frac{4}{3}$	0	0
0	1	1	0	$\frac{1}{3}$	0	200
0	60	0	30	60	1	60,000
			↑ p	↑ q		

Mine I should be operated 30 days and mine II 60 days to achieve minimum cost subject to the constraints; the minimum cost is \$60,000.

REVIEW EXERCISES (pp. 282-283)

1. The maximum is 41 at (2,7); the minimum is 16.5 at $(\frac{11}{2},0)$.
3. No maximum; the minimum is 12 at (6,0).
5. Maximum income is \$2400, obtained by manufacturing 12 of each table, or by manufacturing 13 of A and 9 of B, or by manufacturing 14 of A and 6 of B.
7. Maximize $f(x,y) = 2x - 5y$ subject to $x,y \geqslant 0$, $-4x + 2y \leqslant 1$, $3x - 2y \leqslant 2$.
9. f has maximum value 20 at (10,0).
11. Maximize $f(x,y) = 8x + 3y$ subject to $x,y \geqslant 0$, $x - 4y \leqslant 5$, $-x + 2y \leqslant 3$, $x + 3y \leqslant 2$.
13. g has minimum value 11 at $p = 1$, $q = 0$, $r = 2$.
15. The minimum cost, 32¢, is obtained by combining 6 ounces of food A with 2 ounces of food B.

CHAPTER TEST (p. 284)

1. (See next page)
2. Ten Lani leis and 5 Kehena leis yield maximum daily income of \$42.50.
3. f attains its maximum of 120 at (0,30,0).
4. g has minimum value 33 at (3,7).
5. See 2 above.

1.

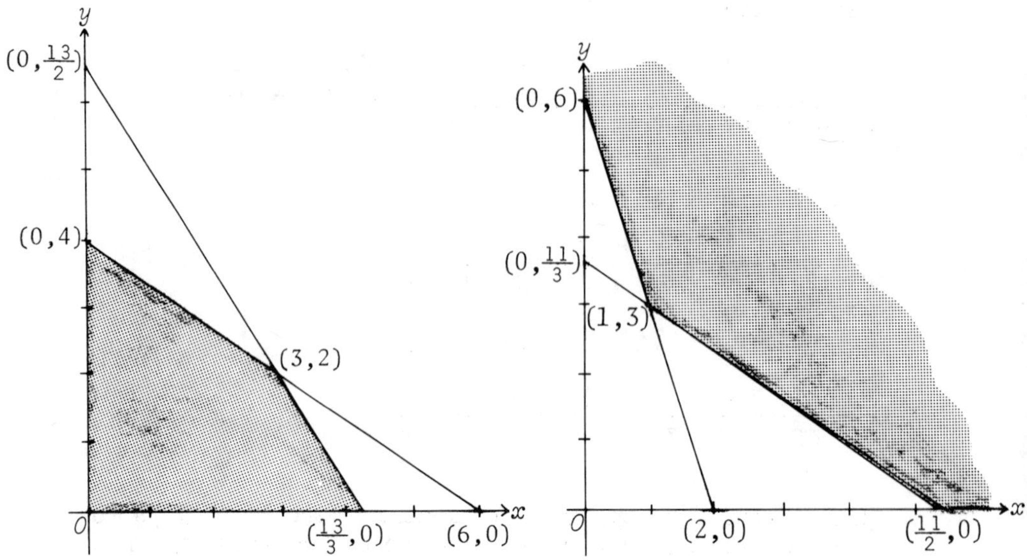

(a) The corner points are (0,0), (0,4), (3,2), $(\frac{13}{3},0)$. $f(x,y) = 2x + 5y$; f is a maximum at (0,4), where it equals 20.

(b) The corner points are (0,6), (1,3), and $(\frac{11}{2},0)$. $f(x,y) = 4x + 2y$ is a minimum at (1,3), where it equals 10.

2,3,4,5. (See preceding page.)

CHAPTER 6

SECTION 29

Section Exercises (p. 289)

1. (a) \$350, \$850 (b) \$3000, \$28,000 (c) \$252, \$1152
   (d) \$562.50, \$2062.50
3. (a) 4 yr (b) 3.18 yr (c) 3.33 yr (d) 37.78 yr
   (e) 13.33 yr (f) 20 yr

Test A (p. 290)

1. $I = PRt = 2300(0.06)(2.5) = \$345$; $A = P + I = 2300 + 345 = \$2645$
2. $(1 + Rt)P = A$; $P = \dfrac{A}{1 + Rt} = \dfrac{1000}{1 + (0.125)(4)} = \$666.67$
3. $P = P(0.08)t$; $t = \dfrac{1}{0.08} = 12.5$ yr

4. $250 = 700R(3)$, $R = \frac{250}{3(700)} = 0.1190$ or 11.90%

SECTION 30

SECTION EXERCISES (pp. 298-299)

Note: A Sharp EL-506H was used for all computations, and answers were rounded off to the nearest cent or hundredth.

1.

AFTER	SIMPLE INTEREST	YEARLY COMPOUNDING
1 year:	5000 + .06(5000) = \$5300	5000 + .06(5000) = \$5300
2 years:	5300 + .06(5000) = \$5600	5300 + .06(5300) = \$5618
3 years:	5600 + .06(5000) = \$5900	5618 + .06(5618) = \$5955.08
4 years:	5900 + .06(5000) = \$6200	5955.08 + .06(5955.08) = \$6312.38

3. Compounding once a year: $800(1.18) = \$944.00$
four times a year: $800(1.045)^4 = \$954.02$
six times a year: $800(1.03)^6 = \$955.24$
monthly: $800(1.015)^{12} = \$956.50$

5. $100(1.05)^9 = \$155.13$

7. twice a year: 9.20%
three times a year: 9.27%
monthly: 9.38%

9. twice-a-year compounding: $\frac{1000}{(1.06)^{12}} = \$496.97$
quarterly compounding: $\frac{1000}{(1.03)^{24}} = \$491.93$
monthly compounding: $\frac{1000}{(1.01)^{72}} = \$488.50$

11. 17.94 years 13. 9.20% 15. \$6644.11

TEST A (p. 299)

1. $r = \frac{12\%}{4} = 0.03$; $n = 7 \times 4 = 28$; $A = 400(1.03)^{28} = \$915.17$

2. $r = \frac{8\%}{2} = 0.04$; $n = 20 \times 2 = 40$; $P = \frac{2000}{(1.04)^{40}} = \$416.58$

3. $\left(1 + \frac{0.06}{12}\right)^{12} = 1.06168$; the effective rate is 0.0617 or 6.17%.

4. $3P = P(1.05)^n$, $n = \frac{\log 3}{\log 1.015} = 73.97$ periods, or 18.45 years.

SECTION 31

SECTION EXERCISES (pp. 308-309)

1. (a) \$1753.85 (b) \$5545.47 (c) \$3928.78 (d) \$74.19

3. \$4403.44 5. \$1099.53 7. \$2705.53 9. \$221.46
11. \$25,677.63

TEST A (p. 309)

1.

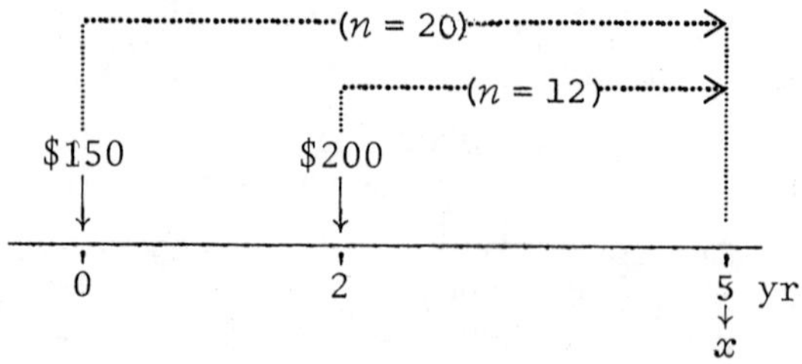

$$x = 150(1.03)^{20} + 200(1.03)^{12} = \$556.07$$

2.

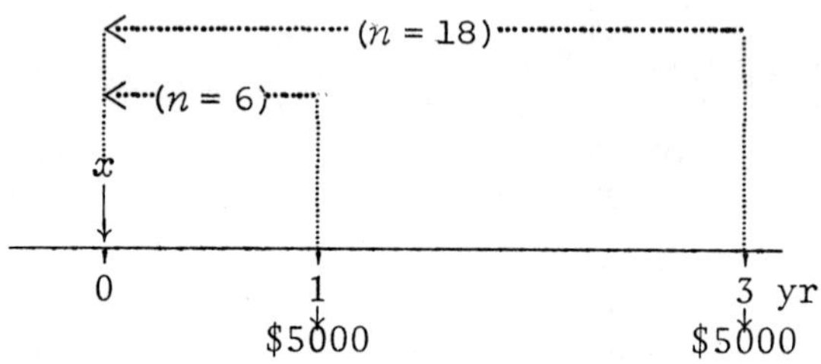

$$x = 5000(1.01)^{-6} + 5000(1.01)^{-18} = \$8890.30$$

3.

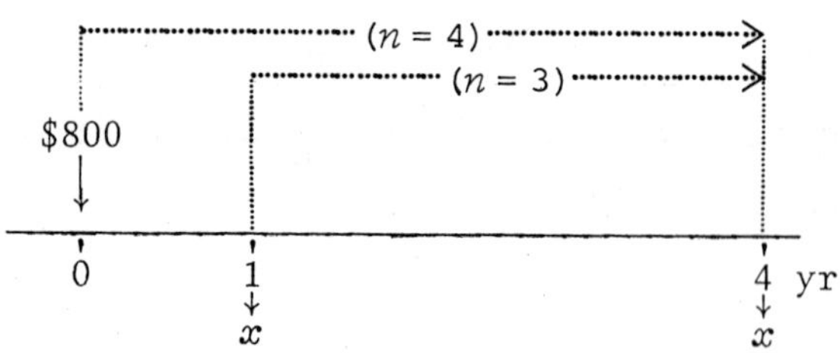

$$800(1.1)^4 = x(1.1)^3 + x \text{ and } x = \$502.48$$

SECTION 32

SECTION EXERCISES (pp. 314-315)

1. (a) \$11,819.00 (b) \$14,147.07
3. \$2082.27 5. \$5436.91

TEST A (p. 315)

1. $55s_{\overline{60}|0.015} = 55\frac{(1.015)^{60} - 1}{0.015} = 55(96.214561) = \$5291.81$

2. $\frac{250}{s_{\overline{20}|0.0075}} = \frac{250(0.0075)}{(1.0075)^{20} - 1} = 250(0.046531) = \$11.63$

SECTION 33

SECTION EXERCISES (pp. 324-325)

1. $10,240.47
3. $361.52; $3014.72
5. (a) $987.59 (b) $162,021.60 (c) $13,786.37
   (d) $33,487.08 (e) $644.36
7. The outstanding loan at any time is the indebtedness at that time. The alternative procedure calculates how much the original loan ($750) amounts to with interest after payment #4, then subtracts the amount of a 4-payment annuity (P = $84.62) after payment #4. The answer is equal to the present value (just after payment #4) of the remaining 8-payment annuity due.

TEST A (p. 325)

1. $125a_{\overline{20}|0.03} = 125\frac{(1.03)^{20} - 1}{(0.03)(1.03)^{20}} = \$1859.68$

2. (a) $\frac{600}{a_{\overline{10}|0.0125}} = \frac{600(0.0125)(1.0125)^{10}}{(1.0125)^{10} - 1} = \$64.20$

   (b) $(64.20 \times 10) - 600 = \$42$

REVIEW EXERCISES (pp. 327-328)

1. $33 3. $757.58 5. (a) 12.36% (b) 12.683%
7. 11.6 yr 9. $1843.89 11. $2978.70 13. $4965.26
15. $477.03 17. $52,942.61 19. $3266.07 21. $173.34

CHAPTER TEST (p. 329)

1. $784 2. $1639.34 3. (a) $1300 (b) $1343.92 (c) $1348.85
4. (a) 12.551% (b) 12.683% 5. $2797.74
6. (a) 11.1 yr (b) 7.9 yr (c) 7.7 yr

7. $7845.45 8. $8406.91; $17,714.88 9. $3761.58
10. $3372.84 11. $830.63
12.

BANK	INTEREST RATE	TERM	MONTHLY PAYMENT	TOTAL INTEREST
A	12%	25 yr	$400	$80,000
B	13%	20 yr	$468.63	$72,471.20
C	14%	15 yr	$466.66	$43,998.80

Note that bank A offers the lowest monthly payments but charges the greatest total amount of interest on the loan.

Index

THIS IS THE END OF THE BOOK